Measures of biological variation have long been associated with many indices of social inequality. Data on health, nutrition, fertility, mortality, physical fitness, intellectual performance and a range of heritable biological markers show the ubiquity of such patterns across time, space and population. This volume reviews the current evidence for the strength of such linkages and the biological and social mechanisms that underlie them. A major theme is the relationship between the proximate determinants of these linkages and their longer term significance for biologically selective social mobility. This book therefore addresses the question of how social stratification mediates processes of natural selection in human groups. Data like this pose difficult and sensitive issues for health policy, and recent developments in this area and in eugenics are reviewed for industrialized and developing countries.

T0291575

SOCIETY FOR THE STUDY OF HUMAN BIOLOGY
SYMPOSIUM SERIES: 39

Human Biology and Social Inequality

PUBLISHED SYMPOSIA OF THE
SOCIETY FOR THE STUDY OF HUMAN BIOLOGY

Numbers 1–9 were published by Pergamon Press, Headington Hill Hall, Headington. Oxford OX3 0BY. Numbers 10–24 were published by Taylor & Francis Ltd, 10–14 Macklin Street, London WC2B 5NF. Further details and prices of back-list numbers are available from the Secretary of the Society for the Study of Human Biology.

Human Biology and Social Inequality

39th Symposium Volume of the
Society for the Study of Human Biology

EDITED BY

S. S. STRICKLAND

University College, London and London School of
Hygiene and Tropical Medicine

and

P. S. SHETTY

London School of Hygiene and
Tropical Medicine

CAMBRIDGE
UNIVERSITY PRESS

CAMBRIDGE UNIVERSITY PRESS
Cambridge, New York, Melbourne, Madrid, Cape Town, Singapore, São Paulo, Delhi

Cambridge University Press
The Edinburgh Building, Cambridge CB2 8RU, UK

Published in the United States of America by Cambridge University Press, New York

www.cambridge.org
Information on this title: www.cambridge.org/9780521104012

First published 1998
This digitally printed version 2009

A catalogue record for this publication is available from the British Library

Library of Congress Cataloguing in Publication data

Human biology and social inequality / edited by S. S. Strickland and P. S. Shetty
     p.   cm. – (Society for the Study of Human Biology symposium series : 39)
     Includes index.
     ISBN 0–521–57043–3 (hardcover)
1. Social medicine – Congresses.   2. Equality – Congresses.
3. Human biology – Congresses.   I. Strickland, S. S. (Simon Slade),
1946–   . II. Shetty, Prakash S.   III. Series.
RA418.S6423   1998
306.4′61—dc21   97–23261   CIP

ISBN 978-0-521-57043-5 hardback
ISBN 978-0-521-10401-2 paperback

Contents

Contributors

Yoav Ben-Shlomo
Department of Social Medicine, University of Bristol, Whiteladies
Road, Bristol BS8 2PR, UK

A. H. Bittles
Department of Human Biology, Edith Cowan University, Joondalup
Campus, Perth, Western Australia 6027, AUSTRALIA

Athanasios Chasiotis
Institut für Psychologie, Universität Osnabruck, Seminarstrasse 20,
D-49069 Osnabruck, GERMANY

Y-Y. Chew
Department of Human Biology, Edith Cowan University. Joondalup
Campus, Perth, Western Australia 6027, AUSTRALIA

Mark Nathan Cohen
Department of Anthropology, State University of New York,
Plattsburgh, New York, NY 12901, USA

Emile Crognier
UPR 221 du CNRS, Pavilion de Lanfant 346, Route des Alpes, 13100
Aix-en-Provence, FRANCE

Stefan A. Czerwinski
Department of Anthropollogy, State University of New York at
Albany, 1400 Washington Avenue, Albany, NY 12222, USA

Ottó G. Eiben
Department of Physical Anthropology, Eötvös Loránd University,
H-1088 Budapest, Puskin utca 3, HUNGARY

Sally Macintyre
MRC Medical Sociology Unit, 6 Lilybank Gardens, Glasgow, G12
8RZ, UK

Robert M. Malina
Department of Anthropology, University of Texas, Austin, TX 78712, USA

Michael G. Marmot
Department of Epidemiology, University College London, Gower Street, London WC1E 6BT, UK

C.G. Nicholas Mascie-Taylor
Department of Biological Anthropology, University of Cambridge, Downing Street, Cambridge CB2 3DZ, UK

Sharon Matthews
Department of Epidemiology and Biostatistics, Institute of Child Health, Guilford Street, London WC1N 1EH, UK

Kerin O'Dea
Department of Human Nutrition, Deakin University, Geelong, Victoria 3217, AUSTRALIA

Jessica A. Ogden
Department of Public Health Policy, London School of Hygiene and Tropical Medicine, Keppel Street, London WC1E 7HT, UK

John D. H. Porter
Department of Clinical Sciences, London School of Hygiene and Tropical Medicine, Keppel Street, London WC1E 7HT, UK

Chris Power
Department of Epidemiology and Biostatistics, Institute of Child Health, Guilford Street, London WC1N 1EH, UK

Lawrence M. Schell
Department of Anthropology, State University of New York at Albany, 1400 Washington Avenue, Albany, NY 12222, USA

Prakash Shetty
London School of Hygiene and Tropical Medicine, Keppel Street, London WC1E 7HT, UK

Carolyn Stephens
Department of Public Health and Policy, London School of Hygiene and Tropical Medicine, Keppel Street, London WC1E 7HT, UK

Sara Stinson
Department of Anthropology, Queen's College, City University of New York, Flushing, NY 11367, USA

Simon Strickland
London School of Hygiene and Tropical Medicine, Keppel Street,
London WC1E 7HT, UK

Eckart Voland
Johnder Straase 1, D-37127 Scheden, GERMANY

Richard G. Wilkinson
Trafford Centre for Medical Research, University of Sussex, Brighton,
Sussex BN1 9RY, UK

Acknowledgements

The editors thank the Wellcome Trust, the London School of Hygiene and Tropical Medicine and University College London, for their support. They also thank Dr Nick Norgan and Dr Stanley Ulijaszek who shared with the editors the chairing of symposium sessions, and Carol Aldous for her patience and understanding.

1   *Human biology and social inequality*

SIMON STRICKLAND AND PRAKASH SHETTY

Introduction

It has long been commonplace to seek to justify an idea by appeal to antiquity. If this were to be attempted for the theme of this book, inquiry would centre on the lineage of three linked ideas: the separation of biological from social domains; the concept of inequality; and the distinction between facts (or science) and values (or moral judgements). Tracing some of the historical development of these ideas cannot, to be sure, justify them in a validatory sense. But no apology need be offered for proceeding in this way, for it highlights important and unresolved issues which form an intellectual background common to the contributions in this book.

Biological vs. social domains

The separation of biological from social domains of inquiry can be seen in the many historical attempts to distinguish natural properties of man from the frequently evaluative world of human experience. Discussions of early Greek thought have often focused upon the distinction between *phusis* (φύσις) and *nomos* (νόμος) as it was deployed by Aristotle (e.g. Barker, 1948: xx–xxiii; Popper 1962: I: 60ff). Lloyd's (1991: 425–6) broader analysis of this terminology, rendered by 'nature' as distinct from 'customs, conventions, and laws' in studies of human behaviour, argues forcefully that this antithesis was interpreted variously and polemically by Greek authors. On one view, the contrast was used to secure the domain of *phusis* for science and that of *nomos* for sociology or political philosophy. Others held that *nomos* was merely the kind of custom or convention which the powerful could ignore or manipulate to their advantage. Still others interpreted *nomos* as the law and the guarantor of justice which, therefore, was precisely what discriminated the human species from other animals; while thinkers like Plato resisted too sharp a contrast. The scope of the idea of *phusis* likewise varied; and rather than discovering 'nature', the Greek

1

theorists in effect invented their own distinctive and divergent ideas about it through competitive debate.

Against such a background, it is worth noting certain properties of Aristotelian thought on *phusis* because these have been influential in later work down to contemporary times. Barker (1948: xxiii) noted that the latinate rendering of *phusis* and *phusikos* by 'nature' and 'natural' lends to these terms connotations of birth or innateness which alter their exactly intended sense. For Aristotelian usage includes the process of growth and the condition of being 'grown', as well as the beginnings of 'growing'. Thus, 'The "nature" of things consists in their end or consummation; for what each thing is when its growth is completed we call the nature of that thing, whether it be a man or a horse or a family' (Barker, 1948: 1253a). In this respect, Lloyd's (1991: 427–30) discussion highlights the evolutionary character of Aristotle's idea of what is natural, and this underlies his claim that humans are political animals – man has a natural desire to reproduce, and this entails the basic form of household association, but lack of self-sufficiency results in agglomerations which culminate by nature (*phusei*) in the *polis* or city-state which is their end (*telos*). The form of explanatory argument, which is common to Aristotle's natural science and his political philosophy, is therefore teleological or functionalist – his use of the concept of nature represents, on the one hand, what is true always or usually so, and, on the other, the goal, end (*telos*), good or ideal which is served. Thus, while Aristotle's concept of nature may often appear to be merely descriptive, it is also often and explicitly normative.

To begin this book in such a way serves two purposes. Firstly to highlight, by contrast with the classical period in so far as it is possible to generalize, the somewhat arbitrary but now institutionalized disciplinary boundaries which segregate scientific endeavours in contemporary research. In bringing together studies of the relationships between biological and social variation in human groups, the contributions in this book illustrate how unhelpful such disciplinary boundaries can sometimes prove to be. The second purpose is to address the particular theme of human inequality.

Inequality vs. inequity

Two issues are raised to address the theme of human inequality. The first is the weight to be attached to discriminating the descriptive and impartial sense of this term from its evaluative, normative or juridical senses. It has been seen that Aristotle's usage was not always conducive to a sharp distinction of this kind, but clarity of argument is served if this is borne in mind. The contrast is represented by Munro (1969: 1–2), for example, as referring, on the one hand, to the common attributes with which all people are born and, on the other, to the claim that all are of similar worth or deserve to be treated equally. In that respect, Munro (1969: 55, 65) argues that the classical philosophical traditions of Europe and of China differ in at least two important ways which have had lasting significance down to contemporary times: firstly, in the weight historically placed on the search for truth as distinct from that for knowledge of how to behave ethically; and secondly, in the emphasis given to a social hierarchy based on presumptively innate inequalities as distinct from one dependent on the cultivation of proper behaviour in a generally open meritocracy.

The second, and related, distinction sometimes found in the literature is that made by Whitehead (1988) between the extent of natural variation and the extent of unacceptable or unfair inequalities. A third issue, also related to discrimination, is that introduced by Stephens in this book between the descriptive and analytical senses, where the latter sense applies to control over the benefits and disadvantages of inequality.

In the contributions in this book (with the exception of Stephens'), the impartial, descriptive sense is generally intended by '[in]equality' and its derivatives; by contrast, the evaluative sense is specified either by its appropriate qualification or by using the term '[in]equity' and its derivatives. A presumptive distinction between facts and values underlies this nomenclature and is considered further below.

The second point above raises the issue of the extent to which human inequality has been understood to be 'natural'. With notable exceptions, the tendency in European thought has for long been to presume an innate inequality. For example, the Black Report (Townsend *et al.*, 1988) remarked on the tendency to assume a human variation which is 'eternal and unalterable'. This was clear from Plato's allegory in the *Republic* associating an ideal hierarchy of occupations with base and precious metals in the composition of each inhabitant of the state. In this passage, Socrates advocates that the ideal community should accept as true a particular fable:

> It is true, we shall tell our people in this fable, that all of you in this land are brothers; but the god who fashioned you mixed gold in the composition of those among you who are fit to rule, so that they are of the most precious quality; and he put silver in the auxiliaries, and iron and brass in the farmers and craftsmen. Now, since you are all of one stock, although your children will generally be like their parents, sometimes a golden parent may have a silver child or a silver parent a golden one, and so on with all the other combinations. So the first and chief injunction laid by heaven upon the Rulers is that, among all the things of which they must show themselves good guardians, there is none that needs to be so

carefully watched as the mixture of metals in the souls of the children. If a child of their son is born with an alloy of iron or brass, they must, without the smallest pity, assign him the station proper to his nature and thrust him out among the craftsmen and farmers. If, on the contrary, these classes produce a child with gold or silver in his composition, they will promote him, according to his value, to be a Guardian or an Auxiliary. They will appeal to a prophecy that ruin will come upon the state when it passes into the keeping of a man of iron or brass. Such is the story; can you think of any device to make them believe it?

(Plato *Republic*, p. 414).

For Plato, it was a short step from these deliberations to the claim that marriage and reproduction should follow essentially the same principles of sound breeding practice which were applied in animal husbandry; and that consequently the braver and better youth should have the greater opportunities to reproduce. This ostensibly 'eugenicist' line of argument presupposes a strongly hereditarean outlook which has contemporary resonance. In a Christian theological context, however, natural inequality was sometimes linked to doctrines of predestination and original sin (Munro, 1969: 19–21). These are ideas which have influenced how responsibility (for ill health or other experiences) is attributed to the individual or to external circumstances. It is worth quoting a passage from Aristotle (*Politics:* 1254b–55a) which argues a claim in some respects similar to that of Plato, but with clearer biological force:

Ruling and being ruled [which is the relation of master and slave] not only belongs to the category of things necessary, but also to that of things expedient; and there are species in which a distinction is already marked, immediately at birth, between those of its members who are intended for being ruled and those who are intended to rule... This characteristic is present in animate beings by virtue of the whole constitution of nature, inanimate as well as animate; for even in things which are inanimate there is a sort of ruling principle, such as is to be found, for example, in musical harmony... But it is nature's intention to erect a physical difference between the body of the freeman and that of the slave, giving the latter strength for the menial duties of life, but making the former upright in carriage and (though useless for physical labour) useful for for the various purposes of civic life – a life which tends, as it develops, to be divided into military service and the occupations of peace. The contrary of nature's intention, however, often happens: there are some slaves who have the bodies of freemen – as there are others who have a freeman's soul. But if nature's intention were realized – if men differed from one another in bodily form as much as the statues of the gods [differ from the human figure] – it is obvious that we should all agree that the inferior class ought to be the slaves of the superior. And if this principle is true when the difference is one of the body, it may be affirmed with still greater justice when the difference is one of the soul; though it is not as easy to see the beauty of the soul as it is to see that of the body. It is thus clear that,

> just as some are by nature free, so others are by nature slaves, and for
> these latter the condition of slavery is both beneficial and just.

Possible inconsistency between bodily properties and occupational rank is
clear in this account, as in that of Plato; and Aristotle (*Politics*: 1255b)
recognized that inequalities could also be based on relations of power and
legal authority. Yet he maintained the general principle that the distinction
between master and slave was natural in the sense of ideal – as with
distinctions between right and left, and between male and female, the first
of each pair was treated as the *telos* or end, the second in each pair as for
the sake of that end. Following this logic, therefore, Aristotle did not treat
the master/slave polarity as belonging to the category of *nomos* (Lloyd,
1991: 428–9; see also Lloyd, 1983).

Turning to contemporary accounts of human inequality, Dahrendorf
(1969) followed Jean-Jacques Rousseau and rejected the Aristotelian claim
of natural inequality on the grounds that the natural rank of all men is one
of equality. However, this author seems to have been motivated in part and
significantly by a desire to establish a new disciplinary field – on his
argument, inequalities in wealth, repute and power could only be understood through a form of sociological inquiry that was intellectually auton-
omous and independent of the study of natural inequalities. 'Thus', he
wrote, 'the difficult philosophical question of the natural rank of man can
be set aside here as irrelevant to the truth or falsity of sociological explanations of social stratification. We rule out only explanations based on the
assumed congruence, or tendency to congruence, of the natural and social
rank orders' (Dahrendorf, 1969: 21–2, footnote 5). In asserting this *a
priori*, and without examining the nature and extent of correlations between rank orders, Dahrendorf was swinging to another arguably unjus-
tifiable extreme. A major purpose of this volume is to present recent studies
which set out explicitly to quantify the degree of congruence, and nature of
the interaction, between social and biological variation in human groups.

Scientific facts vs. value judgements

The distinction between facts and values refers back to the earlier distinction between descriptive and evaluative senses of 'equality', or that of
'equality' from 'equity'. For underlying the apparently straightforward
terminological distinction between types of equality, however, are the
problem of differentiating facts from values and the related question of
whether a value-free science is possible. Reasoning from a utilitarian
viewpoint, J.B.S. Haldane (1932) argued that scientific quantitative
methods should be applied to ethical problems, and he put his view thus:

'It is true that science from its nature can only say what is, was, or will be, and not what ought to be ... But our views as to the status of good action are profoundly affected by our views of the universe'.

Few scientists would probably disagree with this claim, although many would be wary of the weight which Haldane placed on eugenics as an ethical principle (Haldane, 1932). That the fact–value contrast poses a seemingly intractable problem, however, and one which remains current particularly in the field of medical genetics, is illustrated well by an example from human nutrition. Thus, in a recent review of concepts of 'adaptation' used in nutritional research, Waterlow (1990) contends that it is essential to disentangle the biological responses to nutritional stresses from the value judgements which are inevitable in a human context – there are established statistical methods for determining empirically whether a relationship between stress and response is linear or threshold in character, and not even the latter pattern necessarily legitimates objectively a judgement that would otherwise be arbitrary or value-laden. By contrast, in an extended essay on the nutrition and economics of well-being and destitution, Dasgupta (1993) rejects the distinction between facts and values, arguing that the 'commonality of the human experience' means that the determinants of well-being are quantifiable regardless of people's own conceptions of what is good. It is such divergent approaches as these that underlie disparate ideas of the significance of human inequality for future action, and thus the kinds of prescription for inquiry and policy which are deemed to be appropriate.

In defence of a middle way, one might quote Lloyd's (1991: 366–7) remark on this disjunction:

> If the fact-value contrast has occasionally been invoked in a highly simplistic fashion, that too does not mean that it was not, is not, useful and important. Indeed though the fact–value contrast has been criticized and to some extent eroded, in some contexts it is still the case that insufficient, not excessive, attention has been paid to it.

Thus, Lloyd (1991: 370–1) invites us to reflect on our anthropocentricity while preserving our humanity, and to ponder 'the unwisdom of the notion that it is better for *no* link [between science and morality] to remain, for rather, clearly, the *connection* is of the utmost concern'. The studies presented here, and the lively debate which they occasioned at the Society for the Study of Human Biology symposium, illustrate the many and varied forms which this dialogue can take.

Relationships between biological and socio-economic characteristics of populations

The contributions gathered in this volume represent several of the themes that were already present in the classical philosophical literature cited above, but three are particularly prominent. The first of these is the degree of congruence between measurable biological and social economic properties of human populations. In her comparative overview, Macintyre presents evidence for the consistent and ubiquitous associations between indices of physical health – mortality, self-reported ill health and growth – and social stratification by occupation, income or educational level. Educational attainment tends to be the most widely applicable measure in international comparisons. Different causes of death may show distinct social class gradients, while measures of long-term health status show more socio-economic differentials than do short-term measures of health state. The magnitude of differentials varies across countries – the proportional increase in mortality as one moves from the top to the bottom of an occupational hierarchy is narrowest in Norway and Denmark, greatest in France. Relationships of health and longevity to socio-economic status follow a stepwise gradient, not a threshold pattern, regardless of the measure of inequality applied. Evidence of mortality differentials between occupational groups increasing since the 1950s was suggested by the Black Report (Townsend *et al.*, 1988); and Macintyre adds to this the considerable evidence of increasing socio-economic disparities in the West over the past 20 years. This has been associated with declining overall mortality but accompanied by increasing mortality differentials between social groups. While absolute poverty can threaten health, the question is raised of the relation between processes that produce socio-economic disparities in health and those which result in increased life expectancies. It may be surprising that successive improvements in living standards would be health enhancing, but the generally linear trend is consistent with the accumulation of health risks (resulting from material and social disadvantages) over the lifetime.

Four basic explanations were posited by the Black Report (Townsend *et al.*, 1988) on inequalities in health for social class health gradients in the UK. These were:

(1) Artefact: the consequence of methodological inadequacies, for example resulting from the problem of class attribution at different periods, and changes in the occupational structure of the population over time as represented by the health benefits of

upward social mobility or the benefits of good health for upward social mobility.

(2) Natural or social selection: the phenomenon of within-generational social mobility being influenced by presumptively innate health status (physical strength, vigour or vitality).

(3) Material or structural circumstances: economic and social factors including processes of impoverishment, and the loose concept of 'stress' which could be argued to mediate between unemployment and ill health.

(4) Cultural or behavioural patterns: individual choices, autonomously (and therefore irresponsibly) made, which determine patterns of behaviour resulting in onset of disease and associated mortality.

The Report argued that the most persuasive answer lay in materialist-type explanations, with perhaps different forms of explanation applying to distinct stages of the life-cycle (Townsend *et al.*, 1988). However, the independence of these four types of explanation is questionable. Whitehead (1988) argued persuasively that behavioural choices are related to or constrained by living conditions so the separateness of the last two claims is open to criticism. The first two points are linked in that they both concern biologically selective social mobility – they are differentiated principally in the extent to which the biological properties in question are presumed to be intrinsic ('innate') or acquired, and in whether the focus is on upward as distinct from downward social mobility.

In subsequent work, the extent of the contribution made by selective social mobility to health inequalities has been vigorously disputed – Illsley (1986) argued that data on social mobility at marriage indicate artefact and bias, resulting from selective class recruitment and loss, to be the major factor underlying inequalities in perinatal mortality. Thus he emphasized the effects of premarital growth and development on later maternal nutrition and health during the child-bearing years. By contrast, Wilkinson (1986b) emphasized the postmarital health environment. He claimed that, if marriage results in a class differential in relative stature of 20% and of perinatal mortality in 116%, then the differential attributable to other factors than selective social mobility must in this case approximate $116-20\% = 96\%$.

It is, however, questionable whether the property of stature and risk factors for perinatal mortality are biologically independent to the extent presumed by Wilkinson in this case. From a broader perspective, quantitative assessment of the magnitude of cumulative health risks is required. It is represented in this book by the analysis of longitudinal British data from birth to young adulthood in the contribution by Power and Matthews (Chapter 3). The analysis by these authors uses the National Child Development Study data set from 1958 through to the latest round in 1991, supplementing earlier studies in this series (see also Power *et al.*, 1986). It shows that health gradients occur regardless of the measures of inequality applied, though as in some other studies education shows the greatest disparities. An entirely artefactual explanation is therefore unpersuasive. While there is health-related social mobility, some mechanisms of which are illustrated in the contributions by Porter and Ogden (Chapter 6) and by Schell and Czerwinski (Chapter 7; see also Wadsworth, 1986), health gradients remain after allowing for baseline health status and for mobility itself. Thus selective processes do not represent a major explanation for the observed gradients. With respect to the last two explanatory factors, it is again made clear that these are inter-woven and cannot be easily separated. In this data analysis, social gradients in height at age 23 years were affected by early childhood and genetic factors (mid-parent height), and by effects of social class at birth and birth weight which were additive. Contributions made by circumstances of the intervening childhood and adolescent years were major sources of health inequalities. Further work is required to disentangle the social and physical factors responsible at each stage of life.

Related to the question of how different measures of inequality are associated with health differentials is that of the importance to be attributed to *absolute* material circumstances, as distinct from *relative* difference in socio-economic status. Underlying this is a broader issue concerning how inequalities can appropriately be measured, whether in terms of population-based rates or in terms of the sum of individual-based risk factors. The contributions by Wilkinson (see also Wilkinson, 1986a,b, 1994), Stephens, and Ben-Shlomo and Marmot (Chapters 4, 16 and 17) emphasize the significance of relative differentiation. Wilkinson argues that associations between socio-economic and health inequalities are stronger within societies than they are when examined across societies or with an historical perspective – thus the empirical relationships between individual income and health (for example) do not result in a society-level association between economic growth and population health.

Effects of the epidemiological transition, in which the social distribution of 'diseases of affluence' changes with economic development, perhaps offer a partial explanation for this. However, Wilkinson argues that, above a certain threshold, absolute living standards seem to have little effect on physical well-being and the pronounced health inequalities observed then reflect the penetrating influence of the social environment. This influence is exerted through violence and through varied psycho-social processes (discussed also by Porter and Ogden, Stephens, and by Ben-Shlomo and Marmot). For example, recent economic growth in Western nations has resulted in increased real incomes, but has been linked to increasing rather than declining mortality differentials between classes. On Wilkinson's argument, this underlines the importance of egalitarianism and social cohesion in determining the link between a population's income distribution and its experience of mortality, contrary to the emphasis placed by others on material conditions or individual behavioural variation (Phillimore et al., 1994). Thus, factors such as social cohesion and the supportive nature of the entire community and cultural attitudes related to a stable family structure may contribute to these differences as exemplified by the Roseto studies (Egolf et al., 1992).

A picture broadly consistent with this is provided by Eiben's review of population studies in Hungary and Poland during liberalization between the 1960s and early 1990s. These data also indicate rising levels of socioeconomic inequality (see also Macintyre, Chapter 2). However, there are differences from, as well as similarities with, the Western European experience (see also Power, 1994). Thus, rising inequality in Eastern Europe has been associated with an increasing incidence of mortality from cardiovascular and ischaemic heart disease; and there has been a resurgence of tuberculosis, faltering of the long-term secular trend of improved linear growth performance in children, and persistent differences in anthropometric status across the urban–rural divide.

Variation in the resurgence of tuberculosis, and its relationship to patterns of infection with HIV (Human Immunodeficiency Virus), is the subject of the contribution by Porter and Ogden. These authors show that the study of inequality continues to be as relevant to epidemiological practice today as it has been in the past. Unemployment, homelessness, and ethnicity are all linked to the re-emergence of tuberculosis in Europe and America over the past 15 years. Infectious diseases can be classified as 'democratic' (for example, influenza) or 'undemocratic' (cholera, tuberculosis) depending on the extent to which vulnerability to them varies across socio-economic gradients within or between populations or regions of the world – the landscape of vulnerability to the 'undemocratic' diseases differs markedly between the societies of the North and those of the South. A significant shift of focus from individual risk analysis to assessment of social vulnerability has altered ways of thinking about the transmission of infectious diseases. Thus the classical biomedical model of disease transmission as a dynamic event in a static social structure cannot well accommodate the social dimensions of disease and its transmission – epidemiologists and those professionally engaged in public health need to consider how individuals are related to their communities and to the wider social order.

In cross-sectional studies, explanations for the varying congruence between biological and social differentials have generally been sought in proximate environmental factors. Many of these have been reviewed elsewhere comparatively on the basis of population studies from around the world (e.g. Strickland and Tuffrey, 1997). In the present volume, insight into some underlying processes is afforded by studies of what the Black Report called 'transmitted deprivation' (Townsend *et al.*, 1988). The contribution by Schell and Czerwinski shows how social factors – primarily operating through residential segregation by ethnicity, income and race – differentially distribute exposure to toxic pollutants such as lead in America. Lead passes across the placenta, so pathological effects can be intergenerational. Schell and Czerwinski speculate that the impact of lead exposure on cognitive development could result in subsequent behavioural patterns which also increase adverse health risks and the possibility of long-term downward social mobility. This is therefore a selective-mobility hypothesis of the kind which in some circumstances may help to explain why health inequalities persist.

Cognitive development is also thought to be influenced by the nutritional environment during growth, and this provides another instance of a selective-mobility hypothesis. Using a range of studies, Stinson reviews the substantive evidence for effects of nutrition on cognitive function in the 'real world' as distinct from the world of the IQ test. She concludes that the impact of nutrition can be mediated by patterns of motor development and interaction between the child and the environment. In general, there is strong evidence for such effects, although the degree to which they endure and contribute to socio-economic status in the long-term is a subject requiring further research.

While nutrition may vary with socio-economic conditions, these can also influence choice of, and access to, diverse types of physical activity. Their close interaction and biological consequences are represented by two contributions. The first, by O'Dea (Chapter 9), describes the impact of sudden affluence, and reversion to traditional living patterns, on risk of non-insulin dependent diabetes (NIDDM) in populations of the developing world. This study indicates the importance of physical activity level, diet, weight and fatness in the aetiology of the condition. Low fertility is a correlate of NIDDM, and data on its changing prevalence suggest that processes of natural selection have operated which arguably lend support to Neel's (1962) 'thrifty genotype' hypothesis. However, also to be considered is the related – 'thrifty phenotype' – argument, which effectively encompasses the former. This holds that foetal programming may determine the properties of carbohydrate metabolism which result in impaired glucose tolerance in later life. Thus, some populations may be doubly at risk through effects of westernization across the life-span. This study illustrates clearly how biologically heritable characteristics can interact with the proximate environment to result in pronounced inequalities in chronic degenerative morbidity.

The second contribution of relevance here is that by Malina (Chapter 10). At the level of elite sports, opportunities are unequally distributed because recruitment is highly exclusive and often begins early in childhood, for example in gymnastics. The study of sports physiology can sometimes provide models for understanding how extremes of physical activity modulate physiological functions. Malina shows that sport-specific selective factors are implicated in the variables associated with comparatively late maturation in female gymnasts, divers, and ballet dancers. Mechanisms may involve foetal programming and stresses operating on the hypothalamic–pituitary–ovarian axis in women. However, there appear to have been no studies of possible long-term impact on fertility outcomes. While there is evidence of effects of physical exertion in men on the characteristics of sperm, there are no data on the extent to which these effects are reversible. Apart from limited data on cardio-vascular related mortality and morbidity before reproductive age, there is as yet little evidence that high levels of physical fitness in athletes have consequences for their reproductive fitness.

Social mobility and its influence on patterns of inequality

The possibility that biologically selective social mobility influences patterns of inequality is the second theme which was represented, somewhat problematically for Plato and Aristotle, in the classical works cited earlier. Although Power and Matthews (Chapter 3) found that this phenomenon contributed comparatively little to health gradients in the British data, the subject is of wider importance. It concerns the study of the genetic structure of populations, and the extent to which unequally distributed characteristics result from heritable or non-heritable sources of variation. Thus it involves properties with a significant genetic component, and more broadly the analysis of differential reproductive performance. Against this background, the findings of Power and Matthews could be speculated to result in part from the impact of comparatively aggressive health care provided by the welfare state, and partly from procedures for classifying physical conditions in terms of morbidity. In other social environments, or with respect to different biological properties, one might conjecture that the nature and outcome of selective social mobility processes could be quite various. Thus, Franz Boas (1938: 81) argued that natural selection in humans operates primarily through social stratification; Haldane discussed the relevance of innate inequalities for eugenics, and possible correlations between wealth and heritable components of intelligence (1932: 24, 106–7); and still earlier Charles Darwin had explored the general idea at length in *The Descent of Man* (1871).

In this volume, Mascie-Taylor reviews the role of social and geographical mobility in altering the genetic stucture of populations, and considers critically the arguments over determination of IQ and its unequal social distribution connected with the work of Cyril Burt. In general, IQ is significant in both intra- and inter-generational social (occupational) mobility, but non-specific factors are as important, or indeed more so. Evidence for socio-economic class differences in polymorphisms and blood groups is weak in the UK. Migration studies show clearer disparities between stable and mobile populations, for example with respect to anthropometric traits. As with the patterns of chronic degenerative disease in newly affluent populations, such evidence needs to be considered in the light of recent work on epigenetic inheritance mechanisms acting by genomic imprinting *in utero* to modulate expression of capacities for physical growth (Golden, 1994; Maynard Smith and Szathmáry, 1995: 247–50).

As pointed out by Illsley's (1955) classic study of stillbirths and perinatal mortality in Scottish women, social mobility can also occur through marital arrangement. This is reflected by the somewhat inconsistent data on health inequalities among women in the contributions by Macintyre and Power and Matthews. Through processes of assortative mating, marital arrangements are sensitive to social inequality; and the extent to which there is consequent inequality in reproductive success is the subject of the contributions by Voland and Chasiotis (Chapter 12) and by Crognier (Chapter 13). Voland and Chasiotis use rich data on the historical demography of rural Germany in the eighteenth and nineteenth centuries. These authors address the question how far the mating and reproductive decisions of women, and their social mobility on marriage, can be understood to have caused social inequality in the reproductive fitness of men. By comparing prosperous farmers and landless workers, they are able to indicate what approximate to 'manipulative' reproductive strategies of farmers and 'opportunistic' strategies of the landless. This study provides substantial evidence that land ownership was a component of the process of natural selection, and that differential reproductive success accumulated to yield pronounced and long-lasting social status differences in Darwinian fitness. These authors emphasize that the psychological traits characteristic of male social economic competitiveness seem to have resulted from sexual selection by females, and argue that the over-reproduction of social elites is a largely female-driven phenomenon which is itself also contingent upon male social status.

In this respect, there is a general lack of substantive and comparable evidence from non-Western population studies. The contribution by Crognier uses case-study material from rural Morocco to examine how agro-ecological zones and socio-economic status differences are associated with a number of dimensions of reproductive activity. His findings indicate that the impact on reproductive success (numbers of live-births, and of offspring survival to sexual maturity) of slight socio-economic variations was more pronounced than was that of agro-ecological region. Higher frequencies of childlessness and widowhood resulted in poorer outcomes in the less well-to-do groups, and, while the agro-ecological environment affected infant survival chances, there were socio-economic means of compensating for this to a greater extent than there were such means for coping with socio-cultural forces. In this case, indeed, reproductive success was itself a measure of status more important than comparatively minor socio-economic disparities, a point which has not generally been prominent in studies of low-fertility Western populations.

The significance of fertility for socio-economic inequality is emphasized on a larger temporal and spatial scale in Cohen's review of archaeological evidence. This indicates a clear correlation between population density, size and degree of technological and hierarchical complexity. Cohen's view is that this association has been driven since the Palaeolithic by persistent and in general increasing reproductive success. 'Civilization' is identified by the emergence, at various times and in various regions of the world, of class-based stratification. This is permanent, largely unrelated to individual capabilities and inflexible. It depends on coercive power in the form of the prevailing ethic, as shown by the ability to command human labour in building massive structures like the pyramids and on exclusive ownership of essential resources. Socio-economic inequalities occur, then as now, both within and between communities linked by trading and political networks. In spite of some advantages, the weight of the archaeological and comparative evidence suggests that civilization has conspicuously failed to deliver enhanced health; rather, it has tended to harm that of the majority of its populations. Improvements in population health since the nineteenth century have been unequally distributed between regions of the world and between classes within its nations. The evolutionary implication is that, in terms of population health, the human species has been a victim of its own reproductive success.

Implications for health policies in the management of health inequalities and inequities

Cohen (Chapter 14) ends his contribution by pointing out the 'moral obligation to invest in the health and nutrition of those whom class stratification and civilization otherwise harm'. This opens up the third facet of the topic to which the classical thinkers animadverted, namely the inferences which can legitimately be drawn, from studies of the preceding kinds, for health policy and the management of health inequalities. It is at this juncture that the principles linking 'facts' with 'values', 'science' and 'morality', need to be scrutinized with care.

It was argued by the Black Report (Townsend *et al.*, 1988) and by Whitehead (1988) that the observed extent of health inequalities was unacceptable and, at least in principle, largely avoidable. Consequently their policy recommendations centred on measures for reducing inequalities through generally redistributive economic mechanisms. This was coupled with longitudinal health monitoring, and the overall aim was to achieve equal access to treatment and equal opportunity for attaining a uniform level of health. In retrospect, despite the Scandinavian evidence that health inequalities can be largely eliminated (see Macintyre, Chapter 2), the avoidability of *all* biological inequalities must inevitably be idealistic, and their possible interaction – as that between, for example, IQ, occupational status and chronic disease in adulthood – suggests that there will always be some association between social and biological inequality.

The problem of linking 'facts' to 'values' arises particularly clearly in the case of eugenicist policies. In their contribution, Bittles and Chew (Chapter 15) review the origins of the eugenics movement since the time of Galton in the nineteenth century, and proceed to describe the current aims of policies which have been established recently in Singapore and China. These policies were initially linked to anti-natalist programmes, as with China's One-child policy, which were intended to restrict the impact of population growth. In Singapore, a eugenically selective pro-natalist line has been adopted since the mid-1980s, and in China an explicitly eugenicist Law on Maternal and Infant Health Care was enacted in 1995. In practice, the efficacy of these policies remains to be determined. Evidence from male-biased secondary sex ratios suggests that the values of the Chinese legislature have not inevitably been shared by the populace, and that there has persisted a Confucian ethic favouring sons over daughters. Concern about eugenics, which is today widely felt in the West, reflects in part the historical experience of Europe and America since the inter-war period, and the worries of medical geneticists that innovative techniques of diagnosis will be open to ideologically-driven misuse. It arguably also testifies to varying attitudes to authoritarian rule, for these have since classical times characterized divergent philosophical arguments over human nature and inequality in East and West (Munro, 1969; Lloyd, 1996).

Within a wider global context, Stephens examines health inequalities and health policies in the developing world. She highlights the economic and environmental relations between the countries of the North and the South, stressing the need for an understanding of international social processes. Development theory has tended in the past to focus on ideas of 'absolute poverty', of which the symptoms are construed as physical deprivations of water, food, shelter, sanitation, diseases and prophylaxis ('basic needs'). There has been much debate over the relative efficacy of 'trickle-down' versus redistributive economic policies in resolving these deficiencies (see also Townsend *et al.*, 1988). By contrast with this focus, however, equality of health has been a development goal (therefore presuming health *equity*) since the Alma Ata declaration of 1979, which was endorsed again in 1984. Progress towards this has been weakened by structural adjustment policies applied to the management of international debt since the 1980s. Case-study material on urban health problems illustrates the importance of this rapidly expanding sector of the world's population. Urban populations have been hitherto relatively unrepresented in health data compared to rural dwellers. Yet urban health inequalities tend to be more marked – they tend to show the characteristics of an 'epidemiological transition'; they present signs of psycho-social stress resulting in violence, and they are fertile grounds for communicable diseases and for diseases of affluence. In sympathy with Porter and Ogden, and with Wilkinson, therefore, Stephens argues that policy attention needs to focus to a greater extent on health inequalities *per se*, rather than on a 'basic needs' approach linked to simply defined criteria of 'poverty'.

In the final contribution in this volume, Ben-Shlomo and Marmot examine policy options for health inequalities in industrial and post-industrial societies. The evidence indicates that inequalities in health are not inevitable, they can sometimes be attenuated, if not eliminated completely. However, explicit and empirically determined criteria are needed to assess the potential role of policy in achieving this end. Such criteria are essentially rules which link 'facts' (for example, about the importance of an area in public health, about morbidity inequalities or about aetiology) with 'values' (about the significance attached to inequality or particular types of morbidity). Measures designed to influence smoking behaviour, alcohol consumption, living conditions and the working environment show these rules at work and the extent to which they can be effective. Occupation-related morbidity is particularly important in industrial and post-industrial societies and it is linked to lack of control over work, imbalance between effort and reward, 'job insecurity' or 'labour market flexibility', the concept of the effective community and the significance of social cohesion. Thus the potential significance of social policies seems to outweigh that of medical interventions in management of inequalities in health.

Conclusion

This introductory chapter has sought to convey three ideas that have been intertwined in the history of European thought, and which surface repeatedly and to varying degrees of clarity in the contributions in this volume. These were: the separateness of the biological from the social domains of inquiry; the concept of natural human inequality; and the distinction between facts and values. If such a volume can be said to support a single conclusion, it must surely be that these ideas remain as central to the interpretation of human biology and concepts of health at the end of the twentieth century as they were to the philosophical arguments of the classical world. Indeed, the general idea of intrinsic human inequalities finds expression in a far broader mythological context, illustrated by the mystical 'Vedic Hymn to the Primaeval Man, Prajâpati', from whose body derived the four distinct classes of Indian society (Basham, 1954; Macdonald, 1975):

> *When the gods made a sacrifice*
> *With the Man as their victim,*
> *Spring was the melted butter, Summer the fuel,*
> *And Autumn the oblation.*

> *From that all-embracing sacrifice*
> *The clotted butter was collected.*
> *From it he made the animals*
> *Of air and wood and village.*

> *From that all-embracing sacrifice*
> *Were born the hymns and chants,*
> *From that the metres were born,*
> *From that the sacrificial spells were born.*

> *Thence were born horses,*
> *And all beings with two rows of teeth.*
> *Thence were born cattle,*
> *And thence goats and sheep.*

18   S. Strickland and P. Shetty

When they divided the Man,
Into how many parts did they divide him?
What was his mouth, what were his arms,
What were his thighs and his feet called?

The brahman was his mouth,
Of his arms was made the warrior,
His thighs became the vaisya,
Of his feet the sudra was born.

The moon arose from his mind,
From his eye was born the sun,
From his mouth Indra and Agni,
From his breath the wind was born.

From his navel came the air,
From his head there came the sky,
From his feet the earth, the four quarters from his ear,
Thus they fashioned the worlds.

With Sacrifice the gods sacrificed to Sacrifice –
These were the first of the sacred laws.
These mighty beings reached the sky,
Where are the eternal spirits, the gods.

References

Aristotle. *Politics.*Translated by E. Barker (1948). Oxford: Clarendon Press.
Barker, E. (1948). *See* Aristotle.
Basham, A. L. (1954). *The Wonder That Was India*. London: Sidgwick and Jackson.
Boas, F. (1938). *The Mind of Primitive Man*. New York: Macmillan Press.
Dahrendorf, R. (1969). On the origin of inequality among men. In *Social inequality*, ed. A. Beteille, pp. 16–44. Harmondsworth: Penguin Books.
Darwin, C. (1888). *The Descent of Man, and Selection in Relation to Sex*, 2nd edn. London: John Murray.
Dasgupta, P. (1993). *An Inquiry into Well-Being and Destitution*. Oxford: Clarendon Press.
Egolf B., Lasker J., Wolf S. & Potvin L. (1992) The Roseto effect: A 50-year comparison of mortality rates. *American Journal of Public Health* **82**, 1089–92.
Golden, M. H. N. (1994). Is complete catch-up growth possible for stunted malnourished children? *European Journal of Clinical Nutrition* **48** (Suppl 1), S58–S71.
Haldane, J. B. S. (1932). *The Inequality of Man and Other Essays*. London: Chatto

and Windus.

Illsley, R. (1955). Social class selection and class differences in relation to stillbirths and infant deaths. *British Medical Journal* **24**, 1520–4.

Illsley, R. (1986). Occupational class, selection and the production of inequalities in health. *Quarterly Journal of Social Affairs* **2**, 151–65.

Lloyd, G. E. R. (1983). *Science, Folklore and Ideology. Studies in the Life Sciences in Ancient Greece.* Cambridge: Cambridge University Press.

Lloyd, G. E. R. (1991). *Methods and Problems in Greek Science. Selected Papers.* Cambridge: Cambridge University Press.

Lloyd, G. E. R. (1996). *Adversaries and Authorities. Investigations into Ancient Greek and Chinese Science.* Cambridge: Cambridge University Press.

Macdonald, A. W. (1975). On Prajâpati. In *Essays on the Ethnology of Nepal and South Asia*, pp. 1–13. Kathmandu: Ratna Pustak Bhandar.

Maynard Smith, J., & Szathmáry, E. (1995). *The Major Transitions in Evolution.* Oxford: W.H. Freeman.

Munro, D. J. (1969). *The Concept of Man in Early China.* Stanford: Stanford University Press.

Neel, J. V. (1962). Diabetes mellitus: a 'thrifty' genotype rendered detrimental by 'progress'. *American Journal of Human Genetics* **14**, 353–62.

Phillimore, P., Beattie, A. & Townsend, P. (1994). Widening inequality of health in northern England, 1981–91. *British Medical Journal* **308**, 1125–8.

Plato. *Republic.* Translated by F.M. Cornford (1941). Oxford: Clarendon Press.

Popper, K. R. (1962). *The Open Society and its Enemies.* Vol. I *Plato.* London: Routledge.

Power, C., Fogelman, K., & Fox, A. J. (1986). Health, social mobility during early years of life. *Quarterly Journal of Social Affairs* **2**, 397–413.

Power, C. (1994). Health and social inequality in Europe. *British Medical Journal* **308**, 1153–6.

Strickland, S. S. & Tuffrey, V. R. (1997). *Form and Function.* London: Smith-Gordon.

Townsend, P., Davidson, N., & Whitehead, M. (1988) *Inequalities in Health.* Harmondsworth: Penguin Books.

Wadsworth, M. E. J. (1986). Serious illness in childhood and its association with later-life achievement. In *Class and Health*, ed. R. G. Wilkinson, pp. 50–74. London: Tavistock Publications.

Waterlow, J. C. (1990). Nutritional adaptation in man: General introduction and concepts. *American Journal of Clinical Nutrition* **51**, 259–63.

Whitehead, M. (1988). The health divide. In *Inequalities in Health*, ed. P. Townsend, N. Davidson & M. Whitehead, pp. 221–381. Harmondsworth: Penguin Books.

Wilkinson, R. G. (Ed.). (1986a). *Class and Health: Research and Longitudinal Data.* London: Tavistock Publications.

Wilkinson, R. G. (1986b). Occupational class, selection and inequalities in health: A reply to Raymond Illsley. *Quarterly Journal of Social Affairs* **2**: 415–22.

Wilkinson, R. G. (1994). Divided we fall: The poor pay the price of increased social inequality with their health. *British Medical Journal* **308**, 1113–14.

2 Social inequalities and health in the contemporary world: comparative overview

SALLY MACINTYRE

Introduction

Socio-economic inequalities in a large number of indicators of health (e.g. age at death, physique, physical functioning, disability, mental health, and both chronic and acute physical ill health) have been systematically observed in the UK for the last 150 years. Concern with social class differences in health is sometimes perceived to be a particularly British phenomenon, but socio-economic gradients in health have been documented in a wide range of countries including both market and non-market industrialized economies, societies with both well and less well developed welfare provision, and societies in various stages of development. It is important to stress the ubiquity (over both time and space) of the observed pattern of systematically poorer health and a shorter life span being associated with each successively lower position in any given system of social stratification whether that is measured by occupational social class or by other indicators such as prestige, education, or access to material resources. This contribution illustrates this point with data from a range of countries, using a variety of measures both of health and social position. It presents a comparative overview of patterns of health in the contemporary world and their relationship to measures of social class, structured around eight general observations about social inequalities and health.

Overview

Firstly, socio-economic differentials in longevity have been observed since reliable record keeping started, which in Britain and many other countries was in the first half of the nineteenth century. In his monumental report on sanitary conditions among the labouring poor, Chadwick collated data from many different areas of Britain (Chadwick, 1842). Table 2.1 shows

20

Table 2.1. *Longevity of families belonging to various classes: average age at death*

District or town	Gentry and Professional	Farmers and tradesmen	Labourers and artisans
Rutland	52	41	38
Bath	55	37	25
Leeds	44	27	19
Bethnal Green	45	26	16
Manchester	38	20	17
Liverpool	35	22	15

Based on Chadwick, 1842; cited in Wohl, 1983, p. 5.

the average length of life for members of families in three different social strata – gentry and professional; farmers and tradesmen; labourers and artisans – in various towns and counties in England, demonstrating successively shorter life expectancy as one moves down the social scale. It also shows that although the gradient in length of life by social stratum was apparent in all these localities, it was set at different levels in different places. For instance, in Bath, a relatively prosperous small town, it was set at a higher level than in the industrial cities of Manchester and Liverpool (Wohl, 1983).

Indeed, in mid-nineteenth century Britain, the close relationship between ill health and poverty was a commonplace observation. Elsewhere in Chadwick's report he notes that:

> To obtain the means of judging of the references to the localities in the sanitary returns from Aberdeen, the reporters were requested to mark on a map the places where the disease fell, and to distinguish with a deeper tint those places on which it fell with the greatest intensity. They were also requested to distinguish by different colours the streets inhabited by the higher, middle and lower classes of society. They returned a map so marked as to disease, but stated that it had been thought unnecessary to distinguish the streets inhabited by the different orders of society, as that was done with sufficient accuracy by the different tints representing the degrees of the prevalence of fever.
>
> *(Chadwick, 1842; quoted in Flinn, 1965)*

Variation in death rates between the poorer and richer areas of Paris was observed by Villermé around the same period (Flinn, 1965). Table 2.2 shows districts in Paris ranked by the proportion of properties exempt from taxation because of poverty, and demonstrates that death rates followed a roughly similar rank order, being lower in the richer areas and higher in the poorer areas.

Table 2.2. *Mortality rates from 1817 to 1821 in arrondissements in Paris ranked by proportion of properties exempt from taxation*

Arrondissement	Proportion exempt	Deaths at home 1 in:
Montmartre	0.07	62
Chausse d'Antin	0.11	60
Roule, Tuileries	0.11	58
Luxembourg	0.15	51
Pt. St. Denis	0.19	54
Faubourg St. Denis	0.22	53
St. Avoie	0.22	52
Monnaie, Invalides	0.23	50
Isle St. Louis	0.31	44
Ste Antoine	0.32	43
Jardin du Roi	0.38	43

Based on Villermé; cited in Flinn, 1965, p. 237.

Although social inequalities in health have been documented in Europe for at least 150 years, public and scholarly interest in them has been episodic and cyclical rather than constant – every few years they tend to be 'rediscovered' and treated as though they were a new observation (Blaxter, 1981). The early 1990s saw the 'rediscovery' of inequalities in the UK, the USA and several European countries, but this recent flurry of interest should not lead us to ignore the substantial body of work on the topic over the last 150 years (Macintyre, 1997).

Secondly, although it is often suggested that interest in inequalities in health is a particularly British obsession arising from the class stratification of British society, it is clear that socio-economic differentials exist in all the industrialized societies for which data are available. The three most commonly used indicators of socio-economic position are occupational social class, achieved educational level, and income. Occupational social class as an indicator tends only to be available in the UK and Western Europe, therefore education is the most commonly used measure for international comparisons because it is available in most countries (Berkman and Macintyre, 1997). Table 2.3 shows mortality rates among men in 1986 in the USA (Pappas *et al.* 1993). The death rates decreased with increasing levels of education and with increasing income for both whites and blacks at most education and income levels, however, the rates for blacks was higher than those for whites. Table 2.4 gives the pattern for white and black women, showing relationships similar to those among men between education and income on the one hand and mortality rates on the other, with the rates being higher among blacks in most subgroups (Pappas *et al.* 1993).

Table 2.3. *Death rates in 1986 among 25 to 64-year-old men by education and income by race, USA (age adjusted death rates per 1,000)*

Education (years completed)	White	Black
School		
0–11	7.6	13.4
12	4.3	8.0
College:		
1–3	4.3	5.0
≥ 4	2.8	6.0
Income ($):		
< 9,000	16.0	19.5
9,000–14,999	10.2	10.8
15,000–18,999	5.7	9.8
19,000–24,999	4.6	4.7
≥ 25,000	2.4	3.6

Based on Pappas *et al.* 1993.

Table 2.4. *Death rates in 1986 among 25 to 64-year-old women by education and income by race, USA (age adjusted death rates per 1,000)*

Education (years completed)	White	Black
School		
0–11	3.4	6.2
12	2.5	3.9
College:		
1–3	2.1	3.2
≥ 4	1.8	2.2
Income ($):		
< 9,000	6.5	7.6
9,000–14,999	3.4	4.5
15,000–18,999	3.3	3.7
19,000–24,999	3.0	2.8
≥ 25,000	1.6	2.3

Based on Pappas *et al.* 1993.

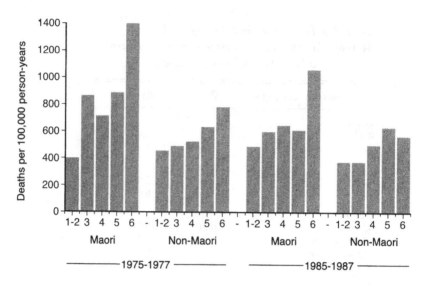

Figure 2.1. New Zealand Male Age Standardized Mortality Rates per 100,000 person years by ethnicity and class. (*Source:* Pearce *et al.* 1993. Reproduced by permission of the authors.)

In New Zealand an occupational social class scale (the Elley-Irving scale), similar to the British one, is used (Pearce *et al.* 1985, 1993). Figure 2.1 shows higher mortality among men in lower occupational social classes (e.g. 5 and 6) than higher social classes (e.g. 1 and 2) both in the mid-1970s and mid-1980s; it also shows a less consistent social class gradient for Maori than for non-Maori men, but higher mortality among Maoris than non-Maoris.

Although it has been difficult to obtain data on socio-economic differentials in health from Eastern Europe countries formerly in the communist bloc, it is clear that before the fall of communism in the early 1990s, there were clear socio-economic gradients in mortality and morbidity in many of these countries. Table 2.5, for example, shows infant mortality rates in Hungary by mothers' level of education. It shows the familiar pattern – those with higher education having only one-third the rate of infant mortality as those with the least education (Orosz, 1990).

Thirdly, inequalities tend to be observed in almost all measures of health and using all measures of socio-economic position. Measures of inequality of health commonly included in the analyses of social class can be divided into those based on death, on self-reported health and on growth and functioning (see Table 2.6). Nearly all of these show socio-economic gradients in health in the expected direction, although the magnitude of socio-economic differentials vary according to the measure used.

Table 2.5. *Infant mortality per 1000 live births by mother's level of education, 1987, Hungary*

Number of grades completed	Rate
0–7	36.1
8	19.0
9–12	14.4
13+	10.3
average	17.3

Based on Orosz, 1990

Table 2.6. *Health Measures*

Death-based
All cause age/sex standardized mortality rates/ratios
Specific cause age/sex standardized mortality rates/ratios
Years of potential life lost
Disability free life years
Infant/perinatal mortality

Self-reported health
Self-perceived general health
Self-reported longterm illness
Bed days/restricted activity days
Symptoms

Growth and function
Height
Birth weight
Blood pressure
Respiratory function

Different causes of death have different social class gradients. Table 2.7 gives data from Hungary which shows rate ratios for major causes of death by years of education in relation to a reference group of those with the highest education (Orosz, 1990). The biggest difference is for infectious and parasitic diseases and respiratory disease, while there is little difference in cancers or circulatory diseases. Certain cancers, for example breast and some blood cancers, tend to have no social class gradient or an inverse social class gradient (Davey-Smith *et al.* 1991).

Measures of health *status* (a relatively long-term property of the individual) tend to show more marked social class differences than measures of health *state* (relatively short-term properties of the individual such as current symptoms) (Blaxter, 1989). Whether mortality or morbidity

Table 2.7. *Rate ratios of age standardized specific cause mortality rates: those with the highest educational qualification; males, Hungary 1980*

Cause of death	Years of education				
	0–5	6–7	8	9–12	13+
Infections and parasitic	5.13	2.54	1.58	1.44	1.00
Neoplasm	1.49	1.08	0.98	1.06	1.00
Circulatory	1.83	1.30	1.08	1.21	1.00
Respiratory	6.33	2.26	1.51	1.32	1.00
Digestive	1.65	1.49	1.32	1.15	1.00
Accident, poisoning and violence	3.60	2.01	1.64	1.38	1.00

Based on Orosz, 1990.

measures produce steeper gradients by socio-economic status varies with the age group being considered. Self-reports of health may be confounded by different thresholds for recognizing and reporting illness among different social class groups; this has led to some stress on the importance of direct measures, such as height, which cannot be affected by self-reporting biases.

Measures such as life expectancy free of disability (Graham and Davis, 1990), mortality amenable to medical treatment (Marshall *et al.*, 1993) and years of potential life lost (Blane *et al.*, 1990) have shown that lower socio-economic status groups not only die earlier but are more disabled while they are alive. Diseases amenable to medical treatment and years of potential life lost produce steeper social class gradients than standardized mortality rates (SMRs: these present rates of death for a particular group as a ratio of rates of death for all groups). Thus, while the relationship between social class and health or death is nearly always in the same direction, its magnitude varies with the measures of health and of class used.

Fourthly, the magnitude of socio-economic differentials varies between countries. There have been a number of attempts to compare their magnitude systematically, mainly using occupational class or education as the basis for cross-country comparisons (Valkonen, 1989, 1993; Kunst *et al.*, 1992; Kunst and Mackenbach, 1994). Table 2.8 compares an inequality index (the proportional mortality increase as one moves from the top to the bottom of an occupational hierarchy) across six European countries (Kunst and Mackenbach, 1994). In the 1970s, mortality differentials were the smallest in Norway followed by Denmark. Compared to Norway, inequalities were about 1.5 times greater in Sweden, twice as great in England and Wales, five times as great in Finland and six to eight times as

Table 2.8. *Inequality index* for males 35–64 years, ca. 1970–80,* in six countries

	Basic estimate	Manual v non-manual occupations
Norway	0.18	0.21
Denmark	0.23	0.25
Sweden	0.29	0.30
England and Wales	0.40	0.36
Finland	0.95	0.96
France 1976–80	1.66	1.14
1980–89	`1.64	1.21

* Proportional mortality increase as one moves from the top to the bottom of the occupational hierarchy.
Based on Kunst & Mackenbach 1992.

great in France. Although it is sometimes assumed that England and Wales have relatively high levels of socio-economic inequalities in health compared to other industrialized countries, it appears that they occupy the middle range for Europe. To check for the effect of different occupational scales the authors also made a simpler comparison between manual and non-manual occupations and found a very similar pattern and rank order (Kunst and Mackenbach, 1994).

A similar measure was recently used to compare inequalities in health by educational level in six European countries and the USA (Elo and Preston, 1996). Table 2.9 shows that the USA and England and Wales had similar inequality co-efficients for both men and women – for men these were medium to low compared to other countries, while for women they were high (Elo and Preston 1996).

The pattern for a range of self-reported health measures differs for males and females (Kunst *et al.*, 1992). For males, inequalities were found to be smallest in Norway, Sweden, UK and Spain; middling in Netherlands, Denmark, Finland and Japan; and large in Germany, Italy, Canada and the USA. For females there was less consistency among the health measures used (perceived general health, health complaints, short-term and long-term disability, chronic conditions and height), but the largest inequalities were found in Italy, Spain, Canada and the USA (Kunst *et al.*, 1992).

Fifthly, the main causes contributing to inequalities in health vary between countries. Table 2.7 showed big gradients in infections and parasitic disease, but not in circulatory diseases, in Hungary (Orosz, 1990). In the 1980s it was found that the causes responsible for excess deaths among unskilled compared to other men were mainly: accidents in Denmark,

Table 2.9. *Inequality co-efficients
(percentage reduction or mortality per one
year increase in education) by sex and
country, 35–54 years old, ca. 1970–1980*

	Males	Females
Denmark	8.1	3.8
Finland	9.2	6.1
England & Wales	7.4	7.8
Hungary	8.2	2.2
Norway	8.6	5.8
Sweden	8.0	4.8
USA (1979–85)		

Based on Elo and Preston 1996.

Norway and Finland; cardiovascular, cancer and respiratory diseases in
England and Wales; and cancer, accidents and cirrhosis in France (Leclerc
et al., 1990).

Sixthly, as shown in Tables 2.1–2.9, it seems that health and longevity
tend to have a stepwise, not a threshold, relationship to socio-economic
status. That is, health or length of life improves with each successive
improvement in socio-economic status, rather than everyone below a
threshold of poverty being equally at risk and everyone above it being
equally healthy and long lived (Macintyre, 1994). This tends to apply
whether one measures the gradient by income, occupational class or type
of area of residence. Figure 2.2, for example, shows SMRs at age 40–64 for
postcodes of residence in Scotland at the 1991 census, the postcodes being
classified into seven categories of deprivation using the Carstairs Morris
index based on census variables (Carstairs and Morris, 1991). There is a
steady increase in SMRs with increasing scores on this area deprivation
index (Mcloone, 1994).

Similar observations have been reported from the USA. Among both
white (see Table 2.10) and black men screened for inclusion in the multiple
risk factor intervention trial, the risk of subsequent death (adjusted for risk
factors such as smoking) rose steadily from the highest income neighbourhoods to the lowest (Davey-Smith *et al.*, 1996a,b). The occupational class
gradient for white men in New Zealand shows a similar linear pattern (see
Figure 2.1). This graded increase in health and longevity with increasing
socio-economic advantage has been seen by some as somewhat of a puzzle,
since it seems plausible that there might be threats to health from absolute
poverty but less plausible that each successive improvement in living

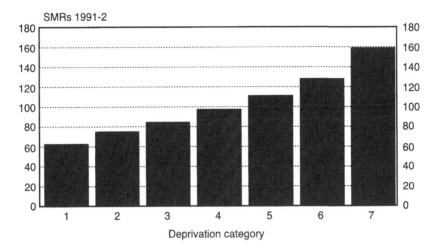

Figure 2.2. Standardized mortality rates for (SMRs) ages 40–64. Age and sex standardized mortality ratios within each deprivation category. All Scotland = 100. (Based on McLoone, 1994.)

standards is health enhancing (Adler *et al.*, 1994; Macintyre, 1994). However, this linear pattern is consistent with general secular trends in health and life expectancy, which show continuing improvement at a population level as living standards rise. It is also consistent with a model of the generation of socio-economic differentials in health being the outcome of a cumulation of material and social advantages and disadvantages over the life course (Davey-Smith *et al.*, 1994).

Seventhly, there is evidence that socio-economic differentials within countries have increased over the last decade or two, despite continuing declines in overall mortality. In Scotland there is evidence of increased differentials between postcode sectors ranked by deprivation between the 1981 and 1991 censuses (McLoone and Boddy, 1994); the same has been shown within Glasgow (McCarron *et al.*, 1994) and in the North of England (Phillimore *et al.*, 1994). In Finland socio-economic differentials increased between 1971 and 1985 among middle-aged and elderly men. Differentials among women were smaller than those among men and did not increase over this period (Valkonen, 1993). Class differences in New Zealand among men increased slightly between the mid-1970s and mid-1980s despite an overall mortality decrease of 15% (Pearce *et al.*, 1993). In the USA, disparities in death rates by income and education increased between 1960 and 1980 in all sex and race groups (although the increase was less for women than for men) (Pappas *et al.*, 1993). These increasing disparities seem mainly due to faster declines in mortality rates among high

Table 2.10. *All cause mortality by level of median family income by zip code of residence: white men screened for MRFIT (305,099 men followed-up for 12 years)*

Income ($)	Age and risk factor adjusted relative risk
< 7,500	1.79
7,500– 9,999	1.50
10,000–12,499	1.56
12,500–14,999	1.49
15,000–17,499	1.44
17,500–19,999	1.29
20,000–22,499	1.29
22,500–24,999	1.20
25,000–27,499	1.19
27,500–29,999	1.12
30,000–32,499	1.11
≥ 32,500	1.00
	(reference group)

MRFIT: multiple risk factor intervention trial.
Adapted from Davey-Smith *et al.*, 1992.

socio-economic status groups. However, in the UK there is evidence of absolute rise in mortality among some poorer groups, both in the 1971–81 and 1981–91 periods. In Scotland, death rates among 20 to 29-year-olds in the more deprived areas increased between the censuses (McLoone and Boddy 1994), and in the North of England mortality rates increased among young men in the poorest areas over the same period (Phillimore *et al.*, 1994). Continued or increasing social differentials in mortality in a period during which mortality rates have steadily declined, raise intriguing questions about the relationship between the processes producing social inequalities in health and the processes producing increasing life expectancy. The co-existence of these two trends suggests that different processes may be at work and that human health and life expectancy are extremely sensitive to social conditions.

Eighthly, and relatedly, it appears that strong and consistent linear socio-economic gradients in health are found more frequently for men than for women, and for ethnic majority populations than for ethnic minorities. There is less consistency in patterns of socio-economic inequality in health, both between countries and between health indicators, for women than for men. Several studies show linear gradients for men but slight deviations from linearity among women. The Whitehall Study for instance, found fewer consistent differences in morbidity by grade among women than among men (Marmot *et al.*, 1991). Other studies of women have found higher rates of poor health in the top socio-economic category compared to the second top category. For example, by 1992 perinatal mortality rates in the UK were slightly higher in women in the highest social class groups than those in the next highest (Office of Population Censuses and Surveys, 1995), and in Finland in the mid-1980s, rates of stillbirth, perinatal deaths, and preterm birth were lowest in the second highest socio-economic group, not the highest; a finding that persisted after adjusting for age, parity, urbanization, marital status and time of first antenatal visit (Hemminki *et al.*, 1990, 1992). This suggests that women are less sensitive than men to socio-economic conditions, that they are exposed to different aspects of socio-economic position or that our measures of social stratification are less appropriate for women than for men (Koskinen and Martelein, 1994).

Similarly, socio-economic gradients in health are less consistently found among ethnic minority populations in many countries, suggesting that the socio-economic measures may have different meanings and consequences for them. Class gradients are less apparent or reversed for ethnic minority groups such as South Asians in the UK, and residential segregation and discrimination seems to make income or educational scales less appropriate as predictors of mortality and morbidity for blacks than for whites in the USA (Pappas *et al.*, 1993). Marmot *et al.* (1984) showed varying social class gradients for country of birth in the UK in the early 1970s, with steep and linear gradients for those born in Ireland, flat or inconsistent gradients for those born in the Indian subcontinent and Europe, and reverse or inconsistent gradients for those born in the Caribbean or African commonwealth (see Figure 2.3). Again, this suggests that our occupational social class measures may not accurately measure exposures to socioeconomic conditions in such groups, and/or that complicating selective migration and other processes are at work.

Conclusions

This broad overview of social inequalities in health and longevity suggests that:

(1) Socio-economic inequalities in health have been observed since statistics have been available.
(2) They are observable in all industrialized countries.
(3) They are observable in most measures of health and longevity, and using most measures of socio-economic status.
(4) Their magnitude varies between countries.
(5) Main causes contributing to inequalities vary between countries.

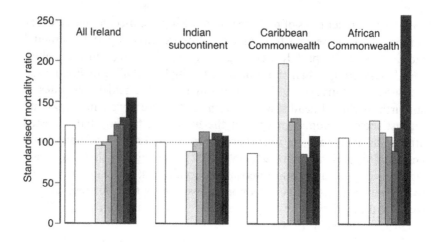

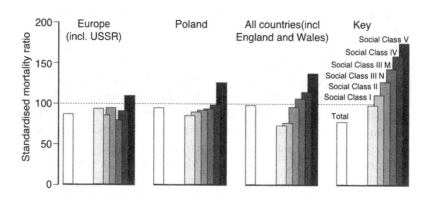

Figure 2.3. Standardized mortality ratios for male immigration to England and Wales aged 15–64 by country of birth and social class, 1970–72. (Source: Immigrant Mortality in England and Wales 1970–78 (1984), Crown Copyright 1984. Reproduced by permission of H.M.S.O. and of the Office for National Statistics.)

(6) Health and longevity tend to have a stepwise, not threshold, relationship with socio-economic status.

(7) Socio-economic differentials in mortality have tended to increase in recent decades.

(8) Socio-economic differentials in health are less strong and consistent for women and for minority populations.

Although none of these points are unknown or controversial within those areas of medical sociology or social epidemiology concerned with socio-economic inequalities in health, they may be less well known, and if not controversial then at least surprising, to people from other disciplines, to lay people, and to policy makers. Social inequalities in health tend to be rediscovered periodically in the academic and policy worlds, and have indeed been the subject of increasing interest in the mid-1990s in many countries, including the USA where they were formerly not given much attention (Macintyre, 1997). However, they seem to be an ubiquitous feature of industrialized societies, and show clearly the importance of interactions between social and biological factors in determining human health and life expectancy.

References

Adler, N., Boyce, T., Chesney, M., Cohen, S., Folkman, S., Kahn, R. L. & Syme, S. L. (1994). Socioeconomic status and health; the challenge of the gradient. *American Psychologist* **49**, 15–24.

Berkman, L. & Macintyre, S. (1997). The measurement of social class in health studies; old measures and new formulations. In *Social Inequalities and Cancer*, ed. M. Kogevinas, N. Pearce, M. Susser & P. Boffeta. I.A.R.C. Scientific Publications No. 138. Lyon: International Agency for Research on Cancer.

Blane, D., Davey-Smith, G. & Bartley, M. (1990). Social class differences in years of life lost; size, trends, and principal causes. *British Medical Journal* **301**, 429–32.

Blaxter, M. (1981). *The health of the children; a review of research on the place of health in cycles of disadvantage*, vol. 3. *Studies in Deprivation and disadvantage*. London: Heinemann Educational Books.

Blaxter, M. (1989). A comparison of measures of inequality in morbidity. In *Health inequalities in European countries*, ed. J. Fox, pp. 199–230. Aldershot: Gower.

Carstairs, V. & Morris, R. (1991). *Deprivation and Health in Scotland*. Aberdeen: Aberdeen University Press

Chadwick, E. (1842). *Report of an Enquiry into the Sanitary Conditions of the Labouring Population of Great Britain*. London: Poor Law Commission.

Davey-Smith, G., Leon, D., Shipley, M. & Rose, G. (1991). Socioeconomic differentials in cancer among men. *International Journal of Epidemiology* **30**, 39–45.

Davey-Smith, G., Neaton, J. D., Samler, L. & Wentworth, D. (1992). Mortality differentials by income during 12-year follow-up of 305,099 middle-aged men in the MRFIT screening study. *BSA/ESMS Conference, September 1992*.

Davey-Smith, G., Blane, D. & Bartley, M. (1994). Explanations for socioeconomic differentials in mortality: evidence from Britain and elsewhere. *European Journal of Public Health* **4**, 131–44.

Davey-Smith, G., Neaton, J., Wentworth, D., Stamler, R. & Stamler, J. (1996a). Socioeconomic differential in mortality risk among men screened for the multiple risk facor intervention trial. I. White men. *American Journal of Public Health* **86**, 486–96.

Davey-Smith, G., Wentworth, D., Neaton, J., Stamler, R. & Stamler, J. (1996b). Socioeconomic differentials in mortality risk among men screened for the multiple risk facot intervention trial. II. Black men. *American Journal of Public Health* **86**, 497–504.

Elo, I. & Preston, S. (1996). Educational differences in mortality: United States, 1979–85. *Social Science and Medicine* **42**, 47–57.

Flinn, M. W. (Ed.). (1965). *Report on the Sanitary Conditions of the Labouring Population of Great Britain by Edwin Chadwick*. Edinburgh: Edinburgh University Press.

Graham, P. & Davis, P. (1990). Life expectancy free of disability: a composite measure of population health. *Community Health Studies* **14**, 138–45.

Hemminki, E., Malin, M. & Rahkonen, O. (1990). Mother's social class and perinatal problems in a low problem area. *International Journal of Epidemiology* **19**, 983–90.

Hemminki, E., Merilainen, J., Malin, M., Rahkonen, O. & Teperi, J. (1992). Mother's education and perinatal problems in Finland. *International Journal of Epidemiology* **21**, 720–4.

Koskinen, S. & Martelein, T. (1994). Why are socioeconomic mortality differences smaller among women than among men? *Social Science and Medicine* **38**, 1385–90.

Kunst, A. & Mackenbach, J. (1994). International variations in the size of mortality differences associated with occupational status. *International Journal of Epidemiology* **23**, 742–50.

Kunst, A., Geurts, J. & van den Berg, J. (1992). International variation in socio-economic inequalities in self-reported health; a comparison of the Netherlands with other industrialised countries. Voorburg/Heerlen: Netherlands Central Bureau of Statistics.

Leclerc, A., Lert, F. & Fabien, C. (1990). Differential mortality: some comparisons between England and Wales, Finland and France based on inequalities measures. *International Journal of Epidemiology* **19**, 1001–10.

Macintyre, S. (1994). Understanding the social patterning of health: the role of the social sciences. *Journal of Public Health Medicine* **16**, 53–9.

Macintyre, S. (1997). The Black report and beyond; what are the issues? *Social Science and Medicine* **44**(6), 723–46.

Marmot, M. G., Adelstein, A. & Bulusu, L. (1984). *Immigrant Mortality in England and Wales 1970–78: Causes of Death by Country*. Office of Population Census and Surveys, Studies on Medical and Populations Subjects, no. 47. London: HMSO.

Marmot, M. G., Davey-Smith, G., Stansfield, S., Patel, C., North, F., Head, J., White, I., Brunner, E. & Feeney, A. (1991). Health inequalities among British civil servants: the Whitehall II study. *Lancet* June 8th, pp. 1387–93.

Marshall, S. W., Kawachi, I., Pearce, N. & Borman, B. (1993). Social class differences in mortality from diseases amenable to medical intervention in New Zealand. *International Journal of Epidemiology* **22**, 255–61.

McCarron, P., Davey-Smith, G. & Womersley, J. (1994). Deprivation and mortality in Glasgow: changes from 1980 to 1992. *British Medical Journal* **309**, 1481–2.

McLoone, P. (1994). *Carstairs Scores for Scottish Postcode Sectors for the 1991 Census*. Glasgow: Public Health research Unit, University of Glasgow.

McLoone, P. & Boddy, F. A. (1994). Deprivation and mortality in Scotland, 1981 and 1991. *British Medical Journal* **309**, 1465–70.

Office of Population Censuses and Surveys (1995). *Mortality Statistics; Perinatal and Infant: Social and Biological Factors*, Series DH3 no. 26 ed. London: HMSO.

Orosz, E. (1990). The Hungarian country profile: inequalities in health and health care in Hungary. *Social Science and Medicine* **31**, 847–57.

Pappas, G., Queen, S., Hadden, W. & Fisher, G. (1993). The increased disparity in mortality between socioeconomic groups in the United States 1960 and 1986. *New England Journal of Medicine* **329**, 103–9.

Pearce, N., Davis, P. B., Smith, A. H. & Foster, F. H. (1985). Social class, ethnic group and male mortality in New Zealand 1974–8. *Journal of Epidemiology and Community Health* **39**, 9–14.

Pearce, N., Pomare, E., Marshall, S. & Borman, B. (1993). Mortality and social class in Maori and non-Maori New zealand men: changes between 1975–77 and 1985–87. *New Zealand Medical Journal* **106**, 193–6.

Phillimore, P., Beattie, A. & Townsend, P. (1994). Widening inequality of health in Northern England 1981–91. *British Medical Journal* **308**, 1125–8.

Valkonen, T. (1989). Adult mortality and level of education: a comparison of six countries. In *Health Inequalities in European Countries*, ed. J. Fox, pp. 142–62. Aldershot: Gower.

Valkonen, T. (1993). Problems in the measurement and international comparisons of socio-economic differences in mortality. *Social Science and Medicine* **36**, 409–18.

Wohl, A. S. (1983). *Endangered Lives: Public Health in Victorian Britain*. London: J M Dent.

3 Accumulation of health risks across social groups in a national longitudinal study

CHRIS POWER AND SHARON MATTHEWS

Introduction

The existence of social inequalities in health is a well established phenomenon. That is, people in relatively low social positions tend to have more morbidity and disability and have a higher rate of premature mortality than those in higher social positions (Macintyre, Chapter 2; Davey-Smith et al., 1990). Even though health inequalities are pervasive, their magnitude varies over time and by place (Kunst and Mackenbach, 1994). Britain in the 1980s witnessed widening inequality (Davey-Smith et al., 1990) evidenced by differences in all cause mortality in Northern England (Phillimore et al., 1994) and Scotland (McLoone and Boddy, 1994) and for childhood injury deaths (Roberts and Power, 1996). The causes of these inequalities are widely debated, particularly since the publication of the Black report in the early 1980s (Townsend and Davidson, 1992). The Black report (Department of Health and Social Security, 1980) put forward alternative explanatory causes for health inequalities, namely artefact, selection, lifestyle and material circumstances. Other explanations have been suggested, including the distribution, accessibility and quality of medical care (Mackenbach et al., 1989; Marmot et al., 1995), factors in early life (Barker and Osmond, 1992), psychosocial factors (Madge and Marmot, 1987) and the physical environment (Jozan, 1989). Several of these explanations have been investigated for the different causes of mortality and morbidity, stages of the life course and for gender (see Macintyre).

Despite the vast literature on the causes of health inequalities, only a few longitudinal studies take account of the cumulative effects of different risks or duration of exposure to a particular risk over the life course. There are exceptions, such as the Whitehall study (Marmot et al., 1991) although

36

this commences in mid-life. Nevertheless, there is now acceptance that co-occurrence of different types of risk factors over time is likely to be responsible for the emergence and maintenance of health inequalities. This view is emphasized in a recent Department of Health report commenting on systematic variations in mortality and morbidity across social groups within the UK (Department of Health, 1995). The report concludes, 'it is likely that cumulative differential exposure to health damaging or health promoting physical and social environments is the main explanation for observed variations in health and life expectancy, with health related social mobility, health damaging or health promoting behaviours, use of health services, and genetic or biological factors also contributing'.

The British birth cohort studies have contributed to the appreciation that health inequalities commence early in life and that differential risk continues throughout the life course. For example, the 1958 birth cohort has shown social differences in birth weight (Butler and Bonham, 1963), height (Goldstein, 1971; Goldstein and Peckham, 1976), obesity (Power and Moynihan, 1988), social adjustment (Davie, 1973) and accidents (Essen and Wedge, 1982; Wedge and Essen, 1982). Over the previous decade the 1958 birth cohort has been studied intensively in order to improve understanding of how adulthood health inequalities are generated (Power *et al.*, 1986, 1990, 1991, 1996; Fogelman *et al.*, 1989; Power, 1991; Power and Manor, 1992; Manor *et al.*, 1997). The present chapter draws together some of this recent work on social inequalities in health, focusing especially on ill health as established from the latest sweeps of the study, at ages 23 and 33. However, before these findings are presented it is necessary to address two important methodological issues – i.e. the measurement of health status and of social position.

Measurement of health status

Health status can be measured with a range of 'objective' and 'subjective' (or self-reported) measures. Although mortality is commonly used, it is of limited value for age groups with low death rates. Measuring the health of younger adults is therefore problematic, as they are a relatively healthy group with a low prevalence of chronic disease (Blaxter, 1990). Although biological markers can be useful, in general they may be limited to a narrow definition of 'health'. The analysis which follows draws on symptom-based indicators although we predominantly focus on self-rated health – an overall or general assessment of an individual's health as rated by the individual. Because self-rated health is subjective it is likely to be affected by response set, whereby at a given level of objective ill health, different individuals report different levels of subjective ill health. For accurate assessment of health inequality it is important that social groups report their health in a similar manner (Blane *et al.*, 1996). There is some evidence that at a given level of objective ill health, the higher income groups report worse health than lower social classes (O'Donnell and Propper, 1991). However, it has also been found that the predictive effect of self-rated health on mortality is similar in both manual and non-manual groups (Wannamethee and Shaper, 1991). Self-rated health is of particular interest because it has been shown to be associated with other measures of current health status and fitness (*Allied Dunbar National Fitness Survey*, 1992), as well as predicting later mortality (Kaplan and Camacho, 1983; Idler and Angel, 1990; Wannamethee and Shaper, 1991; Cox *et al.*, 1993; Appels *et al.*, 1996; Moller *et al.*, 1996). Despite their limitations, self-report measures are particularly useful for this generally healthy stage of life.

Measurement of social position

There are also measurement problems in respect of social status. Numerous measures are utilized, most commonly occupation, education and income. Some theoretical distinctions are drawn between measures, with education being regarded as 'cultural capital' and occupation and income indicating working and housing conditions (Dahl, 1994). However, each measure has limitations especially in relation to theoretical grounding, internal heterogeneity, women's employment and economic inactivity and non-hierarchical classification. Within the UK, there has been a continuing debate on the validity of occupation-based measures, notably the Registrar General's social classification and the socio-economic groups schema (Rose, 1995). Thus, non-occupation based measures are commonly used as proxy measures, including housing tenure (Blaxter, 1990; Arber and Ginn, 1991), education, car access (Macintyre and West, 1991) and combinations of these (Goldblatt, 1990). Nevertheless, relationships between health and social position are generally consistent with different social measures. It is noteworthy however, that the extent of health inequality varies with different social measures (Macran *et al.*, 1994; Rahkonen *et al.*, 1995; Manor *et al.*, 1997). Alternative social measures are presented below for the 1958 study. However, most analyses are based on occupational class so as to relate to the large literature using this measure.

The 1958 birth cohort study

The 1958 birth cohort includes all children born in England, Wales and Scotland during one week (3–9 March) in 1958 (see Table 3.1). The study

Table 3.1. *Summary of the 1958 British Birth Cohort*

	All living in Britain born 3–9 March, 1958 (including immigrants 1958–1974)					
	PMS 1958 (Birth)	NCDS1 1965 (Age 7)	NCDS2 1969 (Age 11)	NCDS3 1974 (Age 16)	NCDS4 1981 (Age 23)	NCDS5 1991 (Age 33)
Target sample	17,773	16,883	16,385	16,915	16,457	16,455
Data sources	Parents	Parents School Tests Medical	Parents School Tests Medical Subject	Parents School Tests Medical Subject Census	Subject Census	Subject
						Spouse/ partner Children
Achieved sample	17,414	15,458	15,503	14,761	12,537	11,407

Source: Shepherd, 1995.

originated in the Perinatal Mortality Study, whose aim was to determine the social and obstetric factors associated with stillbirth and death in early infancy. Information was collected on 98% of births totalling 17,414. Subsequent follow-up of survivors was undertaken at ages 7, 11, 16, 23 and 33 years, with 11,407 subjects included in the most recent sweep (Ferri, 1993). Despite sample attrition, those remaining in the study are generally representative of the original sample (Power *et al.*, 1991; Ferri, 1993). Information at ages 23 and 33 was obtained through an interview with the study subject, unlike at previous sweeps when parents and schools (teachers and doctors) provided information, as well as the cohort members.

Socio-economic measures

Several occupation-based measures are used. Social origins are represented by father's occupational class at the time of the respondent's birth (using the Registrar General's 1950 classification) and also when the cohort member was aged 16 or at age 11 if 16-year data were not available. Two social class measures define the social destinations of the cohort members, namely own social class at ages 23 and 33 using the 1980 and 1990 Registrar General's classification respectively. Both men and women were allocated to a social class on the basis of their own current or most

recent occupation. Analyses are based on four social classes: classes I and
II combined, class III non-manual, class III manual and classes IV and V
combined. In addition, two alternative measures of social position are
presented, namely (1) housing tenure as reported at age 33, categorized as
owner occupier and local authority renter, and (2) educational qualifications achieved by age 33, categorized into five groups: above A level, A
level or equivalent, O level or equivalent, less than O level and no qualifications.

Health measures

Health measures at age 33 are the primary focus here, although several
measures from age 23 are also presented. The measures include self-rated
health, long-standing illness, psychological health, respiratory status,
height, obesity, joint pain and arthritis, back pain, stomach problems,
hypertension, eczema, hayfever, and menstrual and gynaecological problems (see Appendix 3.1 for further details).

Explanatory variables

Variables representing the specific explanatory areas of 'inheritance' at
birth, socio-economic circumstances, education, health in childhood/ado-
lescence and health-related behaviour are also described (see Appendix
3.2).

Findings

Social differences in health at age 33

By age 33, social class differences are evident for most health indicators
included in the study, with symptom prevalence increasing from higher (I
& II) to lower (IV & V) social class (Table 3.2). Trends are especially
pronounced for poor/fair health rating (ranging from around 9% in classes
I & II to 20% in IV & V), high Malaise scores (4.1 to 11.5% for men; 6.5 to
18.6% for women), limiting long standing illness for men (4.5 to 9.5%) and
respiratory symptoms (16 to 30% and 12 to 23% for men and women
respectively). The difference between the extreme social groups is also
pronounced for arthritis/joint pain (in men), stomach trouble (in women),
hypertension, and for women, menstrual problems. Only two health indicators show the reverse of this general pattern across social classes, i.e.
hayfever and eczema, although a statistically significant trend ($p < 0.05$) in
this direction is evident only for men.

Table 3.2. *Morbidity and symptom prevalence (%) at age 33 by social class at 33 years of age*

Health/morbidity[a]	Sex	Social class at age 33 I & II	IIIN	IIIM	IV & V	All	Total sample	χ^2 trend
Percentage in each class	M	39.8	10.7	33.3	16.3	—	5275	
	F	32.6	36.5	7.4	23.6	—	5305	
Poor/fair self-rated health	M	8.8	13.2	14.7	20.0	13.1	5223	71.10***
	F	9.8	12.0	15.9	19.5	13.4	5228	63.34***
Poor/fair self-rated health last	M	2.8	6.1	5.2	9.5	5.0	5224	48.33***
12 months	F	6.8	8.3	10.7	12.3	8.9	5227	29.43***
Long-standing illness	M	15.9	15.8	16.5	19.8	16.7	5250	4.58*
	F	12.8	12.8	15.4	16.2	13.8	5275	8.33**
Limiting long-standing illness	M	4.5	5.9	5.8	9.5	5.9	5250	20.75***
	F	5.4	5.0	6.7	7.1	5.8	5276	4.66*
High Malaise score	M	4.1	6.4	7.2	11.5	6.6	5243	49.54***
	F	6.5	10.7	13.6	18.6	11.4	5269	105.56***
Arthritis/joint pain ever	M	12.2	14.6	16.0	17.5	14.6	5253	17.95***
	F	15.2	14.7	17.8	17.8	15.8	5268	5.11*
Back pain (last 12 months)	M	27.2	29.7	33.2	32.4	30.3	5237	15.97***
	F	25.4	26.1	29.8	32.6	27.7	5260	21.12***
Migraine	M	12.9	12.1	11.5	13.0	12.4	5244	0.26
	F	23.9	26.1	23.7	28.7	25.8	5264	7.08**
Stomach trouble	M	19.5	17.4	19.3	22.3	19.7	5242	1.71
	F	12.2	13.0	12.1	17.1	13.7	5252	13.33***
Hypertension	M	4.0	4.4	4.0	7.6	4.6	5239	9.22**
	F	5.0	6.0	6.2	8.0	6.2	5258	10.32**
Respiratory symptoms	M	15.8	14.2	24.3	29.6	20.7	5129	86.09***
(one or more)	F	12.4	14.3	23.5	23.3	16.5	5152	74.16***
Asthma/wheezing	M	25.9	26.0	28.1	32.4	27.7	5254	11.28***
	F	26.5	25.5	32.3	32.6	28.0	5273	18.00***
Asthma/wheezing on	M	6.3	6.6	5.8	7.1	6.3	5254	0.09
medication	F	7.7	8.0	10.3	9.6	8.4	5273	4.8*
Eczema	M	17.2	15.6	12.3	13.4	14.8	5244	15.64***
	F	20.3	19.9	19.3	18.7	19.7	5263	1.21
Hayfever	M	24.6	21.3	16.2	16.1	20.1	5240	47.95***
	F	22.0	21.0	19.5	19.6	20.9	5264	2.83
Body Mass Index – obesity	M	9.4	9.9	12.4	12.7	11.0	5175	11.50***
	F	15.0	13.3	22.5	18.5	15.8	5185	12.85***
Menstrual problems	F	16.5	16.7	16.4	21.9	17.8	5210	14.03***
Other gynaecological problems	F	16.3	17.0	17.7	16.2	16.6	5212	0.002

*** p < 0.001; ** p < 0.01; * p < 0.05.
[a] See Appendix 3.1 for definitions.

Explanations for social differences in health

Some of the alternative explanations for social differences in health mentioned in the introduction have been investigated with the 1958 birth cohort data. These are considered here in relation to the Black report categories of artefact, selection, lifestyle and material circumstances (Townsend and Davidson, 1992).

Artefact

According to the artefact explanation, health differences are a product of measurement inadequacies, such as numerator-denominator biases in the decennial supplements. For instance, this arises when there is a different assignment of social class on the death certificate (numerator) than at the census (denominator). Recently such explanations have been given less credence, since studies overcoming such methodological difficulties, notably the Office of National Statistics (formerly OPCS) Longitudinal Study, confirm that social gradients in mortality are evident in the national figures (Fox and Goldblatt, 1982).

Furthermore, numerous studies (Blaxter, 1990; Goldblatt, 1990; Arber and Ginn, 1991; Macintyre and West, 1991) reproduce social differences in health using several alternative measures of social position which are not dependent on occupational class. This has contributed to the decline in support for artefactual explanations. An illustration of health differences by housing tenure and education is presented here using data from the 1958 birth cohort. Figure 3.1 confirms that the existence of health inequalities is not an artefact associated with a particular social measure. However, the extent of health inequality varies according to the measure of social position used. Within the cohort study, education (as represented by highest qualification achieved) demonstrates greater inequality than social class of origin (Manor et al., 1997)

Selection

A second explanation suggests that health differences are the result of a continuous process of health-related social mobility. That is, those moving up the social scale are relatively healthy and those moving down are relatively unhealthy. With data on social position and health at several ages, the 1958 birth cohort is ideal for exploring both inter-generational (that is from class of origin to class of destination) and intra-generational mobility (Power et al., 1986, 1990, 1991, 1996). Table 3.3 illustrates the

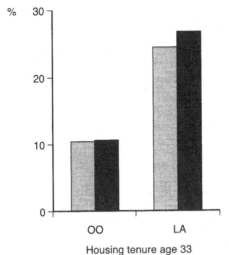

Housing tenure age 33
OO : owner occupier   LA ; local authority renter

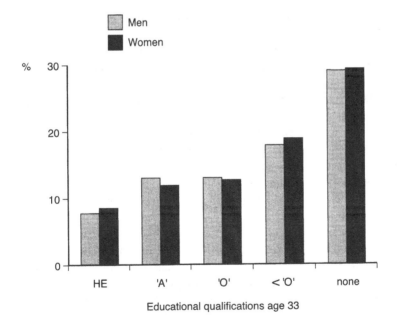

Educational qualifications age 33

HE : higher education   'A' : A level or equivalent   'O' : O level or equivalent
< 'O' : less than O level   none : no qualifications

Figure 3.1. Social differences in self-rated health at age 33 years.

Table 3.3. *Inter-generational mobility from social class IIIM* and health at age 23*

	Men			Women		
	Upwards (N = 194) %	Stable (N = 595) %	Downwards (N = 231) %	Upwards (N = 210) %	Stable (N = 631) %	Downwards (N = 264) %
Poor/fair self-rated health	4.1	9.1	13.4	5.2	8.6	16.3
High Malaise score	1.5	3.7	7.4	6.2	9.8	16.7
Psychological morbidity	3.6	2.7	6.5	3.8	5.2	10.6
Short stature	5.2	11.9	12.3	10.0	9.4	14.8

* Includes those in class IIIM at birth and age 16.
Source: Power et al., 1991, p. 70.

relationship between health and inter-generational mobility. Cohort members whose own social class at age 23 differed from that of their father's during their childhood are identified. This comparison is restricted to those fathers who were in class III manual when the subject was born and who were still in the same social class when the subject was 16. This is to hold constant the effect of father's social class. As can be seen from Table 3.3, those men and women who had been upwardly mobile by age 23 reported better health than those who had moved down. Those who did not deviate from their social origins (the stable category) tended to be intermediate. For example, 4.1% of men who were upwardly mobile rated their health as fair/poor, 9.1% did so among those who were stable and 13.4% among downwardly mobile men. The only exception to this trend was for psychological morbidity of men only.

Table 3.4 shows the magnitude of health-related intra-generational mobility seen between the ages 23 and 33 in the birth cohort study. In this instance, health status at age 23 appears to influence the pattern of subsequent social mobility. Men and women rating their health as fair or poor were more likely to be downwardly mobile by age 33, particularly for those starting in classes I & II at age 23 (odds ratio: 3.14 for men and 2.03 for women). Men with a fair/poor health rating were also consistently less likely to be upwardly mobile; whereas among women the odds ratios for upward mobility were not significant and were also less systematic. This raises the prospect that explanations for social inequalities in health may vary depending on gender.

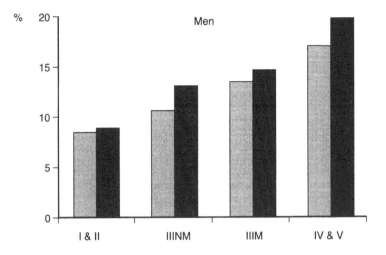

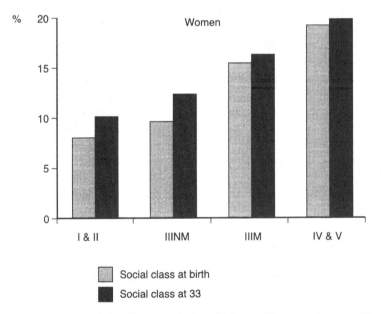

Figure 3.2. Social class gradients in fair/poor self-rated health at age 33 years.

Table 3.4. *Odds ratios of upward and downward social mobility, fair/poor health at age 23 relative to excellent/good health*

Social mobility ages 23 and 33	I & II	IIIN	IIIM	IV & V	I & II	IIIN	IIIM	IV & V
		(Men)				(Women)		
Downward	3.14*	1.96	1.54	—	2.03*	1.69*	1.80	—
Stable	1.00	1.00	1.00	1.00	1.00	1.00	1.00	1.00
Upward	—	0.70*	0.67	0.62*	—	1.13	1.77	0.75
n/N	51/966	57/774	141/1739	80/761	53/1012	202/2282	47/388	108/777

Social class at age 23

* $p < 0.05$; n/N = numbers with fair-poor health/total sample.
Source: Power et al., 1996.

Analyses of the 1958 birth cohort confirms that both inter- and intra-generational social mobility is influenced by health status. However, it should be noted that the numbers who are mobile with a fair/poor health rating, especially those moving downwards, are relatively small in comparison with the total sample size. The numbers of individuals who are socially mobile due to their health is pertinent when considering health selection as an explanation for health gradients. In previous work we have demonstrated that despite the existence of health-related social mobility, the latter does not have a major impact on health gradients (Power et al., 1991, 1996). Evidence for this can be seen in Figure 3.2 where social gradients attenuate slightly, but do not disappear when using class at birth (origin) as the social stratifier. That is, health gradients are still apparent even after removing any effect of subsequent social mobility. These results for age 33 are similar to those for self-rated health at age 23 (Power et al., 1991).

An alternative approach for assessing the impact of health-related social mobility on health gradients is to examine the gradients in the sub-sample identified as 'healthy' from a baseline measure. Table 3.5 illustrates this with social differences in self-rated health at age 33, presented for both the total sample and for a sub-sample that excludes those with poor/fair health at age 23. When the effect of ill health at baseline is removed, the familiar health gradient remains and is of similar magnitude to that seen in the full sample of cohort members. It would appear therefore, that whilst social mobility is related to health, this in itself does not represent a major determinant of inequalities in health. Hence other factors need to be considered.

Table 3.5. *Class gradients in prevalence (%) fair/poor health at age 33*

Social class at age 33	Men with data at ages 23 & 33	Men excluding fair/poor health at age 23	Women with data at ages 23 & 33	Women excluding fair/poor health at age 23
	(N = 4240)	(N = 3911)	(N = 4459)	(N = 4049)
I & II	8.5	6.7	9.4	7.3
IIIN	13.3	10.1	11.4	8.5
IIIM	14.4	11.9	15.5	13.5
IV & V	17.7	13.8	18.8	14.7
All	12.5	9.9	12.8	9.8

Source: Power *et al.*, 1996.

Material circumstances and health related behaviour

The Black Report (Townsend and Davidson, 1992) identified material circumstances and cultural/behavioural explanations for health inequalities. To a degree these explanations are interwoven because lifestyles are strongly influenced by socio-economic circumstances. Most investigations of material circumstances and health-related behaviour are cross-sectional or they relate subsequent morbidity and mortality to a single previous measure (Haan *et al.*, 1987; Madge and Marmot, 1987; Kooiker and Christiansen, 1995; Marmot *et al.*, 1995). Thus, few studies are able to take account of cumulative effects of different risk factors or of the duration of exposure to a specific risk factor, such as hazardous working environment or smoking behaviour. However, when health differentials were examined in the 1958 birth cohort at age 23, the cumulative effects of both material circumstances and health behaviours appeared to be of great importance. This is illustrated in Table 3.6 in which several factors occurring at different life stages are presented. Social differences in self-rated health are expressed as odds ratios of poor/fair rated health in classes IV & V versus I & II. The unadjusted odds ratios are then adjusted to examine the effect of earlier life circumstances, as represented by: social class at birth; housing tenure at age 11; social adjustment as measured by the Rutter behaviour score at age 16 (Rutter, 1967); school absence through ill health; smoking at age 16; end of school qualifications; unemployment; and, for women, maternal age at first child birth. These particular variables had been selected from an earlier stage of analysis to represent 'inheritance' at birth, earlier socio-economic circumstances, educational achievement, health and health related behaviour (Power *et al.*, 1991). The ordering of the

Table 3.6. *Relative odds (classes IV & V versus I & II) 'poor/fair' health rating at age 23*

	Men (N = 2652)
Unadjusted	3.04 (95% CI 1.83, 5.05)
Adjusted for:	
Social class at birth	2.90*
Housing tenure at 11	2.91*
Behaviour score at 16	2.25*
School absence through ill-health[+]	2.09*
Smoking at 16	1.96*
End of school qualifications	1.86*
Unemployment	1.71*
	Women (N = 2827)
Unadjusted	3.28 (95% CI 2.12, 5.05)
Adjusted for:	
Social class at birth	3.00*
Housing tenure at 11	2.66*
Behaviour score at 16	2.16*
School absence through ill-health[+]	2.05*
Smoking at 16	2.00*
End of school qualifications	1.62*
Unemployment	1.58
Age at first child	1.52

[+] Age 15–16; * 95% confidence interval (CI) excludes 1.
Source: Power *et al.*, 1991, p. 139.

variables in Table 3.6 broadly reflects the time sequence of each factor. For instance, social class at birth, signifying social inheritance, was considered first, then housing tenure at age 11, representing socio-economic circumstances in childhood. The effect of each variable on social differences in self-rated health, is indicated by its impact on the odds ratio after adjusting for all previously listed variables. For instance, the effect of social adjustment (which is a behaviour score at age 16) is observed after adjusting for social class at birth and housing tenure at age 11 (Table 3.6).

For men the unadjusted odds of poor/fair health in classes IV &V were 3.04 times those in classes I & II; for women the odds ratio was 3.28. After adjusting for all variables, class differentials are substantially reduced, although remaining significant for men. Also, each variable in general

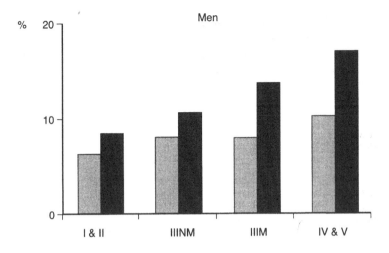

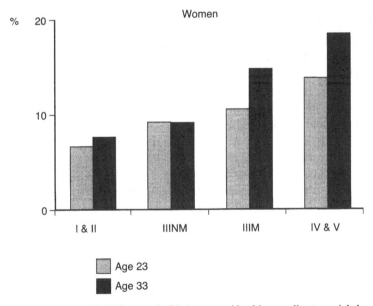

Figure 3.3. Differences in fair/poor rated health according to social class at birth.

contributes to the reduction in the odds ratio. This is especially noted for variables entered at the later stages of the analysis. For instance, adjusting for male unemployment resulted in a small reduction in the odds ratio (from 1.86 to 1.71), but this was achieved following adjustment for six other variables. Of particular interest is the impact of social adjustment

Table 3.7. *Odds of poor/fair health at age 33 by social class, relative to class I & II*

Social class age 33	Unadjusted	Adjusted*
Men		
I & II	1.00	1.00
IIIN	1.61	1.44
IIIM	2.00	1.46
IV & V	2.54	1.71
Women		
I & II	1.00	1.00
IIIN	1.28	1.04
IIIM	1.84	1.35
IV & V	2.32	1.55

* Adjusted for social class at birth, 16 and 23.

and qualifications achieved by the time of leaving school. These analyses suggest that social differences in health are 'explained' by a cumulation of factors, both material and behavioural, throughout the life-course.

In further analyses of the cohort data, we were examining whether this conclusion holds true for health differences 10 years later. The particular focus of this work was to establish whether the gradient in self-rated health at age 33 is affected by similar factors to those identified at age 23, or whether more recent circumstances and events occurring in the intervening years predominate. As a first step, we examined whether the magnitude of health inequality was similar at the two ages (Power *et al.*, 1997). Figure 3.3 presents the data for fair/poor health rating at ages 23 and 33 according to social class at birth. It is apparent that social differences in self-rated health were maintained by age 33. Furthermore, the gradient in poor/fair rated health appears to increase, for men at least, though not quite significantly. By using a prior social class measure (i.e. social class at birth) for both ages, we excluded the possibility that change in the gradient is related to change in social position. For men, it is possible that circumstances between the ages 23 and 33 may be contributing to the self-rated health gradient at age 33.

Preliminary analysis has also been undertaken using social class at earlier stages of life (i.e. at birth, ages 16 and 23). It should be emphasized that these social class measures are used as proxies for a diverse range of factors in the social and physical environment occurring at each life stage. Table 3.7 presents the unadjusted odds of poor/fair health rating at age 33 in each social class compared with those in classes I & II. Odds ratios for

Table 3.8. *Relative odds of short adult stature (classes IV & V versus I & II)*

	Men (N = 3664)
Unadjusted	2.36 (95% CI 1.65, 3.37)
Adjusted for:	
Midparent height	1.63*
Social class at birth	1.67*
Birth weight	1.55*
Height at 7	1.11
	Women (N = 3818)
Unadjusted	2.31 (95% CI 1.66, 3.23)
Adjusted for:	
Midparent height	1.59*
Social class at birth	1.47*
Birth weight	1.38
Height at 7	0.97

* 95% confidence interval excludes 1.
Source: Power *et al.*, 1991, p. 144.

classes IV & V are 2.54 for men and 2.32 for women; when the odds ratios were adjusted for social class at ages 23, 16 and at birth, health differentials were substantially reduced (Table 3.7). This preliminary analysis suggests that social differences in self-rated health are due to cumulative socio-economic circumstances throughout the life-course.

The results presented thus far largely focus on self-rated health at ages 23 and 33. A different pattern could emerge with other health indicators. For comparative purposes therefore, we include the results of similar analyses for adult short stature (that is, heights below the lowest decile) at age 23. As seen from Table 3.8, social differences in short stature are affected largely by early life and genetic factors (mid-parent height, social class at birth, birth weight and height at age 7). Social class differences are not entirely accounted for by differences in parental height, since there appears to be an additional effect of other early life factors represented by social class at birth and birth weight. That these analyses conform to expectations as to the factors responsible for social differences in short stature, provides support for the general approach used in these attempts to investigate determinants of social inequalities in health.

It is important to recognize that there are methodological problems, specifically in relation to confounding and causality, that arise when examining life course influences on later health. In the analysis described above, the use of a temporal sequence of factors at different life stages may ensure that particular circumstances or events precede health outcomes, but this approach does not necessarily establish causality. Indeed, it is possible that the causal relationships have been obscured by taking account of earlier circumstances with which later 'causal' characteristics may be associated. It is therefore plainly evident that causes of social class differences in health are difficult to unravel (Power *et al.*, 1991).

Conclusions

Social inequalities in health have been shown to be well established by early adulthood according to self-reported health measures in the 1958 birth cohort. In general, the relationship between health and social position is graded, such that improvements in health occur at each increasing level of the social hierarchy. This pattern of health inequality seen quite early in life is a portent of that for chronic disease and mortality in later life. The graded relationship suggests that health variations are not simply reflecting differences between poor and non-poor sections of society. Further, it appears that artefactual explanations or health selection do not provide an adequate explanation for health inequalities, as demonstrated by the data from the 1958 birth cohort. It can also be argued that the causes of socio-economic differences in health have earlier origins than studies of mid-life factors might suggest. Evidence from the 1958 birth cohort indicates that human resources or 'capital' (as indicated by social adjustment and educational achievements) may be particularly important. Further attention should therefore be directed at how such influences accumulate at the different stages of life.

Acknowledgements

Chris Power is supported by the Canadian Institute for Advanced Reseach as a Weston Fellow. Sharon Matthews is supported by the Economic and Social Research Council (ESRC award number R000235189).

Appendix 3.1. *Health measures*

(1) Self-rated health at ages 23 and 33: subjects assessed their health

as excellent, good, fair or poor overall and in the past 12 months. These were dichotomized into excellent/good and fair/poor categories.

(2) Long-standing illness and long-standing illness that limits daily activities (at age 33).

(3) Psychological health status (ages 23 and 33) was derived from the Malaise Inventory, a 24-item checklist of symptoms such as anxiety, irritability and depressed mood; both psychological and somatic items are included (Rutter *et al.* 1970). Scores of seven or more were used to identify a high risk group. Internal consistency and external validity of the scale have been demonstrated for this age group (Rodgers and Power, unpub. data).

(4) Psychological morbidity (age 23) – psychological and emotional problems experienced since age 16 for which specialist help had been sought. These include anorexia nervosa, drug and alcohol dependency and personality disorders.

(5) Respiratory symptoms, based on the MRC questions (Medical Research Council, 1970) on morning and day/night-time cough and phlegm.

(6) Asthma/wheeze – subjects reported at age 33 whether they had had wheezing or whistling in the chest in the past. Subjects reported if they required prescribed medication in the last 12 months.

(7) Body size, indicated by the Body Mass Index (calculated by dividing weight in kilograms by the square of height in metres). Subjects at age 33 were measured (heights were measured without shoes, using a stadiometer reading to one centimetre; weights were measured in indoor clothing, using Salter portable scales). As recommended by the Royal College of Physicians (1983) men were categorized as obese if their BMI was greater than 30.0 and women if it was greater than 28.6.

(8) Birthweight.

(9) Height –
- at age 7 (measured);
- age 23 (self-reported);
- parental height at birth (self-reported), the analysis used mid-parent heights defined as the average of the mother's and father's heights.
- adult short stature is defined as heights below the lower decile.

Respondents were also provided with a list of physical problems/conditions (age 33) and asked if they had ever suffered from the condition or been told they had the condition/problem. These included:

(1) Arthritis, rheumatism or painful joints, ever and in the last 12 months.
(2) Back pain, lasting more than one day, ever and in the last 12 months.
(3) Migraine.
(4) Stomach trouble.
(5) Hypertension.
(6) Eczema.
(7) Hayfever.
(8) Menstrual problems and other gynaecological problems (women only).

Appendix 3.2. *Explanatory variables*

The explanatory areas examined were drawn from previous analyses as set out in *Health and Class: The early years* (1991).

Socio-economic circumstances
(1) Housing tenure age 11 (renters versus owners).
(2) Unemployment by age 23 (one or more periods versus none) between leaving school and the 23 year interview.
(3) Age at first child (child by 23 versus no child).

Education/attitudes
(4) End of school qualifications (none versus one or more).

Health in childhood/adolescence
(5) School absence through ill health (age 16) during the previous year (≥ 1 month versus < 1 month).

Behaviour
(6) Smoking at 16 (smoker versus non-smoker).
(7) Behaviour score at 16 ('deviant' verus 'normal') based on the Rutter Behaviour Scale (Rutter, 1967).

References

Allied Dunbar National Fitness Survey (1992). London: Sports Council and Health Education Authority.
Appels, A., Bosma, H., Grabauskas, A., Gostautas, A. & Sturmans, F. (1996). Self-rated health and mortality in a Lithuanian and Dutch population. *Social*

Science & Medicine **42**, 681–9.

Arber, S. & Ginn, J. (1991). Gender and later life: a sociological analysis of resources and constraints. *British Sociological Association Conference, Manchester, 25–28 March 1991.* Manchester: British Sociological Association.

Barker, D. J. P. & Osmond, C. (1992). Inequalities in health in Britain: specific explanations in three Lancashire towns. In *Fetal and Infant Origins of Adult Disease*, ed. D. J. P. Barker, pp. 68–78. London: *British Medical Journal*.

Blane, D., Power, C. & Bartley, M. (1996). Illness behaviour and the measurement of class differentials in morbidity. *Journal of Royal Statistical Society* (Series A) **159**, 77–92.

Blaxter, M. (1990). *Health and Lifestyles.* London: Tavistock.

Butler, N. R. & Bonham, D. G. (1963). *Perinatal Mortality.* Edinburgh: Livingstone.

Cox, B. D., Huppert, F. A. & Whichelow, M. J. (1993). *The Health and Lifestyle Survey: Seven Years On.* Aldershot: Darmouth Publishing.

Dahl, E. (1994). Social equalities in ill-health: the significance of occupational status, education and income-results from a Norwegian survey. *Sociology of Health and Illness* **16**, 644–67.

Davey-Smith, G., Bartley, M. & Blane, D. (1990). The Black Report on socioeconomic inequalities in health 10 years on. *British Medical Journal* **301**, 373–7.

Davie, R. (1973). The behaviour and adjustment in school of seven year olds: sex and social class differences. *Early Child Care and Development* **2**, 39–47.

Department of Health (1995). Variations in health: What can the department of health and the NHS do? London: Department of Health.

Department of Health and Social Security (1980). *Inequalities in Health: Report of a Research Working Group chaired by Sir Douglas Black.* London: Department of Health and Social Security.

Essen, J. & Wedge, P. (1982). *Continuities in Childhood Disadvantage.* London: Heinemann Educational.

Ferri, E. (1993). *Life at 33: the fifth follow-up of the National Child Development Study.* London: National Children's Bureau.

Fogelman, K., Fox, A. J. and Power, C. (1989). Class and tenure mobility, do they explain inequalities in health among young adults in Britain? In *Health Inequalities in European Countries*, ed. J. Fox, pp. 333–52. Aldershot: Gower.

Fox, J. & Goldblatt, P. O. (1982). *Longitudinal Study: Sociodemographic Mortality Differentials.* London: HMSO.

Goldblatt, P. (1990). *Longitudinal Study: Mortality and Social Organisation 1971–1981.* OPCS LS No.6. London: HMSO.

Goldstein, H. (1971). Factors influencing the height of seven-year-old children. Results from the National Child Development Study (1958 cohort). *Human Biology* **43**, 92–111.

Goldstein, H. & Peckham, C. (1976). Birthweight, gestation, neonatal mortality and child development. In *The Biology of Human Foetal Growth*, ed. D. F. Roberts & A. M. Thomson, pp. 81–102. London: Taylor Francis.

Haan, M., Kaplan, G. & Camacho, T. (1987). Poverty and health: prospective evidence from the Almeda County Study. *American Journal of Epidemiology* **125**(6), 989–98.

Idler, E. L. and Angel, R. L. (1990). Self-rated health and mortality in the

56   C. Power and S. Matthews

NHANES-1 epidemiologic follow up study. *American Journal of Public Health* **80**, 446–52.

Jozan, P. (1989). A compilation of some aspects of area mortality differentials in some European countries. In *Health Inequalities in European Countries*, ed. J. Fox, pp. 173–98. Aldershot: Gower.

Kaplan, G. A. & Camacho, T. (1983). Perceived health and mortality: a nine-year follow-up of the human population laboratory cohort. *American Journal Epidemiology* **117**, 292–304.

Kooiker, S. & Christiansen, T. (1995). Inequalities in health: the interaction of circumstances and health related behaviour. *Sociology of Health and Illness* **17**, 495–524.

Kunst, A. E. & Mackenbach, J. P. (1994). International variation in the size of mortality differences associated with occupational status. *International Journal of Epidemiology* **23**, 1–9.

Macintyre, S. & West, P. (1991). Lack of class variation in health in adolescence: an artefact of an occupational measure of social class? *Social Science and Medicine* **32**, 395–402.

Mackenbach, J. P., Stronks, K. & Kunst, A. R. (1989). The contribution of medical care to inequalities in health: differences between socio-economic groups in decline of mortality from conditions amenable to medical intervention. *Social Science and Medicine* **29**, 369–76.

McLoone, P. & Boddy, F. A. (1994). Deprivation and mortality in Scotland 1981 and 1991. *British Medical Journal* **309**, 1465–70.

Macran, S., Clarke, L., Sloggett, A. & Bethune, A. (1994). Women's socio-economic status and self assessed health: identifying some disadvantaged groups. *Sociology of Health and Illness* **16**, 182–208.

Madge, N. & Marmot, M. (1987). Psychosocial factors and health. *Quarterly Journal of Social Affairs* **3**, 81–134.

Manor, O., Matthews, S. & Power, C. (1997). Comparing measures of health inequality. *Social Science and Medicine* **45**, 761–71.

Marmot, M., Bobak, M. & Davey-Smith, G. (1995). Explanations for social inequalities in health. In *Society and Health*, B. A. Amick, S. Levine, A. R. Tarlov & D. Chapman-Walsh, pp. 172–210. New York: Oxford University Press.

Marmot, M. G., Davey-Smith, G., Stansfield, S., Patel, C., North, F., Head, J., White, I., Brunner, E. & Feeney, A. (1991). Health inequalities among British civil servants: the Whitehall II study. *Lancet* **337**, 1387–93.

Medical Research Council (1970). *Questionnaire on Respiratory Symptoms.* London: MRC.

Moller, L., Kristensen, T. S. & Hollnagel, H. (1996). Self-rated health as a predictor of coronary heart disease in Copenhagen, Denmark. *Journal of Epidemiology and Community Health* **50**, 423–8.

O'Donnell, O. & Propper, C. (1991). Equity and the distribution of UK National Health Service resources. *Journal of Health Economics* **10**, 1–19.

Phillimore, P., Beattie, A. & Townsend, P. (1994). Widening inequality of health in northern England. *British Medical Journal* **308**, 1125–8.

Power, C. (1991). Social and economic background and class inequalities in health among young adults. *Social Science and Medicine* **32**, 411–18.

Power, C., Fogelman, K. & Fox, A. J. (1986). Health and social mobility during the early years of life. *Quarterly Journal of Social Affairs* **2**, 397–413.

Power, C., Hertzman, C., Matthews, S. & Manor, O. (1997). Social differences in health: life cycle effects between ages 23 and 33 in the 1958 birth cohort. *American Journal of Public Health.* **87**, 1499–1503.

Power, C. & Manor, O. (1992). Explaining social class differences in psychological health among young adults: a longitudinal perspective. *Social Psychiatry and Psychiatric Epidemiology* **27**, 284–91.

Power, C., Manor, O. & Fox, A. J. (1991). *Health and Class: the Early Years.* London: Chapman Hall.

Power, C., Manor, O., Fox, A. J. & Fogelman, K. (1990). Health in childhood and social inequalities in young adults. *Journal of the Royal Statistical Society* (Series A) **153**, 17–28.

Power, C., Matthews, S. & Manor, O. (1996). Inequalities in self-rated health in the 1958 birth cohort: life time social circumstances or social mobility? *British Medical Journal* **313**, 449–53.

Power, C. & Moynihan, C. (1988). Social class and changes in weight-for-height between childhood and early adulthood. *International Journal of Obesity* **12**, 445–53.

Rahkonen, O., Arber, S. & Lahelma, E. (1995). Health inequalities in early adulthood: A comparison or young men and women in Britain and Finland. *Social Science and Medicine* **41**, 163–71.

Roberts, I. & Power, C. (1996). Does the decline in child injury mortality vary by class? A comparison of class specific mortality in 1981 and 1991. *British Medical Journal.* **313**, 784–6.

Rose, D. (1995). *A Report on Phase I of the ESRC Review of OPCS Social Classifications.* London: OPCS.

Royal College of Physicians (1983). Obesity. *Journal of the Royal College of Physicians* **17**, 5.

Rutter, M. (1967). A children's behaviour questionnaire for completion by teachers. *Journal of Child Psychology and Psychiatry* **8**, 1–11.

Rutter, M., Tizard, J. & Whitmore, K. (1970). *Education, Health and Behaviour.* London: Longman.

Shepherd, P. (1995). The National Child Development Study (NCDS). *An Introduction to the origins of the study and the methods of data collection*, 1. Working Paper no. 1. London: SSRU, SLM.

Townsend, P. & Davidson, N. (1992). *Inequalities in Health: The Black Report and the Health Divide*, 2nd edn. Harmondsworth: Penguin.

Wannamethee, G. & Shaper, A. G. (1991). Self-assessed health status and mortality in middle-aged British men. *International Journal of Epidemiology* **20**, 239–45.

Wedge, P. & Essen, J. (1982). *Children in Adversity*, London: Pan.

4   *Equity, social cohesion and health*

RICHARD G. WILKINSON

In this chapter I want to suggest that the reason why national mortality rates are closely related to income distribution is because more egalitarian societies are more highly cohesive and social cohesion is beneficial to health. It will be necessary to prepare the ground by giving a brief summary of the evidence that national mortality rates are indeed related to income distribution.

Let us start out from the so-called ecological fallacy. The concept is used to draw attention to the dangers of inferring that relationships identified using grouped population data necessarily hold at the individual level. It is rarely invoked to point out the equally serious danger that findings which seem to hold at the individual level may not hold for whole populations. Yet it is clear that policy makers in fields such as public health need to address themselves to results which hold true for whole populations. Do the studies of even large numbers of *individuals* provide a good guide to the way changes in the social and economic structure of our society affect health? Although individual studies may tell us that children are safer if they are driven to school than if they walk or cycle, we know that growing car use not only increases the dangers faced by those who do walk or cycle, but also pollutes the air everyone has to breathe. Here the individual data would be a poor guide to the healthiest transport system for a whole society. This is the ecological fallacy in reverse: the healthiest policy for individuals does not add up to the healthiest policy for populations.

A particularly important area in which this dilemma crops up involves our ignorance as to whether various socio-economic indicators are related to health because they are a guide to relative position in society or to absolute standards. Income appears to be related to health within societies more because it is an indicator of where people stand in society relative to others than because the differences in absolute living standards are so important. Although richer individuals, occupations and areas appear to be healthier than poorer ones, it is not safe to assume that richer industrial-

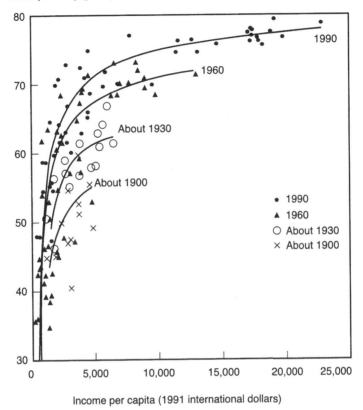

Figure 4.1. Life expectancy and income per capita for selected countries and periods. (Reproduced from Wilkinson, 1996.)

ized societies like the USA are healthier than poorer ones like Greece. The same might be true for education. Do people with more education tend to be healthier than those with less because the absolute level of an individual's educational standards are beneficial to health, or simply because higher educational standards raise people's socio-economic status relative to others in society?

Figure 4.1 is taken from a graph published in the World Bank World Development Report for 1993. It shows the relationship between average living standards and life expectancy in different countries at successive dates during the twentieth century. Living standards are expressed on the horizontal axis in 1991 international dollars. Instead of converting from national currencies using exchange rates as they are arbitrarily determined by speculation on the international currency markets, currencies have been converted according to the domestic purchasing power of each national currency. As well as taking account of national price differences the figures also take account of inflation.

This chapter is concerned primarily with the developed – i.e. industrialized[1] – countries which lie on the lengthening part of the curves as they approach the horizontal near the top (Figure 4.1). The shape of the curves shows that even the cross-sectional relationship between average living standards and average mortality rates is weakened among the industrialized countries. It takes enormous quantities of additional income before life expectancy is appreciably higher. When looking at changes over time, it can be seen that the relationship between average income and life expectancy is even weaker than the cross-sectional relationship suggests – instead of moving out along a given curve as they get richer, the passage of time seems to move countries onto successively higher new curves. In effect, the same level of income buys more life expectancy now than it used to. Statistical analysis of both the cross-sectional data and the changes over time suggest that not much more than 10% of the increases in life expectancy enjoyed in most of the industrialized countries comes from increasing average incomes. Preston's estimate was about 12%, my own – using more recent data – is about 9% (Preston 1975; Wilkinson 1996). This means that around 90% of the improvements in mortality rates come from unknown factors which appear to improve the gearing of the relationship between income and health as time passes. I have discussed possible explanations of this pattern elsewhere (Wilkinson 1996).

This picture of weak relationships *between* income and health between industrialized societies stands in stark contrast to the close relationships which exists between them *within* these societies. The vast body of research on health inequalities testifies to the strength of the relationship within societies. Whether you classify by employment grade, car ownership, housing tenure, education, social class or income, you find the same picture of a strong relationship between health and indicators of socio-economic status within societies. Perhaps the most remarkable demonstration of the relationship – specifically with income – comes from two papers by Davey-Smith *et al.* (1996a,b) using data from the Multiple Risk Factor Intervention Trial (MRFIT). Death rates among over 300,000 American white men were classified by the median income of the ZIP code area in which they lived. They produced an almost perfect rank ordering of mortality across 14 income categories – the mortality of only one income category was slightly out of place in the overall gradient (Davey-Smith *et al.*, 1996a). Mortality rates among 20,000 American black men were

---

[1] These terms, which are essentially synonymous, include the 'post-industrialized' countries; they are distinct from the 'newly industrialized' countries, which are the middle-income countries and are still described as 'developing' or 'less developed' than western nations.

perfectly rank ordered across all 10 of the income categories into which they were classified (Davey-Smith *et al.*, 1996b).

The remarkable contrast between the income–mortality fit within industrialized countries and the lack of one between them illustrates the danger of inferring from individual data to societal relationships. The individual relationship between income and health does not add up to produce a societal relationship between economic growth and improvements in health – even when looking at change over a period as long as 20 years which allows for possible time lags in disease causation.

There are two parts to the most likely explanation of why this is so. Firstly, there are reasons to think that the epidemiological transition marked a stage in economic growth during which the vast majority of the population gained reliable access to the basic necessities of life (see Wilkinson, 1994c). Particularly important are the reversals in the social distribution of coronary heart disease and a number of other conditions including obesity. The epidemiological transition is characterized not only by the decline of infectious diseases as the main causes of death, but also by a tendency for the so-called diseases of affluence to become the diseases of the poor in affluent societies. After this threshold level of income (around $5000 international 1991 dollars per capita in Figure 4.1) has been reached, the absolute standard of living is no longer such an important constraint on standards of health. In addition to the reversal in the social distribution of heart disease and obesity, deaths from stroke, hypertension, duodenal ulcers, nephritis and nephrosis, and lung cancer also become more common lower down the social scale (Koskinen, 1988). Two other pieces of evidence also suggest that the epidemiological transition marks the attainment of some threshold standard of living for a large majority of the population. One is the fact that the proportion of babies weighing less than 2500 g at birth has remained between 6% and 7% of all births since the 1950s despite very substantial increases in living standards (Macfarlane and Mugford, 1984; Power 1994). Lastly, although the age of menarche in girls declined earlier this century, it too has ceased to decline since the 1950s (Roberts, 1994).

The second part of the explanation of why the effects of individual income on individual health no longer add up to a strong international relationship between average income and life expectancy is that the relationship we are seeing within countries is the effect of *relative* income on health.

I have in the past published a number of papers showing that narrower societal income distributions are closely related to higher average life expectancy among industrialized countries, both cross-sectionally and when looking at changes over time. Correlations of 0.6, 0.7 or even 0.8

have been found (Wilkinson 1986, 1992, 1993, 1994a,b,c; McIsaac and Wilkinson 1997). The same relationship has been reported among industrialized and developing countries by a number of other people using other sources of data from different periods (Rodgers, 1979; Flegg, 1982; Le Grand, 1987; Waldmann, 1992; Wennemo, 1993; Duleep, 1995). The relationship between inequality and average mortality rates has also been shown for administrative areas within societies. Mortality within small areas of England was found to be related to the degree of socio-economic inequality within each area (Ben-Shlomo et al., 1996). Most strikingly, mortality rates for the 50 states of the USA have been found to be strongly related to the extent of income inequality within each state (Kaplan et al., 1996; Kennedy et al., 1996).

The association between population mortality rates and income inequality (or socio-economic inequality more generally) is not removed by controlling for average incomes, absolute poverty, levels of government expenditure, smoking, race, expenditure on medical care, etc. It appears across a wide range of circumstances including industrialized and less developed countries; it has been found in data dealing with changes over time as well as in cross-sectional data; and it has been found between countries as well as between regions and areas within countries.

It is hard to imagine a variable which has both a sufficiently powerful influence on mortality and is linked to inequality in such a way that it would give rise to a spurious relationship in all these circumstances. Nor is the idea of reverse causality a plausible explanation. No one has suggested that health has an important influence on income distribution – certainly not one sufficiently powerful to over-ride the influence of taxes and benefits, levels of unemployment and profits, levels of education, industrial structure and international competition. In addition, the issue of reverse causality between health and socio-economic status has been examined repeatedly in the health inequalities literature (Blane et al., 1993; Fox et al., 1985; Power et al., 1990; Lundberg, 1991; Power and Matthews, Chapter 3). If it is not a major factor even when using data classified by occupation and so confined to the economically active population of working age, it is very unlikely to be important when using mortality data for the whole population, including the old, the young, and those who are not economically active and among whom illness is unlikely to affect household income. Given that life expectancy is influenced primarily by the higher death rates in infancy and old age, it is inconceivable that an effect of illness on incomes could give rise to a close correlation between income inequality and average life expectancy.

To clarify, this is not a relationship between the extent of income inequalities and health inequalities within countries (though that almost certainly underlies it), it is a relationship between income inequality and mortality rates for the whole population over which inequality is measured. When first encountering this relationship, some imagine that perhaps the reason why income distribution is important may be simply because it provides a better specification of the absolute income levels of different proportions of the population – it is assumed that the fundamental relationship is still between health and absolute income or material standards. However, this interpretation cannot be sustained. For this to be an effect of individuals' absolute income would depend on there being a curvilinear relationship between income and mortality – if only absolute income levels mattered, how income was distributed would make no difference to the whole population's mortality rate if there were a simple straight-line relationship. Whether the relationship is linear or non-linear is controversial. The data covering some 300,000 people in the MRFIT mentioned above, suggest it is linear (Davey-Smith *et al.*, 1996a). Others have suggested it is at least gently curved (Backlund *et al.*, 1996). But it would have to be very sharply curved for the differences in income distribution between countries to be as strongly related as they are to national mortality rates. On top of that, if the underlying relationship was really between health and absolute material standards, there would be a strong relationship between life expectancy and average income rather than the weak international relationship shown in Figure 4.1. The evidence that the relationship with average absolute income is weak is confirmed – even within countries – by the data from the 50 states of the USA. The correlation between mortality and average income for each state is only 0.3 and disappears altogether ($r = 0.06$) when controlling for income distribution (Kaplan *et al.*, 1996). The relationship with income distribution remained strong not only after controlling for average income but also after controlling for the absolute income levels of the poor in each state (Kennedy *et al.*, 1996a,b). Even on its own this is strong evidence that we are not dealing with curvilinear effects of absolute income.

We can also tackle this from the point of view of changes over time. If there were a curvilinear relationship between absolute income and mortality we would expect health inequalities to narrow during the course of economic growth. However, the evidence suggests that in a number of developed market economies there has been a long-term widening of mortality differentials (Wilkinson, 1989; Mackenbach 1992; Pappas *et al.*, 1993; see also Macintyre, Chapter 2; Power and Matthews, Chapter 3; and Ben-Shlomo and Marmot, Chapter 17). Finally, a study by Waldmann using data from some 70 rich and poor countries showed that if the incomes of the poorest 20% in each country were held constant statistically, higher incomes among the top 5% (which would at least have been expected to reduced the infant mortality among their children), were instead associated with a significant increase in infant mortality rates across the population as a whole (Waldmann, 1992).

Thus the relationship we are dealing with is primarily a relationship between societal mortality rates and the scale of material inequality within each society. Instead of the absolute material standard of living giving rise to a particular standard of health as an expression of a relationship between material circumstances and biology, we are dealing with the health effects of social relationships structured by economic inequality.

To put the same point in another way: imagine looking at a relative frequency distribution of household incomes in a population – it tends to be skewed to the right rather than being normally distributed. What we are saying is that mortality is determined not by the proportions of the population vertically above various levels of absolute income marked on the horizontal axis, but by the horizontal distances under the curve which show how stretched out the income distribution is and how people's incomes differ within the population.

During the course of the epidemiological transition the health effects of the improvements in the material environment brought by continued economic development faded into the background and unmasked the powerful effect which the social environment has on health. The weakening of the relationship between mortality and absolute living standards does not mean that material factors make no difference to health. It is more likely to mean that the material benefits and disbenefits brought by economic growth are almost evenly balanced. This still makes a strong statement about our changed relationship with economic growth which perhaps ties up with the growing concern about its impact on the environment and the quality of life. The interpretation of the epidemiological transition, summarized above, reflects a change which we might expect to occur at some point in the long history of economic development.

To understand the relationship between income distribution and health we need to view it from rather a different perspective. Although we are now much less aware of the concern for equity than we are of the desire to increase living standards, in the forms of society which prevailed during most of human prehistory, the maintenance of a high degree of equality was a primary concern (see Cohen, Chapter 14). Harmonious social relations were the crucial determinant of the quality of life. Competition and conflict between species rarely has the potential to be so damaging as conflict within a species. Though another species may compete for the same food sources, among other human beings there is the potential for conflict for everything we value and need – not only food, but also housing, clothing, tools, valuables and sexual partners. At the same time, our fellow human beings are also our greatest source of help, comfort, pleasure, solace, self-confirmation – not to mention learning, acquisition of skills, language and mutual protection. Getting the relationship right is all important.

Given the potential for the best and the worst, maintaining harmonious social relations was crucial not only to the quality of life, but also to survival. As a result, overt expressions of material self-interest and competition for the necessities of life was virtually outlawed as a precondition for social life in primitive societies. Instead of goods being exchanged via the open self-interest of the market or haggling, hunting and gathering societies were based largely on systems of reciprocal gift exchange and food sharing. Most important was the need to avoid competition for necessities. Thus even in the few societies where some form of primitive money existed, there were different spheres of exchange and its use was confined to the exchange of luxuries. Systems of gift exchange and sharing ensured a high degree of equality. Woodburn (1982) refers to societies of hunters and gatherers as 'assertively egalitarian'. What Firth (1959, p. 290) says of the New Zealand Maori is typical: their system of food distribution and sharing meant that 'starvation or real want in one family was impossible while others in the village were abundantly supplied with food'. Numerous accounts indicate that similar patterns existed in different parts of the world. Although food usually belonged to those who collected it, it was distributed evenly throughout the tribe or village.

Designed to serve social purposes, gift exchange went well beyond anything practically or economically necessary. Sahlins (1974) accounts for its dominance in primitive societies in terms of the need to make and keep the peace between people. 'Whatever the utilitarian value (of reciprocal gift exchange), and there need be none, there is always a moral purpose ... to provide a friendly feeling ... and unless it did this it failed its purpose' (p. 220). 'The striking of equivalence (in reciprocal gift exchange), ... is a demonstrable foregoing of self-interest on each side, some renunciation of hostile intent or of indifference in favour of mutuality' (p. 220). As he points out, gifts make friends and friends make gifts.

Just as equality and gift exchange are the complementary institutions of mutuality, so inequality and the overt self-interest of market exchange express its absence. The need to avoid conflict over access to material goods was so great that primitive societies found it necessary to remove material inequality as an occasion for conflict and conduct social relations through continuous assertions of mutuality. Drawing on Hobbes's conception of relations between people in a state of nature as consisting of 'warre of each against all', Sahlins (1974) argues that in the absence of an over-riding sovereign power capable of keeping the peace, gift exchange and food sharing was the institutional means by which primitive societies turned potential conflict into co-operation.

There is evidence from several sources indicating that the link between income distribution and population mortality rates hinges on social cohesion. Given the potential for conflict and co-operation it is understandable that we should be highly sensitive to the character of social relations. High levels of social cohesion seem to be a common feature of societies which were unusually egalitarian and healthy. As I have discussed examples – such as Britain in the First and Second World Wars, the Italian-American town of Roseto in Pennsylvania and Japan – in more detail elsewhere (Wilkinson, 1996), I shall discuss them only briefly here.

During both the decades which include the world wars (1911–21 and 1940–51) life expectancy at birth increased by between six and seven years for both men and women. This was very nearly three times as fast as the average increase during the rest of the period 1901–91 (Winter, 1988). These dramatic improvements took place while medical resources were overstretched by the need to look after the casualties of war, and while production was diverted from civilian consumption to armaments. The suggestion that food rationing was the cause of the health improvements during the Second World War ignores the fact that the improvements were just as rapid during the First World War – despite the absence of food rationing.

During both wars a number of factors came together to diminish inequality (Winter, 1985, 1988). Firstly, the increased demand for labour virtually ended unemployment. Secondly, labour market forces led to a dramatic narrowing of differences in earnings among those in employment (Milward, 1984). Thirdly, in order to gain increased co-operation in the war effort, the Government pursued more egalitarian policies, not only by making taxes more progressive, but also by taxing luxuries and subsidising necessities to protect the poor from the effects which shortages would have had on prices. Growing equality and the sense of a unity of purpose in the face of a common enemy combined to create the much vaunted sense of cohesion and camaraderie during the wars. In the absence of rising material standards of consumption this provides strong evidence of the importance of equality and social cohesion as determinants of health.

My next example is the Italian-American town of Roseto in Pennsylvania which attracted the attention of epidemiologists during the 1960s when it was found to have much lower death rates, particularly from coronary heart disease, than the surrounding towns. It was found that Roseto's health advantage could not be explained in terms of traditional risk factors (Bruhn & Wolf, 1979). Epidemiologists were however impressed by the very closely knit and egalitarian nature of community life (see also Ben-Shlomo and Marmot, Chapter 17). They said they could not distinguish between the rich and the poor in any way – not through dress, speech or behaviour. Ostentation was frowned on and there were clear boundaries to how far the pursuit of money was acceptable. Bruhn and Wolf (1979, p. 136) noted that:

> the sense of common purpose and camaraderie ... precluded ostentation or embarrassment to the less affluent, and the concern for neighbors ensured that no one was ever abandoned. This pattern of remarkable social cohesion, in which the family, as the hub and bulwark of life, provided a kind of security and insurance against any catastrophe, was associated with the striking absence of myocardial infarction and sudden death ...

Their prediction that the community's health advantage would be lost as values changed and the community became less cohesive was later strikingly confirmed (see Egolf *et al.*, 1992).

The greater cohesiveness of Japanese society compared with developed market economies in the West has often been noted (Dore, 1973; Bayley, 1976). Many of the important differences seem to spring from changes which were initiated in, or immediately after, the Second World War. They put industrial relations and the control of Japanese industry on a different course. Over the decades since then, Japan has developed the narrowest income distribution among the developed market economies and the highest life expectancy in the industrialized world (Marmot and Davey-Smith, 1989). As an indication of the growth of social cohesion, Japanese crime rates decreased, particularly in the inner-city areas, and their relationship with deprivation weakened (Clifford, 1976). The impression is that the public sphere of life in Japan is more clearly part of social life and is governed more by shared moral values than by the market.

The death rates most closely related to international differences in income distribution include alcohol-related deaths, accidents and injuries (McIsaac and Wilkinson, 1997). Although all the major causal groupings – infections, cardiovascular diseases, cancers – are at least weakly related to income distribution, the most closely related causes seem to be indicative of socio-economic stress. Correlations with income distribution across the 50 states of the USA also show associations with most of the major causes of death (Kennedy *et al.*, 1996b). However, there are again strong indications of social stress – the strongest correlations are with homicide rates and violent crime (Kaplan *et al.*, 1996; Kennedy *et al.*, 1996b) and with 'unintentional injuries' (Kawachi *et al.*, 1997). Associations between income inequality and crime within the USA were also reported earlier by Braithwaite (1979) who used data for 192 Standard Metropolitan Statistical Areas. In addition, as well as other reports of an association with homicide within the USA (Blaus & Blaus, 1982; Currie 1985; Crutchfield, 1989; Balkwell, 1990), associations with homicide have also been found using international data covering between 30 and 40 countries (Braithwaite and Braithwaite, 1980; Messner, 1982).

The associations with crime, as well as the particularly strong correlations with homicide, alcohol-related deaths, accidents and injuries are convincing evidence of the increased social stress one might expect to accompany a decline in social cohesion consequent on widening income differences (see Stephens, Chapter 16). A striking confirmation of these relationships comes in a paper by Kawachi et al. (1997) which shows that the extent to which people trust each other is highly correlated both with income distribution and mortality across the USA. The proportion of people in each state agreeing with the statement that 'most people would try to take advantage of you if they got the chance' has a correlation of 0.8 with income distribution on the one hand and mortality on the other. The credibility of this as an indicator of social trust is strengthened by the fact that the proportion of people agreeing with statements such as 'most of the time people try to be helpful' and 'most people can be trusted' showed a similar pattern of strong negative correlations (Kawachi et al., 1997).

It looks then as if inequality damages health by creating social tensions and damaging social cohesion. The epidemiological findings which most obviously throw light on how variations in social cohesion could affect health are those dealing with social support and social affiliations. Social cohesion is sometimes measured in terms of people's membership of voluntary groups and associations (Putnam et al., 1993). Wider social affiliations and involvement in community life are known to be associated with health (Berkman and Syme, 1979). Closer social support, friendships and confiding relationships are also beneficial to health (House et al., 1988; Berkman, 1995).

There is also evidence that bad social relations and anger can be damaging to health (Salonen et al., 1991; Rael et al., 1995).

It is likely that narrower income differences are associated with smaller health differences within societies. Indeed van Dorslaer et al. (1997) reported a 0.9 correlation between income inequality and inequalities in self-reported morbidity across nine European countries. Narrower income differences may simply be a determinant and indicator of smaller differences in social status and so of a smaller health disadvantage associated with low social status. Studies of the physiological effects of low social status in other primates have provided strong evidence of the harmful effects of the chronic social stress associated with low social status (Sapolsky, 1993; Shively and Clarkson, 1994; Shively et al., 1997). The psycho-endocrinological pathways these studies describe have a number of parallels with the physiological differences found between Civil Servants at different levels in the office hierarchy (Brunner, 1996). If a large part of the social inequalities in health are a direct result of psycho-social processes and of chronic social stress associated with social division, social exclusion and differentiation, then it seems likely that the scale of the health disadvantage associated with low social status will be increased or decreased as socio-economic status differences are increased or decreased. In short, psycho-social processes related to the scale of social divisions in a society may explain health inequalities as well as the link between income distribution and mortality.

Although some of the current literature on social cohesion appears to give pride of place to the frequency of voluntary associations like choral societies and bowling leagues (Putnam, 1995), there is another way of looking at the strength of a society's associational life. In Putnam's study of social cohesion in the Italian regions (Putnam *et al.*, 1993) the associational strength of different regions earlier in the twentieth century was primarily a function of the membership of trade unions, co-operatives, mutual aid societies and political parties. In other words, it was related to the strength of institutions which provided practical support to manual workers and which represented their political interests. Using a measure of people's sense of powerlessness, Putnam *et al.* (1993, p. 110) shows that a much higher proportion of the population feel powerless in regions where associational life and social cohesion are weaker. This is particularly interesting in the context of the link between cohesion and health because a low sense of control has been found to be closely associated with poor health (Johnson and Hall, 1988; Karasek *et al.*, 1988) and is obviously related to the 'empowerment' approach to health promotion.

Putnam *et al.* (1993) emphasized the links between greater equality and the strength of civic life. As a result of the survey work on people's attitudes in the regions of Italy, they were able to say, 'Citizens in the more civic regions, like their leaders, have a pervasive distaste for hierarchical authority patterns' (p. 104), and 'Political leaders in the civic regions are more enthusiastic supporters of political equality than their counterparts in less civic regions' (p. 102). They summarized by saying, 'Equality is an essential feature of the civic community' and append a footnote to say that income distribution is more equal in the more civic regions (p. 105). The correlation given between greater income equality and more civic communities is 0.8, which is remarkably close to the correlation reported by Kawachi *et al.* (1997) among the states of the USA.

It is hard to avoid the conclusion that the social environment now exerts a more powerful impact on health in industrialized countries than the direct effects of differences in the material environment. Indeed the primary importance of the material environment now lies in the way in which it structures social relations. Both the *qualitative* evidence from examples of societies which are both egalitarian and healthy, and the correlations between *quantitative* measurements of social cohesion and income distribution suggest that social cohesion is or provides the crucial link between income distribution and mortality. However, as soon as we start to say that the effects of income distribution on mortality are mediated by social cohesion, there is a temptation to imagine that there is another more direct route to social cohesion which does not involve tackling inequality itself. But the evidence is clear – the correlations within Italy and the USA both suggest that as much as two-thirds of the differences in social cohesion or social trust are determined by income distribution. To ignore this would be folly. The scale of that folly can be judged from two findings. One is Kaplan and colleagues' (1996) finding that closely related to the degree of income inequality is a nine-fold variation in homicide rates (from about 2 per 100,000 to 18 per 100,000) among the states of the USA. The correlations reported between income distribution and both homicide and violent crime were around 0.7 (Kaplan *et al.*, 1996). Second is the fact that the belief that larger income differentials can be justified as a stimulus to economic growth is now contradicted by four studies using independent sets of data (Alesina and Perotti 1993; Glyn and Miliband, 1994; Persson and Tabellini, 1994; Birdsall *et al.*, 1995). All suggest that narrower income differentials complement economic growth.

This chapter would not be complete without a warning on a particularly intractable methodological problem of working with international income distribution data. Most countries' data come from official surveys of household income, and the problem is that survey response rates vary dramatically from one country to another. They vary from below 60% in some countries to as much as 95% in others (Atkinson *et al.*, 1995; McIsaac and Wilkinson, 1996). This would not matter if the non-responders were randomly distributed throughout the income distribution, but unfortunately they are not and the bias this introduces into measures of income distribution has not been adequately investigated. The work that has been done on the characteristics of the non-responders suggests that they are concentrated disproportionately among the poor and to a lesser extent among the rich (Atkinson and Micklewright, 1983; Redpath, 1986; Goyder, 1987; Wolf, 1988; Couper and Groves, 1996). This means that high non-response rates reduce the size of both tails of the income distribution, so creating the impression that income is distributed more equally than it really is. So much so that among the national data sets held by the Luxembourg Income Study there is a statistically significant correlation between response rates and the ratio of incomes received by the top and bottom 30% (McIsaac and Wilkinson, 1997). Almost no country reports a high degree of income inequality on the basis of low response rates. This bias seemed sufficient to mask the association between income distribution and mortality. However, when the countries were weighted by the response rates of their income surveys the expected association reappeared. In addition, the studies of non-response report that families with children have unusually good response rates – no doubt partly because someone is more often at home when interviewers call, but perhaps also because there is a greater willingness to co-operate. If income inequality is measured just among households with children, the association with mortality is again clearly apparent, regardless of the overall differences in response rates. These problems of the international comparability of income distribution data create serious pitfalls for the unwary and impose very severe limitations on the value of the data.

References

Alesina, A. & Perotti, R. (1993). *Income Distribution, Political Instability, and Investment*. Cambridge, Mass: NBER Working Paper 4486.

Atkinson, A. & Micklewright, J. (1983). On the reliability of income data in the Family Expenditure Survey 1970–7. *Journal of the Royal Statistical Society* **146**, 33–61.

Atkinson, A. B., Rainwater, L. & Smeeding, T. (1995). *Income Distribution in OECD Countries*. Paris: OECD.

Backlund, E., Sorlie, P. D. & Johnson, N. J. (1996). The shape of the relationship between income and mortality in the United States: evidence from the National Longitudinal Mortality Study. *Annals of Epidemiology* **6**, 12–20.

Balkwell, J. (1990). Ethnic inequality and the rate of homicide. *Social Forces* **69**, 53–70.

Bayley, D. H. (1976). *Forces of Order: Police Behavior in Japan and the United States*. Berkeley: University of California Press.

Ben-Shlomo, Y., White, I. R. & Marmot, M. (1996). Does the variation in the socioeconomic characteristics of an area affect mortality? *British Medical Journal* **312**, 1013–14.

Berkman, L. F. (1995). The role of social relations in health promotion. *Psychosomatic Research* **57**, 245–54.

Berkman, L. F. & Syme, S. L. (1979). Social networks, host resistance and mortality: a nine year follow-up study of Alameda County residents. *American Journal of Epidemiology* **109**, 186.

Birdsall, N., Ross, D. & Sabot, R. (1995). Inequality and growth reconsidered – lessons from East-Asia. *World Bank Economic Review* **9**(3), 477–508.

Blane, D., Davey-Smith, G. & Bartley, M. (1993). Social selection: what does it contribute to social class differences in health? *Sociology of Health and Illness* **15**, 1–15.

Blaus, J. & Blaus, P. (1982). The costs of inequality: metropolitan structure and violent crime. *American Sociological Review* **47**, 121.

Braithwaite, J. & Braithwaite, V. (1980). The effect of income inequality and social democracy on homicide. *British Journal of Criminology* **20**(1), 45–53.
Braithwaite, J. (1979). *Inequality, Crime and Public Policy.* London: Routledge.
Bruhn, J. G. & Wolf, S. (1979). *The Roseto Story.* Norman: University of Oklahoma Press.
Brunner, E. (1996). The social and biological basis of cardiovascular disease in office workers. In: *Health and Social Organisation,* ed. E. Brunner, D. Blane & R. G. Wilkinson, pp. 272–99. London: Routledge.
Clifford, W. (1976). *Crime Control in Japan.* Lexington Books.
Couper, M. P. & Groves, R. M. (1996). Social environmental impacts on survey cooperation. *Quality and Quantity* **30**(2), 173–88.
Crutchfield, R. (1989). Labor stratification and violent crime. *Social Forces* **68**, 589–612.
Currie, E. (1985). *Confronting crime.* New York: Pantheon.
Davey-Smith, G., Neaton, J. D. & Stamler, J. (1996a). Socioeconomic differentials in mortality risk among men screened for the Multiple Risk Factor Intervention Trial I. White Men. *American Journal of Public Health* **86**, 486–96.
Davey-Smith, G., Wentworth, D., Neaton, J. D., Stamler, R. & Stamler, J. (1996b). Socioeconomic differentials in mortality risk among men screened for the Multiple Risk Factor Intervention Trial II. Black Men. *American Journal of Public Health* **86**, 497–504.
Dore, R. (1973). *British Factory – Japanese Factory. The Origins of National Diversity in Industrial Relations.* Berkeley: University of California Press.
Duleep, H. O. (1995). Mortality and income inequality among economically developed countries. *Social Security Bulletin* **58**, 34–50.
Egolf, B., Lasker, J., Wolf, S. & Potvin, L. (1992). The Roseto effect: a 50-year comparison of mortality rates. *American Journal of Public Health* **82**, 1089–92.
Firth, R. (1959). *The Economics of the New Zealand Maori.* Wellington: R. E. Owen, Government Printer.
Flegg, A. (1982). Inequality of income, illiteracy, and medical care as determinants of infant mortality in developing countries. *Population Studies* **36**, 441–58.
Fox, J., Goldblatt, P. & Jones, D. (1985). Social class mortality differentials: artefact, selection or life circumstances? *Journal of Epidemiology and Community Health* **39**, 1–8.
Glyn, A. & Miliband, D. (1994). Introduction. In *Paying for Inequality: The Costs of Social Injustice,* ed. by A. Glyn & D. Miliband, pp. 1–23. London: Rivers Oram Press.
Goyder, J. (1987). *The Silent Minority: Non-Respondents on Sample Surveys.* Boulder, Colorado: Westview Press.
House, J. S., Landis, K. R. & Umberson, D. (1988). Social relationships and health. *Science* **241**, 540–5.
Johnson, J. V. & Hall, E. M. (1988). Job strain, work place social support, and cardiovascular disease: a cross-sectional study of a random sample of the Swedish working population. *American Journal of Public Health* **78**, 1336–42.
Kaplan, G. A., Pamuk, E., Lynch, J. W., Cohen, R. D. & Balfour, J. L. (1996). Income inequality and mortality in the United States: analysis of mortality and potential pathways. *British Medical Journal* **312**, 999–1003.
Karasek, R. A., Theorell, T., Schwartz, J., Schnall, R., Pieper, C. & Michela, J.

(1988). Job characteristics in relation to the prevalence of myocardial infarction in the U.S. HES and HANES. *American Journal of Public Health.* **78**, 910–18.

Kawachi, I., Kennedy, B. P., Lochner, K. & Prothrow-Stith, D. (1997). Social capital, income inequality and mortality. *American Journal of Public Health.* (In press.)

Kennedy, B. P., Kawachi, I. & Prothrow-Stith, D. (1996a). Income distribution and mortality: cross sectional ecological study of the Robin Hood index in the United States. *British Medical Journal* **312**, 1004–7.

Kennedy, B. P., Kawachi, I. & Prothrow-Stith, D. (1996b). Important correction. Income distribution and mortality: cross sectional ecological study of the Robin Hood index in the United States. *British Medical Journal* **312**, 1194.

Koskinen, S. (1988). Time trends in cause specific mortality by occupational class in England and Wales. In *Proceedings of IUSSP Conference* held in Florence June 1985. Florence: IUSSP.

Le Grand, J. (1987). Inequalities in health: some international comparisons. *European Economic Review* **31**, 182–91.

Lundberg, O. (1991). Childhood living conditions, health status, and social mobility: a contribution to the health selection debate. *European Sociological Review* **7**, 149–62.

Macfarlane, A. & Mugford, M. (1984). *Birth Counts: Statistics of Pregnancy and Childbirth.* London: HMSO.

Mackenbach, J. P. (1992). Socio-economic health differences in the Netherlands: a review of recent empirical findings. *Social Science and Medicine* **34**, 213–26.

Marmot, M. G. & Davey-Smith, G. (1989). Why are the Japanese living longer? *British Medical Journal.* **299**, 1547–51.

McIsaac, S. & Wilkinson, R. G. (1996). Income distribution and cause-specific mortality. *European Journal of Public Health* **7**, 45–53.

Messner, S. F. (1982). Societal development, social equality and homicide. *Social Forces* **61**, 225–40.

Milward, A. S. (1984). *The Economic Effects of the Two World Wars.* London: Macmillan.

Pappas, G., Queen, S., Hadden, W. & Fisher, G. (1993). The increasing disparity in mortality between socioeconomic groups in the United States, 1960 and 1986. *New England Journal of Medicine* **329**, 103–9.

Persson, T. & Tabellini, G. (1994). Is inequality harmful for growth? Theory and evidence. *American Economic Review* **84**(3), 600–21.

Power, C. (1994). National trends in birthweight: implications for future adult disease. *British Medical Journal* **308**, 1270–1.

Power, C., Manor, O., Fox, A. J. & Fogelman, K. (1990). Health in childhood and social inequalities in young adults. *Journal of the Royal Statistical Society* (Series A) **153**, 17–28.

Preston, S.H. (1975). The changing relation between mortality and level of economic development. *Population Studies* **29**, 231–48.

Putnam, R. D., Leonardi, R. & Nanetti, R. Y. (1993). *Making Democracy Work: Civic Traditions in Modern Italy.* Princeton: Princeton University Press.

Putnam, R. D. (1995). Bowling alone: America's declining social capital. *Journal of Democracy* **6**, 65–78.

Rael, E. G. S., Stanfield, S. A., Shipley, M., Head, J., Feeney, A. & Marmot, M.

(1995). Sickness absence in the Whitehall II Study, London: the role of social support and material problems. *Journal of Epidemiology and Community Health* **49**, 474–81.

Redpath, B. (1986). Family Expenditure Survey: a second study of differential response, comparing census characteristics of FES respondents and non-respondents. *Statistical News* **72**, 13–16.

Roberts, D. F. (1994). Secular trends in growth and maturation in British girls. *American Journal of Human Biology* **6**(1), 13–18.

Rodgers, G. B. (1979). Income and inequality as determinants of mortality: an international cross-section analysis. *Population Studies* **33**, 343–51.

Sahlins, M. (1974). *Stone Age Economics.* London: Tavistock Press.

Salonen, J. T., Julkunen, J., Salonen, R. & Kaplan, A. (1991). Cynical distrust and anger control associated with accelerated progression of carotid atherosclerosis. *Circulation* **83**, 722.

Sapolsky, R. M. (1993). Endocrinology alfresco: psychoendocrine studies of wild baboons. *Recent Progress in Hormone Research* **48**, 437–68.

Shively, C. A. & Clarkson, T. B. (1994). Social status and coronary artery atherosclerosis in female monkeys. *Arteriosclerosis and Thrombosis* **14**, 721–6.

Shively, C. A., Laird, K. L. & Anton, R. F. (1997). The behavior and physiology of social stress and depression in female cynomolgus monkeys. *Biological Psychiatry* **41**, 871—82.

van Doorslaer, E., Wagstaff, A., Bleichrodt, H. *et al.*, (1997). Socioeconomic inequalities in health: some international comparisons. *Journal of Health Economics* **16**, 93–112.

Waldmann, R. J. (1992). Income distribution and infant mortality. *Quarterly Journal of Economics* **107**, 1283–302.

Wennemo, I. (1993). Infant mortality, public policy and inequality – a comparison of 18 industrialised countries 1950–85. *Sociology of Health and Illness* **15**, 429–46.

Wilkinson, R. G. (1986). Income and mortality In *Class and Health: Research and Longitudinal Data*, ed. R. G. Wilkinson, pp. 88–114. London: Tavistock Press.

Wilkinson, R. G. (1989). Class mortality differentials, income distribution and trends in poverty 1921–1981. *Journal of Social Policy* **18**(3), 307–35.

Wilkinson, R. G. (1992). Income distribution and life expectancy. *British Medical Journal* **304**, 165–8.

Wilkinson, R. G. (1993). Income and health. In *Health Wealth and Poverty. Medical World*, special edn, March, pp. 6–11.

Wilkinson, R. G. (1994a). *Unfair Shares: The Effects of Widening Income Differentials on the Welfare of the Young.* Ilford: Barnardos.

Wilkinson, R. G. (1994b). Health, redistribution and growth. In *Paying for Inequality: the Economic Cost of Social Injustice*, ed. A. Glyn & D. Miliband, pp. 24–43. London: Rivers Oram Press.

Wilkinson, R. G. (1994c). The epidemiological transition: from material scarcity to social disadvantage? *Daedalus* **123**(4), 61–77.

Wilkinson, R. G. (1996). *Unhealthy Societies: the Afflictions of Inequality.* London: Routledge.

Winter, J. M. (1985). *The Great War and the British People.* London: Macmillan.

Winter, J. M. (1988). Public health and the extension of life expectancy 1901–60. In *The Political Economy of Health and Welfare*, ed. M. Keynes. Cambridge:

Cambridge University Press.

Wolf, W. (1988). Verzerrungen durch Antwortausfalle in der Konsumerhebung 1984. *Statistische Nachrichten* (Vienna: OSTAT) **43**(11), 861–7.

Woodburn, J. (1982). Egalitarian societies. *Man* **17**, 431–51.

5 Growth and maturation problems of children and social inequality during economic liberalization in Central and Eastern Europe

OTTÓ G. EIBEN

Introduction

A dramatic decline in the health status of populations was observed in the Eastern and Central European countries by the middle of the 1960s. This process did not change in the late 1980s during the period of economic liberalization and can be evidenced by certain basic demographic indices (ratio of natural increase/decrease, life expectancy, etc.) and by such indices as the budget allocation to health and welfare or public education, and in Gross Domestic Product (GDP) levels. Social inequalities have appeared between urban and rural populations indicated by differences in the ratio of pauperism, unemployment, etc., but also in the educational level. Using the well recognized principle that growth and maturation of children is a sensitive index of health and nutritional status and hence an index of the welfare of a population, this chapter discusses the growth and maturation problems of children and the social inequalities reflected in these societies over the period of economic liberalization in Eastern and Central Europe. Based largely on data from Hungary and Poland, the chapter sketches the process of social and economic changes and outlines the present situation using demographic data. The basic idea of this contribution is that urban and rural conditions of life differ in the respective sets of factors that influence child growth and maturation.

It was Villermé (1829) who very early on described the effects of social factors on the rate of growth of children and on final adult stature. Based on the data of Charles Roberts, Bowditch (1877) demonstrated that, across all ages, the sons of the labouring classes were shorter than those of the non-labouring classes. Pagliani (1879) published similar data – boys and girls of the well-to-do classes (*classe agiate*) were on the average heavier and taller than those of the poverty-stricken ones (*povera*). In general, children living under better socio-economic conditions have consistently exceeded in growth and maturation their counterparts living under worse condition (the phenomenon of 'hysteroplasia', Rietz 1906). Ten years later, Pfaundler (1916) described the phenomenon of 'proteroplasia', i.e. the observation that urban children were taller, grew faster and matured earlier than their rural peers.

The Western World has noticed the decline in health in Eastern Europe over the last few decades. A steep, relative and absolute decline in health status in Central and Eastern Europe has been observed since the mid-1960s and has been regarded as a considerable disadvantage to their joining the new Europe. Richard Feachem's (1994) searching analysis about this phenomenon pointed out: 'despite the rhetoric of the former Communist regimes, the emerging reality is that the health status of Central and Eastern Europe is extremely poor and has fallen far behind that in Western Europe in the past 30 years'. He supported his statements with a comparison of health status in the former socialist economies with that in the established market economies in the 1990s. Feachem (1994) demonstrated strong relationships between per capita GDP and several indicators of health, i.e. life expectancy at birth, risk of death in early childhood (0–14 years of age) and in adulthood, etc. The diverging trends in these indicators of general health since 1945 were conspicuous. At the end of World War II, the health status of the socialist countries was comparable to that of Western Europe. Since the mid-1960s, however, the relative trends have changed dramatically in Europe – in the former socialist countries living conditions stagnated or deteriorated whereas in the OECD countries they improved steadily. Feachem also pointed out the fall in life expectancy between the mid-1960s and the late 1980s, which was the greatest change observed, even in Hungary by 3.5 years. Mortality also rose markedly in the adult population during the same period – death rates for males between 45 and 49 years of age increased by 131% in Hungary (but only by 7% in East Germany), while in the OECD countries they decreased. The leading cause of death in this age range was ischaemic heart disease. It is a regrettable fact that 'the gap in health status between the former socialist and the Western European countries is wide, and is especially wide in the male cohorts of working age' (Feachem, 1994). It is a natural consequence that while the deceased adults drop out of economic production, the medical treatment of sick citizens imposes a further charge on society. Examining the causes for this gap, Feachem pointed out several risk factors, such as tobacco and alcohol abuse, unhealthy diet and obesity, air pollution as well as psychosocial factors as being contributory and added that a lack of physicians or hospital capacity was clearly not the problem (Feachem, 1994). Only in the first half of 1996 did the latter become a problem in Hungary.

At the end of his analysis, Feachem (1994) urged that 'repairing the damage associated with the past decades of communist rule, closing the gap in health status, must be a central objective for human-resources policy throughout the region'. It is clear that reversal of these trends will be neither easy nor quick, the countries in question need many years to close this gap. The picture is far from a merry one.

Let us examine the present situation more closely, taking into consideration the special status of children living in this region of Europe.

Social and economic liberalization in Hungary

In the late 1980s social and economic liberalization occurred in the Eastern and Central European countries, in the former 'socialist people's democracies'. In Hungary and also in the other countries in question, there was no radical revolution and hence the changes were called 'the velvety revolution', as in the case of Czechoslovakia. The sudden change in Hungary took place in 1989. Free elections resulted in the triumph of the Hungarian Democratic Forum, a middle-right conservative party. The changes in Hungary took their course in three steps:

(1) The collapse of the political structure, and this included the collapse of the totalitarian dictatorship, the centralized planned economy, several kinds of censorship and of the Soviet occupation, and a 'constitutional' state was created. The new government signed a pact with the liberals and the new socialists, i.e. with the ex-communists. This pact made it possible for many persons to wriggle out of their political responsibility and this caused a lot of tension which survives to this day.

(2) The change in the political structure with the new regime brought about economic changes which included economic liberalization. The new Hungarian government inherited a miserable economy from the former one party-state, which had accumulated an extremely large national debt. Hungary is deep in debt and the economy has been subjected to the oppressive influence of the International Monetary Fund (IMF) that now dictates economic policy and enforces the repayment of the national debt. Compared to the 1980s the tax burden on the people is heavy. The annual inflation rate exceeds 25–30%. As a consequence of these changes the living standards are deteriorating year by year. The

social strata of society have undergone rapid changes – a small group has become very rich, middle classes are sinking into poverty and the lower class lives in dire straits and in severe distress. Rates of unemployment, morbidity and mortality as well as pollution and crime have increased and life expectancy is on the decline.

(3) A third stage in the transition has been the need for a cultural or mental change in society, which was indispensable in order to ensure that the old way of thinking should not reappear again. It is necessary to note that the 1994 elections resulted in an unexpected victory of the Socialist Party, i.e. the ex-communists who formed a coalition with the Liberal Party and since then a Social-Liberal government has ruled in Hungary.

People's health consciousness has diminished during the past four decades in Hungary, i.e. health was not valuable until one became ill. Life chances, awareness of life and social values began to decay. The latter is characterized by the growing number of divorces. In 1994, there were 54,114 marriages in Hungary, i.e. 5.3‰ of the whole population; there were also 23,417 cases of divorce, 9.8‰ of all existing marriages or 2.3‰ in relation to the whole population (Demographic Yearbook, 1993). The number of suicides has increased and Hungary now has one of the highest rates for suicides (Table 5.1). In the 1980s, the suicide rate was between 41 and 45 per 100,000 inhabitants. In 1988, a slight decrease was observed and in 1994 the figure was 35.3 per 100,000 (Demographic Yearbook, 1993).

The 1994 distribution of suicides by social/professional groups in Hungary gives a characteristic picture – this rate is higher among manual than among white-collar workers. Suicide, as a percentage of all recorded deaths, was 12.6% among agricultural manual workers, 10.1% among the non-agricultural manual workers, 7.5% among the non-manual professional groups with high-school or lower education and 6.4% among non-manual groups of higher education (Demographic Yearbook, 1993). The number of alcohol addicts is estimated to be about 500,000–800,000. This means that the interests of about 1.5 million children are in jeopardy. Concurrently, the incidence of alcoholic cirrhosis has also increased. The estimated frequency of obesity and/or abuse of tobacco is also high, regrettably also in children. Along with all these (or precisely in spite of all these), Hungary is today a free country with a market-oriented economy. However, her inhabitants are mostly unhappy, disrupted, bitter, pessimistic and poor. How do the children grow up in such an unfavourable atmosphere? It is a well-known fact that 'the study of growth reflects the concerns as well as the condition of a society' (Tanner, 1981), so that child growth may serve as a mirror of the biological conditions prevailing in a

Table 5.1. *Number of suicides in Hungary*

Year	Per 100,000 inhabitants	As percentage of all deaths
1930–31	32.7	2.0
1938	29.3	2.1
1941	27.0	2.0
1948–49	23.9	2.1
1959–60	25.4	2.5
1969–70	33.9	3.0
1979–80	44.7	3.4
1989–90	41.1	2.9
1994	35.3	2.5

society. The history of the Hungarian growth studies offers some insight into this problem.

Growth studies in Hungary

The growth studies in Hungary over the past 120 years can be divided into four periods (Eiben, 1988a).

The first period of some 50 years began with the first investigations in 1873 and ended with World War I. During this period Hungary was a part of the Austro-Hungarian empire. The population was then much more stable and change of residence was rare in comparison with that seen today and consequently there was less mixing. Social and economic conditions and the overall standard of living were generally poorer than in the 1980s. Growth studies carried out during this period were mostly attributable to individual interest, were indifferently organized and poorly co-ordinated, and carried out by using unstandardized methods. Such studies were often parts of other, mostly medical, investigations so that anthropological aspects were usually secondary. Yet, it is to the credit of these Hungarian scientists that they recognized the importance of the problem that early, and undertook anthropometric growth studies long before those by many larger countries.

The second epoch, lasting some 20–25 years, was the period between the two world wars. Although the environmental factors influencing child growth and development did not significantly improve during this period, they did change. After World War I, the territory of 'historical' Hungary was reduced to less than one-third of its previous area, and the population decreased to less than one-half of its earlier number. It became more mobile and, with increasing capitalization and industrialization, urbanization became more intense. The (relative) genetic equilibrium of the population was upset. Social differences in the living conditions of the various groups of children became greater, especially with respect to child labour. Nutritional conditions also changed, with a greater variety of foods being consumed in the cities than in the villages where consumption of carbohydrates, primarily cereals, was predominant. The first large collection of data on the height of Hungarian children, reported by Bartucz (1923), took place during this period and was followed by organized investigations both in the capital and in the country. This period witnessed a unification of methods both in investigation and analysis, so that from this time onward the results of physical examinations are worth comparing.

The third epoch embraces about 40 years after Word War II. After the war, the difficult economic conditions greatly hindered growth studies, although the effect of the war on the growth and development of children was a matter of major concern (Véli, 1948). Should scientists of the twenty-first century wish to study the relationship between changes in the body dimensions of children and social changes, they will be most interested in the data collected during this period in Hungary. Post-war Hungary is a fascinating place for such investigations, for during this time famine and food shortage had ended and urbanization made considerable progress. With respect to nutritional intake, both quantitative and qualitative starvation ended. In Hungary, there was no famine, apart from a few extreme cases. Undernutrition was also very rare, at least in terms of calorie intake. Moreover, the diet which had hitherto consisted of carbohydrates predominantly became more varied with increasing consumption of animal proteins. These changes benefited all social classes and social strata, and consequently every possible group of children. Increased industrialization resulted in fast urbanization accompanied by remarkable internal migration. Biological mixing increased, changing the genetic composition of the population, so that recent generations exhibit signs of the heterosis effect. In every respect, socio-demographic changes altered life remarkably.

The early post-war growth studies were followed by numerous systematic regional investigations. Credit for their initiation and development must go to Professor M. Malán and his collaborators. During this period the need for growth studies carried out on large samples of children in different regions of the country was fully realized. The methods of investigation and analysis were up to date by current standards. There were both cross-sectional and longitudinal growth studies. On the whole, these were well co-ordinated, and since the 1960s most of them have been conducted within the context of the International Biological Programme, generally led by the University Departments of Physical Anthropology. In recent years, increasing interest has been shown by various scientists. Social paediatricians, school doctors, sports research workers, teachers of mentally retarded children, and many others have become involved and wished to participate in and contribute to such growth studies.

The start of the fourth epoch can be dated to the mid-1980s. In the early 1980s, a nationwide representative cross-sectional growth (and physical fitness) study was carried out and the first Hungarian growth standards were published (Eiben and Pantó 1986, 1987/88; Eiben *et al.*, 1991). This project was based on a careful design, and in order to ensure a proportionate representation it involved about 120 communities. The sample was stratified for each region. It comprised 39,035 healthy, 3 to 18-year-old boys and girls, representing 1.5% of all Hungarian children and youth in that age group. The anthropometric programme was performed in great detail, including 18 body measurements. Scores in seven motor tests and various kinds of socio-demographic data of the families were also recorded. This set of reference data serves now as a basic standard (*etalon*) for Hungary and (we are convinced) also for the whole Eastern–Central European region.

Other extensive regional growth studies (for details see Eiben, 1988a) and a detailed study of the secular trends (Eiben 1988b) are also worth mentioning here.

Some demographic characteristics of Hungary

Demographic characterisitics of Hungary are discussed and compared with both the Polish and British data. The Polish comparison is based on the reasoning that these two populations (i.e. Hungarian and Polish) are ethnically relatively homogeneous. All the other countries in Eastern and Central Europe – Czechoslovakia, Romania, the former Yugoslavia and the Carpathian part of Ukraine – have ethnically heterogeneous populations. Hungary is a small country occupying a territory of 93,000 km^2, with a (decreasing) population of 10.3 million. Poland is much larger (312,700 km^2) with a large (slightly increasing) population of 38.5 million (Table 5.2). Gender distribution does not show any remarkable difference, but female preponderance is more conspicuous in Hungary (Table 5.3). These data originated from the Demographic Yearbook 1993, and from the Statistical Pocketbook 1994.

Live-birth rate in Hungary has shown a decreasing tendency in the past 25 years, from 14.7‰ to 11.3‰. The death rate has, however, increased from 11.6‰ to 14.3‰. As a consequence, there is a natural decrease in population size, the rate of which has reached − 3‰ in the 1990s. A decrease in the rate of live births is observable also in Poland. In the UK,

Table 5.2. *Territory and population*
increase/decrease

Country	Territory 1000 km^2	Population (million)
Hungary	93.0	
1970		10.32
1980		10.71
1985		10.56
1989		10.40
1990		10.36
1991		10.35
1992		10.32
1993		10.29
1994		10.26
Poland	312.7	
1989		37.96
1990		38.12
1991		38.24
1992		38.36
1993		38.46
1994		38.54
UK	244.1	
1989		57.24
1990		57.41
1991		57.65
1992		57.85
1993		57.65

the same occurred only between 1970 and 1980, since then this index is stable at about 13.5‰. Death rates show a slightly increasing tendency in Poland, but a decreasing one in the UK. In the 1990s, the natural increase of the population in Poland has shown a diminishing tendency, while in the UK it seems to be stable (Table 5.4).

In the last few decades, infant mortality rate in Hungary has decreased from 35.9 to 11.5 per thousand live births. Data on live births by birth weight offer further information – the ratio of neonates with a low birth weight (under 2500 g) has decreased from 10.7% to 8.6% since 1970 (Table 5.5). Foetal losses have also decreased. In the late 1930s, it was 131.4 per 1000, in 1945 it was 169.1 per 1000, and today it is 11.5‰. Also the perinatal death rate has decreased in the 1980s and 1990s (Table 5.6). Live birth order also gives a characteristic picture – more than three-quarters of the children are first or second born ones, and only 15% of them are a third-born children (Table 5.7). The average number of children in Hungarian families is now under two and this seems to be related to the educational level of the mothers (Table 5.8). Infant nutrition shows a

Table 5.3. *Population by sex (%)*

Country	Male	Female	Number of females per 1000 males
Hungary (1992)	48.0	52.0	1085
Poland (1990)	48.7	51.3	1052
UK (1990)	48.8	51.2	1049

Table 5.4. *Selected data of the population*

Country	Live births (‰)	Deaths (‰)	Natural increase/decrease (‰)	Decreased under 1 year per 1000 live-borns
Hungary				
1970	14.7	11.6	3.1	35.9
1980	13.9	13.6	0.3	23.2
1985	12.3	14.0	−1.6	20.4
1990	12.1	14.1	−1.9	14.8
1992	11.8	14.4	−2.6	14.1
1994	11.3	14.3	−3.0	11.5
Poland				
1970	16.8	8.2	8.6	
1980	19.5	9.8	9.7	21.3
1985	18.2	10.3	7.9	18.5
1990	14.3	10.2	4.1	16.0
1992	13.4	10.3	3.1	14.4
1994	12.5	10.1	2.4	15.1
UK				
1970	16.2	11.8	4.4	
1980	13.5	11.7	1.8	12.1
1985	13.3	11.8	1.5	9.3
1990	13.9	11.2	2.7	7.9
1992/93	13.5	10.7	2.8	6.3

special picture. In the spirit of the old principle *Retour a la nature!*, breast-feeding has become common again in recent years and the use of artificial formula is decreasing (Table 5.9).

Another important aspect of the Hungarian population is its age-structure. Children between 0 and 14 years of age and people above 60 years each make up 19.2% of the overall population. In Poland, the proportions are 25.1% and 14.9% while in the UK they are 19.0% and 20.7% (Table 5.10). Progressive senescence overshadows both the Hungarian and UK populations.This problem appears to be most critical in Hungary where the proportion of people younger than 30 years of age has dramatically decreased. Although Hungary's population has doubled during the past

Table 5.5. *Live birth by birth weight in Hungary (%)*

Year	Birth weight	
	under 2500 g	over 2500 g
1950	5.5	94.5
1960	9.2	90.8
1970	10.7	89.3
1980	10.4	89.6
1989	9.2	90.8
1990	9.3	90.7
1991	9.3	90.7
1992	9.0	91.0
1993	8.6	91.4
1994	8.6	91.4

Table 5.6. *Foetal losses and perinatal death in Hungary (per thousand)*

Year	Foetal losses			Perinatal death
1937–38	131.4			
1945	169.1			
1955	61.2			
1960	47.6			
1965	38.8			
1970	35.9			
1975	32.8	male	female	
1980	23.2	25.9	20.3	22.9
1990	14.8	16.4	13.1	14.2
1994	11.5	12.8	10.2	9.3

Table 5.7. *Live birth order in Hungary 1994 (%)*

1st born	43.2	} 76.8	} 91.8
2nd born	33.6		
3rd born	15.0		
4th born	4.6		
5th born	1.9		
All others	1.6		

Table 5.8. *Live births according to the educational level of the mothers in Hungary 1994*

Completed grades (school classes)	Live birth (%)	
0	0.4	⎫
1–3	0.4	⎬ 5.1
4–5	1.3	⎪
6–7	3.0	⎭
8	52.5	
9–12	29.6	
13 +	12.7	
unknown	0.1	

Table 5.9. *Infant nutrition*

Year	Breast-feeding only	Mixed nutrition	Artificial foods
1970	42.8	47.9	9.3
1980	34.5	49.3	16.2
1985	44.0	43.8	12.2
1990	44.5	44.5	11.0
1993	46.8	42.2	10.9

Table 5.10. *Age-structure of the population*

Country	0–14	15–39	40–59	60 +
Hungary (1992)	19.2	36.1	25.5	19.2
Poland (1990)	25.1	38.1	22.0	14.9
UK (1990)	19.0	36.7	23.6	20.7

century, the number of individuals under age 30 has increased only by 29%. In the last decades of the nineteenth century and the first decade of the twentieth, people under 30 years of age made up 60% of the inhabitants of Hungary. In the 1940s, this proportion was 50% and in the 1980s it decreased to 43%. Stated another way, of a total population of 7.6 million in 1910, 4.6 million were under 30 years of age, about the same number as in the 1980s–1990s in a population of about 10.5 million. Thus, during 70 years, the population increase of about 3 million has not affected the absolute numbers of youth (Table 5.11).

The per capita GDP informs us about the economic status of a country. In the last few years, after the economic liberalization in Eastern and

Table 5.11. *Proportion of young people (younger than 30) in Hungary*

Year	Population in millions	Young people	
		in millions	in percentage
1910	7.6	4.6	60
1941	9.3	4.6	50
1982	10.7	4.6	43

Table 5.12. *Index of the Gross Domestic Product*

Country	1989	1990	1991	1992	1993
1985 = 100%					
Hungary	106.3	102.6	90.4	86.5	84.5
Poland	111.0	97.6	90.2	92.5	96.1
UK	117.3	117.7	115.1	114.6	116.9
Preceding years = 100%					
Hungary	100.7	96.5	88.1	95.7	97.7
Poland	100.3	87.9	92.4	102.6	103.8
UK	102.1	100.6	97.7	99.5	102.0

Central Europe, this index has gradually diminished in Hungary. Taking 1985 data as reference, this value was 106.3% in 1989, but in 1993 it was 84.5%. In Poland, 1991 was the worst year, with some slight increase seen more recently. In the UK, the same index seems to have been relatively stable at a high level (Table 5.12). The per capita index of GNP (Gross National Product) expressed in US dollars shows how poor Hungary and Poland are (Table 5.13). Table 5.14 shows the change in GDP per capita based on data of the European Comparison Programme in these three countries. In addition to the uncomforting demographic status and ecological situation, the Hungarian population is suffering from a high rate of unemployment, between 25 and 30% annual inflation rate and a decreasing life expectancy. Table 5.15 shows the increasing rate of unemployment. It is worth noting that communist rhetoric had declared the right of labour for all. There was no unemployment, every man and woman had to work for a very low wage. While everybody knew that hidden unemployment persisted, people were not prepared to face and manage a market-oriented economy. The data in Table 5.15 show the numbers of registered unemployed, but in reality their numbers might be even higher.

Morbidity and mortality show a tendency to increase. Diseases of the circulatory system and ischaemic heart disease qualify as the leading causes of death. *Morbus hungaricus*, i.e. pulmonary tuberculosis, was

Table 5.13. *Gross National Product per capita (US $)*

Country	1989	1990	1991	1992
Hungary	2,630	2,790	2,750	2,970
Poland	1,890	1,680	1,780	1,910
UK	14,670	16,030	16,530	17,790

Table 5.14. *Gross Domestic Product (GDP) per capita (based on European Comparison Programme; US $)*

Country	GDP per capita (US $)	GDP per capita (US $%) USA = 100	
		1990	1992
Hungary	6.240	29	25
Poland	5.020	23	21
UK	15.830	74	71

almost eradicated but at the end of the twentieth century it is re-emerging (Table 5.16). The interrelation between these disadvantageous factors and life expectancy is clear. For females, life expectancy has increased from 38 to 74 years during the twentieth century. In males, however, after a rise from 35 to 66 years, in more recent years there is a decrease (Table 5.17).

Historical truth demands one to state that during the last decade of the communist regime (a period called the 'mild dictature'), Hungary was one of the few countries that was in the limelight of the Western World, which now seems to have faded. Today a special situation has evolved – the 1980s was a transition period with growing ecological awareness and some positive demographic phenomena. But since the end of the 1980s, parallel with a spectacular collapse of the communist regime, all the growth pains of a market-oriented economy have given a taste of their ill-nature. All these are conceivable as being late effects and consequences of the four-decade long 'state-socialism'.

Urban–rural differences

It is a well-known fact that urban populations grow faster than rural ones. This rate of increase in Europe used to be 10–25% per decade but in several parts of the world it can be even 100% per decade. However, basic facilities such as enough housing, adequate schools and hospitals lag behind. As a

Table 5.15. *Unemployment*

Country	1985	1990	1991	1992	1993
Hungary					
Unemployment as percentage of the active population	—	—	—	9.3	11.3
Females as percentage of all unemployed				40.1	
Young unemployed as percentage of all unemployed				27.0	
Poland					
Unemployed as percentage of the active population	—	3.5	11.81	13.6	15.7
Females as percentage of all unemployed	—			53.2	
Young unemployed as percentage of all unemployed	—			36.9	
UK					
Unemployed as percentage of the active population		5.8	8.1	9.8	10.3
Females as percentage of all unemployed	31.2			32.4	
Young unemployed as percentage of all unemployed	37.9			29.8	

Table 5.16. *Number of deaths by causes in Hungary*
(1920–36: percentage; 1970–93: deaths per 10,000 inhabitants)

Causes of death	1920	1930	1936	1970	1980	1990	1993
Circulatory system	7.5	11.9	14.8	62.2	71.8	73.7	75.2
Tumors	3.1	6.6	8.3	21.9	26.1	30.1	31.6
Death by accidents, poisoning and violence	3.2	4.2	4.4	9.2	11.5	12.8	12.0
out of these: suicides	1.1	1.9	2.1	3.4	4.5	4.1	3.5
Digestive system	—	—	—	4.4	6.6	8.7	11.6
Respiratory system	14.3	12.8	10.6	5.6	9.4	6.4	6.9
Others				10.3	8.9	7.9	8.7

result, living standards are lowered and social inequalities become sharper. Table 5.18 shows the ratios of urban to rural population in Hungary, Poland and the UK. A ratio of about 60:40 was characteristic for Hungary and Poland during the 1980s, but today only a slight increase is occurring in the urban population.

The Hungarian National Growth Study (HNGS) provided important data on the fact that urban boys and girls are taller than rural ones. The

Table 5.17. *Life expectancy at birth and average age at death in Hungary (year)*

Year	Male	Female
1900–01	35.56	38.15
1920–21	41.04	43.13
1930–31	48.70	51.80
1941	54.95	58.24
1950	59.88	64.21
1960	65.89	70.10
1970	66.31	72.08
1975	66.29	72.42
1980	65.45	72.70
1985	65.09	73.07
1990	65.13	73.71
1994	64.84	74.23

Table 5.18. *Ratio of urban and rural population (%)*

Country	Urban	Rural
Hungary		
1985	61.0	39.0
1992	63.1	36.9
Poland		
1985	60.1	39.9
1990	61.7	38.3
UK		
1985	92.0	8.0
1990	89.1	10.9

mean urban heights are above the 50th percentiles for both sexes. Rural boys and girls, especially in the years of puberty, fall behind the 50th percentiles. This height difference is about 3–4 cm in 15-year-old boys and girls. There is a significant difference between urban and rural children in all age groups. The pubertal growth spurt in urban boys and girls occurs about one-and-a-half years earlier than in those from rural communities. The same trend can be observed for body mass (Eiben *et al.*, 1996).

HNGS data were analysed also by parental profession. Four professional categories were formed. In general it has been found that children of parents working in agriculture are usually the shortest, children of industrial manual workers are taller, while the children of non-manual (i.e. white collar parents) are the tallest; these features are also true for Hungary. It was found that the sons of industrial and agricultural and other manual workers showed very small differences in height, while the sons of non-manual fathers were the tallest. A similar pattern was found when the mother's profession was used, with sons of non-manual mothers towering above all the other groups. For girls, the tendency is again the same. Daughters of non-manual fathers are taller in all age groups than daughters of manual worker fathers, and the differences increase both during pre-puberty and puberty. Compared by the mother's profession, the differences mentioned here are even larger, and they are evident also in early childhood (Eiben and Pantó, 1988).

Education of the fathers and/or mothers when used as an analytical criterion of HNGS data discriminates the sample remarkably well. Means of height (and other length measurements) in the upper social strata were above the national mean and/or the 50th percentile, while the lower ones were below these values (Eiben, 1989). The sons of a father with an uncompleted basic education, i.e. general school, are the shortest; sons of fathers who completed general school level are taller; those of fathers with a vocational training are taller again; those of fathers with secondary schooling are still taller, while the sons of fathers with high school or university level education are the tallest. The higher the fathers' educational level, the taller their sons. In this group also pubertal growth spurt occurs earlier than in the other groups of boys. With respect to maternal education, boys show a similar picture, and indeed, in sons of mothers of a low educational level, backwardness in growth and development is more marked. This phenomenon is more marked in girls, especially in the daughters of fathers and mothers of a low educational level who are the shortest, while the daughters of fathers and mothers with a university degree are the tallest, particularly after puberty. Thus, the higher the educational level of the parents, the taller their sons and daughters. These differences between the two extreme social groups are significant for both sexes and is between 6–7 cm difference. The onset of puberty shows the same tendency – age at menarche and oigarche in the lower education categories is delayed by 1–2 and 5–6 months respectively compared with the upper educational categories (Eiben, 1989).

Most of the factors that cause differences between socio-economic groups are more or less correlates of each other, as, for instance, level of education and professional status, since the former determines in part the latter. This is why it is so difficult to separate the effect of certain ecological factors. Higher education is associated with better nutrition and with better care of infants and children. Additionally, these parents can usually make better use of social services than others. For the welfare of Hungarian youth, the educational level of the parents is a determinant factor. As demonstrated, the higher the educational level of the parents – both fathers and mothers – the taller are their sons and daughters. These differences in height can be usually seen by early childhood, and during pre-puberty and puberty they usually become even more marked. The profession of the parents also characteristically influences growth and the developmental process of their children. The trend to tallness rises steadily from the agricultural manual workers through industrial and other manual workers to the non-manual classes. However, professions of parents as an organising principle – at least in Hungary – is less suitable for the purpose of describing a family's standard of living, or to characterize the child-centred attitude of the home. The most important environmental factor seems to be the educational level. The author is convinced that the cultural milieu is the most important social factor influencing growth and maturation of youth without taking the urban and/or rural environment into consideration. Consequently, it seems important to point out the determinative role of mothers in creating a better background for the family (Eiben, 1989). Similar trends seem to be observable for Poland.

Some Polish data on height and menarche

Hulanicka (1990) reported on the heights and weights of pre-pubertal and pubertal Wroclaw and Warsaw boys (age 13.5–15.3, $n = 6969$). In both cities, the same sources of variation in stature were detected, related to some socio-economic status (SES) factors, such as parental educational level, occupation, origin, number of children per family, etc. Boys from the extreme categories of SES factors showed the same distance in stature in the two independent samples from Wroclaw and from Warsaw – a difference of about 2.5–3.6 cm. Similar to the former surveys, gradient associations of children's height and weight with the degree the urbanization of the subjects' settlement persisted in 1988 (Hulanicka et al., 1990). In comparison with 1978, their 1988 sample revealed an increase of height in children from all the examined environments, but in contrast to the former period, this increment has not been uniform for all age groups for children from towns and big cities. In the latter, the increments were 1.5–1.7 cm, while in rural boys it was 2.4 cm. Children born at the end of the 1970s and whose early childhood coincided with the economic recession, did not diverge in height from their 1978 coevals (Hulanicka et al., 1990).

Bielicki et al. (1986) reported social-class gradients in menarcheal age in Upper Silesia – Katowice, Bytom and Sosnewiec (Poland) – based on a large sample ($N = 19,000$), studied in 1981 in relation to parental education and father's occupation. Age at menarche tended to increase with decreasing parental education, although the gradient was not steep. Mean age at menarche varied between 12.8 and 13.3 years. Mean menarcheal age for an occupational group was strongly dependent upon the group's socio-econ-omic status, the latter being defined by parental education, family income, family size and dwelling conditions (Bielicki *et al.*, 1986). A similar study was repeated in the same region 10 years later (in 1991) by Hulanicka *et al.* (1993). It is remarkable that – contrary to the authors' expectation – there was no change in mean age at menarche in Katowice between the decades investigated. Both studies cited here reported a mean age at menarche of 13.1 years. Hulanicka realized that maturation of children was heavily influenced by very significant inequalities in social and economic condi-tions in that region. She pointed out that since the Polish population is ethnically homogeneous and their sample size was very large, the influence of social differences on maturation was very obvious. As a comparison, mean age at menarche of the HNGS girls was 12.79 years (Eiben, 1988b).

Hulanicka *et al.* (1990) have published other growth data on 6 to 18-year-old Polish boys and girls, based on their growth study carried out in 1987–88. The earliest maturers were girls from the big cities ($m = 12.96$ years), followed by those from towns ($m = 13.4$ years) and villages ($m = 13.53$ years). Rural girls from a certain region reported menarche occuring on average 1.02 years later than the Warsaw girls. A deceleration in the age trend of menarche was found in the 1988 sample of girls as compared to those in the 1978 cohort, mostly from those environments where a concurrent inhibition of the secular trend in the height of pre-pubertal children was also noted. This was an additional fact in favour of the hypothesis connecting economic recession with demographic data. In other words, the biological effects of an economic crisis were observed here.

Conclusion

In summary, we should realize that in the last decades there has been an obvious change in the mode of life in towns and villages, both in Hungary and Poland. In Hungary in the 1950s there was a political will to eliminate the disadvantageous differences prevailing in rural areas. Even today, differences in urban and rural modes of life persist, affecting the biological changes during childhood. Parents of a higher educational level tend to live in towns. Although the majority of these people in Hungary live from a modest salary, they can better exploit the opportunities that exist. The same holds true for nutrition, medical care and treatment, physical educa-tion and sports or even such extra curricular activities such as the oppor-tunities for learning music or languages offered to the children. Also our earlier results have clearly demonstrated the impact of a well-planned and systematic physical education in schools and even in the kindergarten. More educated parents motivate their children to participate in these regular activities (Eiben *et al.*, 1994).

Better life conditions in towns are characterized by positive, clearly perceptible factors which influence the growth and maturation of children and the children's physical status is highly susceptible to these influences. In the 1980s in Hungary, it has been repeatedly proven that favourable environmental factors promoted growth while the unfavourable factors retarded it (Eiben, 1988b). In the mid-1990s, however, social inequalities became more marked in Hungary. It seems obvious that childhood, including pre-pubertal and pubertal periods, is the life-period most sensitive to environmental effects, such as socio-economic factors presented by the urban or rural mode of life. To study these unrepeatable processes would be of vital importance, however, at present (in the mid-1990s) there is no money for such research in Hungary. For the time-being we must rest content with promises of the current socialist-liberal government which declares itself to be an 'expert government'.

The last question to address is how it would be feasible to create more opportunities for growth and acquisition of physical fitness for urban and rural youth. It is an economic and socio-political rather than a human biological problem. Auxologists have for a long time had the ambition to call the politicians' attention to this problem, to elaborate a more equitable distribution, a better health and welfare policy as well as a fair policy aimed at the youth. In this sense we still have a lot to do, quickly and definitely, sparing no pains or money. It would be a manageable way to realize the principle that a healthy child is a happy child. The responsibility of the adults as well as of the authorities, including governments, is tremendous. We must remember that children grow up only once.

Acknowledgement

This paper was funded by the Hungarian National Foundation for Scientific Research (OTKA grant T 013098).

References

Bartucz, L. (1923). Growth of school children in Hungary. *Anthropologiai Füzetek* **1**, 88–92. (In Hungarian.)

Bielicki, T., Waliszko, A., Hulanicka, B. & Kotlarz, K. (1986). Social-class gradients in menarcheal age in Poland. *Annals of Human Biology* **13**, 1–11.

Bowditch, H. P. (1877). The growth of children. Reprinted from *The 8th Annual Report of the State Board of the Health of Massachusetts*, Boston, pp. 1–63.

Demographic Yearbook 1993. Budapest: KSH.

Eiben, O. G. (1988a). History of human biology in Hungary. *International Association of Human Biologists, Occasional Papers 2/4*, p. 75. Newcastle upon Tyne.

Eiben, O. G. (1988b). Secular growth changes in Hungary. *Humanbiologia Budapestinensis*, Suppl. 6, p. 133. (In Hungarian.)

Eiben. O. G. (1989). Educational level of parents as a factor influencing growth and maturation. In *Auxology '88. Perspectives in the science of growth and development*, ed. J. M. Tanner, pp. 227–34. London: Smith-Gordon.

Eiben, O. G. & Pantó, E. (1986). The Hungarian national growth standards. *Anthropologiai Közlemények* **30**, 5–23.

Eiben, O. G. & Pantó, E. (1987/88). Body measurements in the Hungarian youth at the 1980s, based on the Hungarian national growth study. *Anthropologiai Közlemények* **31**, 49–68.

Eiben, O. G. & Pantó, E. (1988). Some data to growth of Hungarian youth in function of socio-economic factors. *Anthropologie* (Brno) **25**, 19–23.

Eiben, O. G., Barabás, A. & Pantó, E. (1991). The Hungarian national growth study I. Reference data on the biological developmental status and physical fitness of 3 to 18-year-old Hungarian youth in the 1980s. *Humanbiologia Budapestinensis*, **21**, 123.

Eiben, O. G, Barabás, A. & Pantó, E. (1994). Körperliche Entwicklung von Stadt- und Landkinder in Ungarn. *Anthropologiai Közlemények* **36**, 101–27.

Eiben, O. G., Barabás, A., Kontra, G. & Pantó, E. (1996). Differences in growth and physical fitness of Hungarian urban and rural boys and girls. *Homo* **47**, 191–205.

Feachem, R. (1994). Health decline in Eastern Europe. *Nature* **367**, 313–14.

Hulanicka, B. (1990). Physical development of boys at puberty as a reflection of social differences in population of the city of Wroclaw. *Materalyi i Prace Anthropologiczne* **111**, 21–45.

Hulanicka, B., Brajczewski, C., Jedlinska, W., Slawinska, T. & Waliszko, A. (1990). *City, town, village. Growth of children in Poland in 1988*, p. 52. Wroclaw: Monographies of the Institute of Anthropology, Polish Academy of Sciences.

Hulanicka, B., Kolasa, E. & Waliszko, A. (1993). Age at menarche of girls as an indicator of the socio-political changes in Poland. *Anthropologie et Préhistoire* **104**, 133–41.

Pagliani, L. (1879). *Lo sviluppo umano per eta, sesso, condizione sociale ed etnica: studiato nel peso, statura, circonferenza toracica, capacita vitale e forza muscolare.* Milan: G. Civelli.

Pfaundler, M. (1916). *Körpermass-studien an Kindern*, p. 148. Berlin: Springer Verlag.

Rietz, E. (1906). Körperentwicklung und geistige Begabung. *Zeitschrift für Schulgesundheitspflege* **19**, 65–98.

Statistical Pocketbook 1994. Budapest: KSH.

Tanner, J. M. (1981). *History of the study of human growth*, p. 499. Cambridge: Cambridge University Press.

Véli, G. (1948). How did war influence the somatic development of children? (In Hungarian.) *Népegészségügy* (Budapest) **29**, 667–74.

Villermé, L. R. (1829). Memoire sur la taille de l'homme en France. *Annales d'hygiene publique et de médicine légale* **1**, 351–99.

6 Social inequalities in the re-emergence of infectious disease

JOHN D. H. PORTER AND JESSICA A. OGDEN

Introduction

The spectre of social inequality in the emergence of infectious diseases has been well described through epidemiological investigations. Despite these findings and the identification of groups to target for health measures, public health has failed to alter this dynamic. Increases in tuberculosis (TB) and human immunodeficiency virus (HIV) infections are associated with inner city poverty in high income countries and with population growth and widespread population displacement to the cities in low and middle income countries. Epidemiology within public health seems to be able, therefore, to successfully identify the problem, but at present is unable to provide interventions to find ways of changing the unequal socio-economic dynamic in the population.

This chapter discusses the reasons for the emergence and re-emergence of infectious diseases in relation to socio-economic parameters by looking at the inter-relationship of agent, host and environment. TB and HIV infection will be used to demonstrate the aspects of social inequality in the emergence of these diseases. Finally, potential ways of altering and improving the present unequal social dynamic will be outlined.

Background

Like other scientists, infectious disease epidemiologists are 'cultural actors, prone to the blind spots and folk theories of their own society' (Arras, 1988). They have a particular disciplinary structure and perspective by means of which they explain and study the emergence of infectious disease. In short, epidemiology is the study of the distribution and determinants of disease in populations, and is the cornerstone of public health practice. This scientific, biomedical framework looks at the effect of events in individuals' lives ('exposures') to determine whether they are associated with the development of disease. These associations are then used to find interventions which, through changes in policy, will lead to the improvement of the overall health status of a population.

An epidemiologist begins work by collecting descriptive information to try and find associations between particular exposures and disease and then tests these hypotheses using analytical methods. Indications of social status are included in this descriptive data collection, and these factors, as this contribution will demonstrate, are often strongly associated with infectious disease transmission. For example, a person living in overcrowded conditions (the exposure) is more likely to transmit TB (the disease) to household contacts. Studies which investigated this association then led to interventions to help prevent TB, which in the biomedical model in the early part of the twentieth century, were to remove people from their homes and to isolate them in sanatoria. With the benefit of hindsight we might say today that a more appropriate intervention would have been to reduce overcrowding – to tackle the cause and not the symptom. Epidemiologists and other public health practitioners now recognize that this kind of social or community intervention provides longer term solutions than the more individually focused interventions favoured in the past.

Because the broader social and environmental context is so important in the transmission of infectious disease, the study of social inequality becomes relevant to epidemiological practice. This is no less true today as we witness the re-emergence of TB and indeed the emergence of 'new' infectious diseases like AIDS.

What dimensions of social inequality are most important in the re-emergence of infectious diseases? To answer this question, epidemiologists look at the interaction between the host, the agent and the environment, knowing that changes in one of these will lead to changes in the others. While it is important to recognize the influence social factors have on this dynamic, it is nevertheless difficult to untangle the many layers of social context in order to define the most salient dimensions for disease transmission.

Agent

Each infectious disease has its own unique character and method of transmission. Some diseases have stronger associations with people living in poverty than others. Thus epidemiologists talk of 'democratic' versus 'undemocratic' diseases. Democratic diseases, such as influenza, are transmitted easily and widely across class as well as racial and ethnic lines, making it difficult to stigmatize and lay blame for illness on less empowered groups (Arras, 1988). Undemocratic diseases like cholera, TB and AIDS, disproportionately affect marginalized groups, who are then rather convenient to blame should the disease emerge elsewhere.

Just as human environments change, so do microbial environments. Organisms change their characteristics to survive. In high income countries for example, the widespread use and abuse of antibiotics during the past 50 years has led to the development of resistant strains of bacteria. Methicillin resistant *Staphylococcus aureus*, for example, is now a common cause of nosocomial outbreaks of staphylococcus in hospitals. An example of this process in low income countries is a new strain of cholera which is resistant to standard antibiotics. Another way to view this process, apropos the wider philosophical interests of this colloquium, is articulated in a recent novel: 'it is clear that modern medicine has created a serious dilemma. The problem is not only that a single bacterium has become more virulent... But in a sense modern medicine has put natural selection out of commission. Something that has helped one individual over a serious illness can in the long run contribute to weakening the resistance of the whole human race to certain diseases' (Gaarder, 1995: 322–3).

Today it is common to think of the world as divided economically between North and South, with wealth in the North and comparative poverty in the South. The social inequality inherent in this division extends also to infectious disease. Diseases 'choose' communities according to their ability to transmit within them. Thus, for example, the introduction of cholera into a refugee camp, where there is poor sanitation and an inadequate water supply, will rapidly lead to an epidemic, while the introduction of the same agent into an industrialized country where there are good sanitary conditions, is unlikely to have such serious consequences.

In addition to the unequal distribution of infectious agents between the high and low income countries, unequal distribution occurs within countries as well, and this distribution is affected by the social conditions of the particular communities. For example, while relatively rare in industrialized countries, where it does exist, TB 'undemocratically' affects those who dwell in society's margins – the homeless, the problem drinkers and the intravenous or injecting drug users.

Host

What makes one person more susceptible to an infectious agent than another, and how much of this susceptibility is due to social factors? In the complex set of interactions that result in disease emergence (or re-emergence) in a population, the human element – population growth, density and distribution; immunosuppression; and behaviour – plays a critical role (Lederberg, 1992). Increases in the size, density and distribution of populations can bring people into contact with new pathogenic organisms or with vectors that transmit these organisms. Immunosuppression, a weakening of the immune system, can be caused by many different factors including: ageing, prematurity, HIV infection, malnutrition, pregnancy, severe trauma and malignancy. TB, for example, is an infection which can be reactivated by immunosuppression, as occurs with HIV infection. Finally, the behaviour of the host can affect the risk of acquiring an infection. The HIV pandemic is the most devastating outbreak of a sexually transmitted disease since the emergence of syphilis almost 500 years ago in Western Europe. In the case of AIDS, sexual intercourse is one of the most important 'behaviours' affecting people's vulnerability to acquiring HIV infection. An increasing awareness of the wider economic and political factors influencing HIV transmission has led in recent years to a growing dissatisfaction with the language of 'risk'. Because the concepts of 'high risk group' and 'high risk behaviour', with their focus on the individual, inadequately account for these broader dimensions, an alternative discourse of *vulnerability* has evolved, with particular emphasis on the notion of social vulnerability. This semantic shift is seen as an important first step in changing the way in which research and researchers across disciplines think about disease and the mechanisms of, or responsibility for, disease transmission, and how more appropriate approaches to intervention might be framed (see Parker, 1991).

Environment

Both the agent and host relate to the environment. Three important aspects of the environment currently affecting the re-emergence of infectious diseases, are: rapid population growth, disasters and decaying public health infrastructures. All of these relate closely to social inequality.

Since the 1960s, people in low and middle income countries have migrated from rural areas to the cities in ever increasing numbers. Urban population growth has been accompanied by overcrowding, poor hygiene, inadequate sanitation and insufficient supplies of clean water. This kind of urban development has also led to ecological damage through the emergence of slum areas and shanty towns. These factors have created conditions under which certain disease-producing organisms and their vectors have thrived. Once again, therefore, people in low socio-economic groups are disproportionately affected.

Emergency situations – or 'disasters' – whether natural or man-made – also engender mass human migrations. Refugee camps are often characterized by poor nutrition and sanitation. These communities then become ready sites for the rapid transmission of infectious disease. Economic collapse, war and natural disasters have also caused the breakdown of public health measures and the emergence or re-emergence of a number of diseases. In New York city, the re-emergence of TB in the mid-1980s was associated with the financial cuts in the TB control budget (Brudney, 1991; Reichman, 1991).

There are circles within circles of social inequality. From the global perspective it is easy to see the economic and social divide between North and South. These divisions are also apparent within countries and within communities. Social inequality would appear to be a part of the human condition. Yet simply recognizing inequality does not indicate avenues to redress it. The question becomes: on what level should solutions to inequality be sought? The prospect of tackling the global redistribution of wealth may cause us to fall back, throw up our hands and give up. Such perspectives also serve to rob those communities most (disproportionally) affected of their capacity to devise their own solutions. In countries where governments have inadequate capacity to handle these problems nationally, at what point can or should communities themselves take the responsibility for action? Perhaps the pie can be more easily digested if eaten in smaller mouthfuls. We return to this point later.

TB and AIDS (Acquired Immune Deficiency Syndrome) are emergent diseases affecting large populations of people worldwide and usefully demonstrate the intimate relationship between infectious diseases and social inequality.

Tuberculosis

Agent/host/environment dynamic in tuberculosis infection

TB is caused by *Mycobacterium tuberculosis* which is transmitted from person to person in airborne droplets. Hosts are most likely to transmit the infection if they are *sputum positive*, that is producing sputum which contains the organism. This mode of transmission means that the physical and physiological environment in which people live affects their risk of acquiring infection. People who are in close contact with a TB case, who are poorly nourished (McMurray *et al.*, 1990a,b; Heywood, 1993) and have a poorly functioning immune system, are at a significant risk of becoming infected.

In the UK, TB has long been linked with social inequality. In the late eighteenth century, the onset of the Industrial Revolution and the decline of agricultural employment together created the conditions for large-scale migration into the cities. In 1842, Chadwick's Report of the Sanitary Condition of the Labouring Population (see Weber, 1899) showed that the majority of the population were living in appalling, overcrowded conditions and that areas with the greatest poverty suffered the highest mortality rates. The overcrowded living conditions in the urban slums undoubtedly contributed to the transmission of infection. It was recognized in 1899 that 'the most powerful factors in producing tuberculosis are: (1) air contaminated by the so called tubercle bacillus, (2) food inadequate in purity, quality, quantity, (3) confined and overcrowded dwellings, (4) a low state of general health and resisting power of the body' (Weber, 1899).

Tuberculosis in high income countries

The declining incidence of tuberculosis in the industrialized world since the middle of the nineteenth century has been attributed to improved socio-economic conditions and the isolation of infectious cases (Figure 6.1). Poor nutrition, overcrowding and fewer beds in sanatoria during times of war, have led to periodic increases in TB. Yet despite great improvements in both social conditions and therapeutic options, tuberculosis has not been eradicated, and incidence continues to be higher among the poor and marginalized (Lerner, 1993).

It is clear that improvements in housing, hygiene, nutrition and the provision of sanatoria, have all had an impact on the incidence of TB in the richer countries, although views differ as to which improvements were the *most* influential. McKeown argues that improvement of the British diet is the most likely explanation for the decline in TB during the nineteenth century and the early part of the twentieth century (McKeown *et al.*, 1962; 1975). Certainly nutrition has been found to affect susceptibility to infectious disease generally (Heywood and Marks, 1993), with some studies demonstrating that the cell mediated immune response to TB is compromised in individuals suffering from malnutrition (McMurray *et al.*, 1990a). Indeed the increased rates of TB seen in Asian immigrants to Britain has been attributed to a reduction in serum vitamin D levels (Rook, 1988). There is also some support from animal studies showing that resistance to TB is reduced by lack of protein and also lack of vitamins A, C and D (McMurray *et al.*, 1990b).

In the UK, the continued fall in TB following World War II was slowed but not reversed by the arrival of immigrants from countries with a high prevalence of TB (Springett, 1964). Much higher rates have been

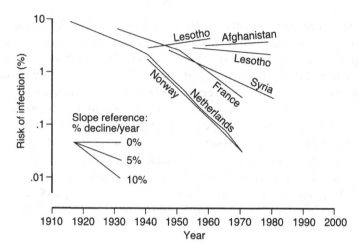

Figure 6.1. Trend in annual risk of infection, selected countries. (*Sources:* Sutherland, 1976; 19: 1–3; Cauthen *et al.*, 1988 WHO/TB/88.154.)

documented in the Indian, Pakistani and Bangladeshi communities since the 1960s (Darbyshire, 1995). One important finding from these surveys has been that notification rates for immigrant groups tended to be highest for recent arrivals, becoming progressively lower with length of stay in the UK (Research Committee of the British Thoracic and Tuberculosis Association, 1975).

Notifications of TB have increased in England and Wales over the past few years, as elsewhere in Europe and the US. An estimated 8000 additional cases occurred between 1982 and 1993 in England and Wales (Darbyshire, 1995). Poverty, unemployment and homelessness are all inextricably linked to this re-emergence of TB. Recent surveys of the homeless in London indicate that 2% of people living in hostels or using day centres have active TB (Citron *et al.*, 1995). Another study in London looked at the association between socio-economic level and notification rates for TB in the period 1982–91. Although the effect was small, the study demonstrated that the average level of notification was correlated with overcrowding, the proportion of migrants in the population and increases in unemployment (Mangtani *et al.*, 1995).

Thus the increase in TB notifications since 1988 may not be simply due to immigration as has been suggested. Socio-economic levels may be one important factor in the incidence of TB, but it is very difficult to untangle the effects of poverty from, for example, ethnicity (Darbyshire, 1995). In the UK the greatest increase in TB between 1980 and 1992 occurred in the poorest 10% of the population. In this group notifications increased by 35% compared with a national increase of 12%.

Tuberculosis in low and middle income countries

The failure to reduce TB in most low and middle income countries in spite of the availability of effective chemotherapy has been attributed to the failure to improve socio-economic conditions, and the evidence from the UK supports this (Darbyshire, 1995). But there are important differences between patterns of infection in the industrializing versus the industrialized (and indeed post-industrial) parts of the world. Because in many parts of the Third World poverty is more widespread than it is in the North, and although TB continues to be associated with poverty and its attendant poor health, the disease and its victims are not pushed to the social margins here as they are in higher income settings. The vast majority of TB patients in countries like India, for example, although poor, are on the whole integrated contributing members of their communities, living otherwise ordinary lives. Household heads, wage labourers, busy housewives and school children are all vulnerable – particularly if they also happen to be poor and living in urban areas. This means that the landscape of vulnerability to infection is very different in these settings, with important implications for appropriate intervention and treatment.

All of the social and economic conditions once so strongly linked to TB in Europe currently characterize many parts of the Second and Third Worlds. In South Africa, the very high incidence of TB in the mines is associated with crowded living and working conditions, as well as the migratory nature of the labour – TB is carried from the mines to the rural areas when men go home on leave or when their tour of duty ends (Strebel and Seager, 1991). There is little reason to suppose the situation is much different wherever industry is based largely on migratory labour. In South Africa mortality rates for TB are higher in the coloured population compared to the white; again reflecting the socio-economic divide (Figure 6.2).

Overcrowding, poor sanitation and malnutrition are, however, pandemic in Third World regions, and so the presence of infectious disease – whether endemic like TB or epidemic like AIDS – does not surprise us. We are nevertheless concerned to make a difference. Certainly no country can control its TB problem in isolation. The world conquest of this disease can only be achieved by international co-operation and by directing funds from the richer nations towards the poorer to reduce the incidence of disease in the developing world (Davies, 1995). But are these the most realistic solutions? Is there potential for generating substantive changes at a more tangible level? Perhaps we should be asking how people manage despite and within the existing conditions of social inequality at the local (household and community) level. Perhaps we need to identify what les-

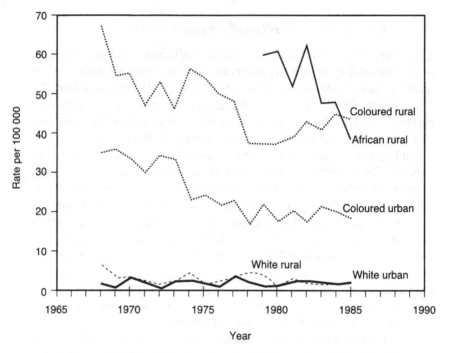

Figure 6.2. Tuberculosis mortality rates for urban and rural areas of the Western Cape Health Region by ethnic group 1968–85. (*Source:* Coovadia and Benatar, 1991.)

sons can be learned, and how existing local capacities can be built upon to generate community-driven change.

HIV

The agent/host/environment dynamic in HIV infection

AIDS is the name given to the syndrome of diseases that results from advanced suppression of the body's immune system following infection by the human immunodeficiency virus (HIV). AIDS is essentially a sexually transmitted disease (STD), with HIV being transmitted mainly through unprotected sexual intercourse. Like other STDs, HIV infection can also be transmitted through blood, blood products, infected needles or donated organs or semen (parenteral transmission), and from a woman to her foetus or infant (perinatal transmission). In 1996 there are estimated to be 21.8 million HIV infected adults and children; 14 million living in Sub-Saharan Africa (Figure 6.3).

HIV in high income countries

AIDS was identified in the USA in 1981. The groups of people initially infected were homosexual men, injecting drug users and haemophiliacs. Thus, in the early days of the epidemic, there appeared to be no particular association with poverty. However, as the epidemic has continued and spread, the epidemiology of the disease has shown the strong link with social inequality and the contribution of poverty to transmission is becoming increasingly apparent.

In the USA in 1990, data based on ethnicity, a proxy for 'class', showed that of newly reported AIDS patients, over 30% were African-Americans and 17% were Hispanic, who represent just over 13% and 8% of the total American population respectively. For women and children the data were more dramatic – among women AIDS patients (15–44 years of age), 57.6% were African-American, and 16.8% Hispanic; among children AIDS patients (younger than 15 years of age), 54.5% were African-American, and 20.6% were Hispanic.

By 1994, the figures showed a similar percentage, but importantly, the median survival time of individuals diagnosed with AIDS was shown to vary by ethnicity. In Connecticut, USA, for example, the median survival for whites was 11.2 months, but only 7.7 and 10.2 months for African-Americans and Latinos respectively (Singer, 1994). These data reflect the broader differences in general health and access to services of these populations.

So, the AIDS epidemic in the USA affects both rich and poor, but increasingly the affected populations are in the inner cities. The agent has shifted to the most susceptible hosts (the poor), and to the environment where transmission through sexual contact and injecting drug use is most likely to occur (the inner city). While poverty is not confined to the inner city, and neither are all poorest ethnic minorities or people of colour, the intersection of urban poverty and socially devalued ethnicity has proven to be a particularly unhealthy combination. As inner city areas have been abandoned by the white middle classes, so AIDS has spread there (Singer, 1994).

Thus the vicious inter-relationship between inequality and disease persists in our own times. The cycle goes as follows: poverty contributes to poor nutrition which, in combination with chronic stress and prior disease, produces compromised immune systems and increasing susceptibility to new infection in individuals. A range of socio-economic problems and stressors increase the likelihood of substance abuse and exposure to HIV which further damages the immune system increasing susceptibility to a host of other diseases including TB.

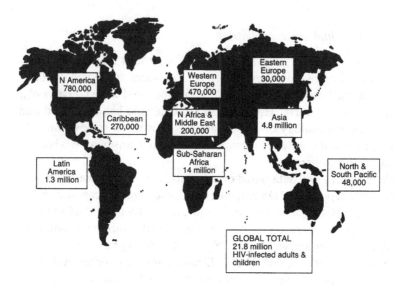

Figure 6.3. Estimated number of persons living with HIV/AIDS – July 1996. (*Source:* UNAIDS, 1996.)

Gender also plays a role. Although early studies suggesting a gender effect in the progression of the disease (indicating faster progression in women than men, e.g. Rothenberg *et al.*, 1987; Bacchetti *et al.*, 1988; White *et al.*, 1989) have not been borne out by more recent research controlling for CD4 count (Melnick *et al.*, 1992) and stage of disease (Creagh *et al.*, 1993), there do seem to be important differences in terms of access to care. Hellinger (1993), for example, has shown that HIV-positive inner city American women are more likely to experience difficulties in obtaining treatment than men. Likewise Bastian *et al.* (1993) indicate that even when these women do get treated, they are treated less aggressively than men. In the light of recent findings about the success of radical anti-retroviral treatments in slowing the progression of the disease, the urgency of removing barriers to care should be more acutely felt, and as DeHovitz (1995: 69) has suggested, such efforts 'must address socio-economic barriers as well as medical ones'. In Africa and elsewhere in the developing world the complex of cultural and economic issues placing poor women at increased vulnerability to infection have been a subject of much anthropological and sociological writing. For an early analysis see Larson (1989); and see Ogden (1996a) for a more recent review of the literature.

Thus socio-economic problems and stressors (such as unemployment, poverty, homelessness, residential overcrowding, substandard nutrition, environmental health risks, infrastructure deterioration, high mobility, family break-up and disruption of social support networks, gangs and street violence, health care inequality) are a group of closely interrelated endemic and epidemic public health conditions, all of which are strongly associated with the transmission of infectious diseases in industrial and post-industrial settings (Wallace, 1990; cf. Mann, 1994).

HIV in low and middle income countries

As the HIV epidemic evolves internationally and as the epidemiology becomes more clear, the most striking feature across the world, is its profound relationship to poverty: those who are most afflicted by poverty are the ones most affected by HIV and AIDS. It is not surprising therefore, that the fastest growing epidemics of AIDS are in Africa, South America and parts of Asia (Evian,1994). In the Americas, the two poorest nations, Honduras and Haiti, and the two largest debtor nations, Mexico and Brazil, report alarming increases in AIDS cases and HIV seroprevalence. In Africa, seven of the eight countries most affected by AIDS are those with the poorest economies – those with US$51 to $350 GNP (Gross National Product) per capita, compared to the African average of US$642 (range $97–$2,481). Research in Zaire suggests that urbanization and the social changes associated with it, best account for the rapid explosion in the rates of seropositivity (Nzilambi, 1988). Despite the compelling simplicity of this urban/rural dichotomy, evidence garnered from the earliest days of the epidemic in Africa indicate that the important causal factor is not necessarily urbanness *per se*, but rather mobility, often a characteristic of cities (e.g. Serwadda *et al.*, 1985; Carswell *et al.*, 1986). Across Asia, the rapid increases in seroprevalence in Thailand are attributable to injecting drug use and the internationally lucrative sex industry (Porter, 1996). In India, commercial sex workers and the homeless create the largest reservoir (Porter, 1996; UNAIDS, 1996). An employment structure built around male migrant labour in parts of southern Africa increases the potential for multiple sexual partner networks, thereby making both men and women more vulnerable to infection.

Although HIV infection in developing countries affects all sections of the population, rich or poor (Evian, 1994), the poor are the hardest hit. But just as HIV *responds* to poverty, AIDS often *engenders* it. The loss of jobs and income, the experience of social stigma, rejection and discrimination, ill health and finally death all contribute to individual and family misfortune, and to the overall, inward turning spiral of poverty (Evian, 1994).

Mobility associated with war and militarism also affects the epidemiology of HIV. Today, low intensity wars in southern Africa encourage the spread of HIV through troop and population movements. The first reported HIV infections in Honduras, South Korea and The Philippines have occurred among commercial sex workers from towns surrounding military bases (Kelley, 1990). War and displacement have profound effects on communities. The population shifts affect social networks and provide a ready climate for activities which render people vulnerable to HIV infection. Crowded conditions, unemployment and inflation fuel the informal economy – of which commercial sexual activity is a part. Poverty itself increases the number of women who use sex to generate income, the number of jobless who sell their blood and their body parts, and in many countries the number of people who sell drugs for livelihood and survival. Malnutrition, which affects epithelial integrity and cellular immunity, and untreated STDs occurring in epidemic numbers within impoverished communities, may contribute substantially to the apparent high transmission rates observed in the Third World.

There is, however, another side to the story. Within many of the countries whose economic and political systems have been devastated by civil war as much as by unequal global distributions, the growth of the informal economy has had some positive effects. There is recent evidence from Uganda, for example, that women are in some ways especially skilled at manoeuvring within and exploiting this 'hidden' economic sector, increasingly finding ways to generate income that do not compromise their health or respectability (Pons and Wallman, 1995, unpub. data; Wallman, 1996; Ogden, 1996b). There is every reason to expect that the increasing economic security of women will lead to an increase in their capacity to renegotiate the social terms which make them vulnerable to HIV.

Conclusion

The analysis of disease by social structure has been used throughout the twentieth century and before, to identify the groups most severely affected by infectious diseases. Epidemiologists collect socio-economic data on all study subjects in order to ensure that bias is not introduced into the outcome of the study from this dimension. But what have the results of these analyses been? Have they led to change? Have they led to improvements in socio-economic conditions for those affected? Some would argue that this method of analysis has indeed been useful and that it continues to be appropriate. However, among public health professionals there continues to be confusion between medical care and health on the one hand, and public health functions on the other, a confusion arising out of a paradigm which considers disease to be a dynamic event occurring within a basically static or fixed society (Mann, 1994). This leads public health practitioners to identify interventions that focus solely on the individual, disregarding the wider social and societal issues that frame the lives of these individuals, and influence both the impact of and responses to the emergence of infectious disease.

The socio-economic model has graphically depicted the fall of TB in western countries. This fall is associated, not with medical interventions, but with improving socio-economic conditions. But what were these improvements and how did each of them contribute to the reduction in TB cases? We do not have this information. There has been very little work to disaggregate the individual parts of the statement 'improved socio-economic conditions have led to the fall in TB cases in England and Wales over the past century' (McKeown *et al.*, 1962). It is difficult, that is why it has not been achieved. It is, however, necessary, because public health interventions based strictly on the biomedical model cannot accommodate these vital social dimensions of disease and disease transmission. For real change to occur epidemiologists and public health practitioners need to recognize the important relationship between individuals and their communities and the ways in which those communities relate to overall political, economic and social structures. In the case of TB control, the public health intervention is case finding and treatment – i.e. for people with a cough to come forward and to be treated by the health service until they are cured. Questions of equity are not addressed. Little work is concentrated on *who* comes to the clinics and the *reasons* why people from the lower socio-economic groups do not attend. At present, the international community is stressing the importance of directly observed therapy (DOT) in the treatment of TB. This strategy was established and tested within the inner cities of the USA, and yet it is being advocated also in developing countries, where a different social/economic and political dynamic prevails. The consequence of a rigid biomedical focus is that the wider, more profound and long lasting interventions such as improved housing, reduction of gender discrimination and ways of building the capacity of communities to manage public health problems on their own terms, are not addressed.

An alternative approach might be more in harmony with the following definition of 'the new public health' offered recently by Professor Jonathan Mann (1994, p. 230):

> the new public health sees both society and disease as dynamic and inextricably linked. Thus, to respond effectively to disease requires societal action. The new public health recognises that the positive impact of traditional public health work will be inherently limited and inad-

equate without a commitment to changing societal conditions which constrain health and create vulnerability to preventable disease, disability and premature death.

Mann proposes an approach to public health that embraces a human right's framework, under the general principle that 'a society in which human rights are promoted and protected, and in which human dignity is respected, is a healthy society.'

Epidemiology/public health needs to look beyond the biomedical model to answer difficult questions about the interrelationships between infectious disease, individuals, communities and the wider political and economic structures in which they exist. This work needs to be conducted with anthropologists, economists, political analysts and other social scientists. The particular perspective provided by each disciplinary focus can and should be constantly informed by the other specialties. The 1996 conference on Human Biology and Social Inequality provided the opportunity for specialists to use the socio-economic model within their own disciplinary framework. Perhaps this constructive co-mingling of paradigms will provide clues for future interventions, and perhaps these multi-specialist fora are an appropriate way forward for public health.

Certainly the problems facing those interested in redressing the inequalities in the re-emergence of infectious diseases are multiple and complex. We feel that the most appropriate response is most likely a multi-disciplinary, multi-level response that reflects this complexity. The tools of epidemiology alone may be insufficient, but they may at least be grist to the collective mill.

References

Arras, J. (1988). The fragile web of responsibility: AIDS and the duty to treat. *Hastings Centre Report* **18**, 10–20.

Bacchetti, P., Osmond, D., Chaisson, R. E. & Moss, A. R. (1988). Survival in patients with AIDS in New York. [Letter] *New England Journal of Medicine* **318**, 1464.

Bastian, L., Bennett, C. L., Adams, J. *et al.* (1993). Differences between men and women with HIV-related *Pneumocystis carinii* pneumonia: experience from 3,070 cases in New York City in 1987. *Journal of Acquired Immune Deficiency Syndrome* **6**, 617–23.

Brudney, K. & Dobkin, J. (1991). Resurgent TB in New York City: human immunodeficiency virus, homelessness, and the decline of TB control programs. *American Review of Respiratory Diseases* **144**, 745–9.

Carswell, J. W., Sewankambo, N., Lloyd, G. & Downing, R. G. (1986). 'How long has the AIDS virus been in Uganda?' *Lancet* **1**, 1217.

Citron, K. M., Southern, A. & Dixon, M. (1995). Out of the shadow: detecting and treating tuberculosis amongst single homeless people. *Crisis*.

Coovadia, H. M. & Benatar, S. R. (Eds.). (1991). *A Century of Tuberculosis; South African Perspectives.* Oxford: Oxford University Press.

Creagh, T., Thompson, M., Morris, A. *et al.* (1993). Gender differences in the spectrum of HIV disease in Georgia. *Ninth International Conference on AIDS, Berlin.* Abstract PO-CO4-2657.

Darbyshire, J. H. (1995). Tuberculosis: old reasons for a new increase? *British Medical Journal* **310**, 954–5.

Cauthen, G. M., *et al.* (1988). WHO/TB/88.154, Geneva, WHO.

Davies, P. D. O. (1995). Tuberculosis and migration. *Journal of the Royal College of Physicians of London* **2**, 113–18.

DeHovitz, J. (1995). Natural history of HIV infection in women. In *HIV Infection in Women,* ed. by H. L. Minkoff, J. A. DeHovitz & A. Duerr, pp. 57–72. New York: Raven Press.

Evian, C. (1994). AIDS and the socio-economic determinants of the epidemic in southern Africa – a cycle of poverty. *Journal of Tropical Paediatrics* **40**, 61–2.

Gaarder, J. (1995). *Sophie's World.* London: Phoenix House.

Hellinger, F. J. (1993). The use of health services by women with HIV infection. *Health Services Research* **28**, 544–61.

Heywood, P. F. & Marks, G. C. (1993). Nutrition and South East Asia. *The Medical Journal of Australia* **159**, 133–7.

Kelley, P. W., Miller, R. N., Pomerantz, R., Wann, F., Brundage, J. F. & Burke, D. S. (1990). Human immunodeficiency virus seropositivity among members of the active duty US army 1985–1989. *American Journal of Public Health* **80**, 405–10.

Larson, A. (1989). The social context of HIV transmission in Africa: a review of historical and cultural bases of East and Central African sexual relations. *Review of Infectious Diseases* **11**, 716–31.

Lederberg, J., Shope, R. E. & Oaks, S. C. (Ed.). (1992). Emerging Infections. In *Microbial Threats to Health in the United States.* Washington, DC: Institute of Medicine, National Academy Press.

Lerner, B. H. (1993). New York city's tuberculosis control efforts: the historical limitations of the 'War on Consumption'. *American Journal of Public Health* **83**(5), 759–65.

Mangtani, P., Jolley, D. J., Watson, J. M. & Rodrigues, L. C. (1995). Socioeconomic deprivation and notification rates for tuberculosis in London during 1982–1991. *British Medical Journal* **310**, 963–6.

Mann, J. (1994). Human rights and the new public health. *Health and Human Rights* **1**(3), 229–33.

McKeown, T., Record, R. G. & Turner, R. D. (1962). Reasons for the decline of mortality in England and Wales during the nineteenth century. *Population Studies* **16**, 94–122.

McKeown, T., Record, R. G. & Turner, R. D. (1975). An interpretation of the decline in mortality in England and Wales during the twentieth century. *Population Studies,* **29**, 391–421.

McMurray, D. N., Bartow, R. A. & Mintzer, C. L. (1990a). Malnutrition induced impairment of resistance against experimental pulmonary tuberculosis. *Progress Clinical and Biological Research* **325**, 403–12.

McMurray, D. N., Bartow, R. A., Mintzer, C. L. & Hernandez-Frontera, E. (1990b). Micronutrient status and immune function in tuberculosis. *Annals of*

New York Academy of Science **587**, 59–69.

Melnick, S., Sherer, R., Hillman, D. *et al.* (1992). HIV-related clinical events and mortality: preliminary observational data from the community programs for clinical research on AIDS (CPCRA). *Eighth International Conference on AIDS, Amsterdam.* Abstract MoCOO31.

Nzilambi, N., De Cock, K. M., Forthal, D. N. *et al.* (1988). The prevalence of infection with human immunodeficiency virus over a 10-year period in rural Zaire. *New England Journal of Medicine* **318**, 276–9.

Ogden, J. (1996a). *Reproductive Identity and the Proper Woman: The Response of Urban Women to AIDS in Uganda.* Unpublished PhD. Thesis, University of Hull.

Ogden, J. (1996b). 'Producing' respect: the proper woman in postcolonial Kampala. In *Postcolonial Identities in Africa*, ed. R. P. Werbner and T. O. Ranger, pp. 165–92. London: Zed Press.

Parker, R. G. (1991). *Bodies, Pleasures and Passions: Sexual Culture in Contemporary Brazil.* Boston: Beacon.

Pons, V. & Wallman, S. (1995). Where have all the young men gone? Demographic shifts in Kampala, Uganda. Paper presented to the *British Association of Population Studies, Durham, September 1995.* (Unpublished.)

Porter, J. D. H., Lea, G. & Carroll, B. (1996). HIV/AIDS and international travel: a global perspective. In *Health and the International Tourist*, ed. S. Clift and S. Page, pp. 68–86. London: Routledge.

Reichman, L. (1991). The U-shaped curve of concern. *American Review of Respiratory Disease* **144**, 741–2.

Research Committee of the British Thoracic and Tuberculosis Association (1975). Tuberculosis among immigrants related to length of residence in England and Wales. *British Medical Journal* **2**, 698–9.

Rothenberg, R., Woefel, M., Stoneburger, R. *et al.* (1987). Survival with the acquired immune deficiency syndrome. *New England Journal of Medicine* **317**, 1297–302.

Serwadda, D., Mugerwa, R., Sewankambo, N., *et al.* (1985). 'Slim disease' – a new disease in Uganda and its association with the HTLV-III infection. *Lancet* **2**, 849–52.

Singer, M. (1994). AIDS and the health crisis of the US urban poor; the perspective of critical medical anthropology. *Social Science and Medicine* **39**, 7, 931–48.

Springett, V. H. (1964). Tuberculosis in immigrants – an analysis of notification rates in Birmingham 1960–62. *Lancet* **1**, 1091–5.

Strebel, P. M. & Seager, J. R. (1991). Epidemiology of tuberculosis in South Africa. In *A Century of Tuberculosis – South African perspectives*, ed. H. M. Coovadia and S. R. Benatar, pp. 58–90. Oxford: Oxford University Press.

Sutherland, I. (1976). Recent studies in the epidemiology of tuberculosis based on the risk of being infected with tubercle bacilli. *Advances in Tuberculosis Research* **19**, 1–63.

UNAIDS (1996). *Status and trends of the global HIV/AIDS pandemic.* Geneva: UNAIDS.

Wallace, R. (1990). Urban desertification, public health and public order: 'planned shrinkage', violent death, substance abuse and AIDS in the Bronx. *Social Science and Medicine* **31**, 801–13.

Wallman, S. (1996). *Kampala Women Getting By: Wellbeing in the Time of AIDS.*

London: James Currey.

Weber, H. W. (1899). On prevention of tuberculosis. *Tuberculosis*, pp. 14–19, 50–55, 101–11.

White, B. M., Swanson, C. E. & Cooper, D. A. (1989). Survival in patients with acquired immune deficiency syndrome in Australia. *Medical Journal of Australia* **150**, 358–362.

7 Environmental health, social inequality and biological differences

LAWRENCE M. SCHELL AND STEFAN A. CZERWINSKI

Introduction

Pollution from industry, transportation, construction and agriculture affects most societies today. However, there are substantial differences among populations in pollutant exposure owing to differences in the prevalence of these four main sources of pollution, and to differences in waste disposal practices. Within a given society also, pollutant exposure may vary considerably. Since pollutants have health consequences, by definition, the distribution of pollutants has consequences for biological variation in human populations as well as for the distribution of infirmity, disability and disease. This presentation will first review some of the evidence for the unequal distribution of pollutants, indicating the types of pollutants so distributed and the means by which this distribution is effected. We will then focus on one pollutant, i.e. lead, and on one set of health consequences, the neurobehavioural and cognitive development of children, and present a model that interrelates pollutant exposure, health effects and social inequality in a positive feedback loop. This model will demonstrate how social factors differentially distribute pollutant exposure on certain social classes, most often the poor, and how this differential exposure translates into neurobehavioural impairment. In the case of many pollutants such as lead, these processes work inter-generationally to perpetuate social stratification. Support for this model will be based on a review of existing literature and results drawn from continuing research.

Social class and the distribution of environmental toxins

Data on the unequal distribution of pollutant burdens within societies are now becoming available. Though a single comprehensive analysis of all

114

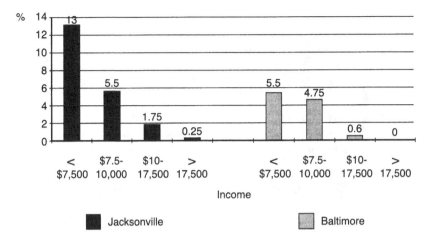

Figure 7.1. Differential exposure to sulphur dioxide by income (US $): percentage of population residing in substandard areas of Jacksonville, Florida, and Baltimore, Maryland. (Adapted from Berry, 1977.)

available data has not been performed, results can be pieced together from a variety of studies conducted since the early 1970s to create a general picture. It is prudent to bear in mind that studies vary in the laboratory methods used for analyzing pollutants, the tools for defining social groups, and also that the actual levels of pollutants have changed during this period.

Pollutants can be measured in environmental media such as water, air, or soil, and in flora and fauna. Measurements of environmental media do not reflect human exposure on a cellular level since they ignore the effects of absorption, metabolism and excretion. Figure 7.1 shows the relationship of exposure to atmospheric pollution from sulphur dioxide according to income level, in Jacksonville, Florida and Baltimore, Maryland during the mid-1970s (Berry, 1977). Sulphur dioxide is a respiratory irritant produced by burning fossil fuels (Waldbott, 1978). In Jacksonville, the group with the lowest income has the largest percentage of the population living where government air quality standards are unmet. As income rises, the percentage decreases. In Baltimore, though levels are lower overall, the negative relationship between exposure and income is evident as well.

Cadmium is another pollutant of concern. It also comes from burning fossil fuels as well as from municipal and medical waste. It is inhaled or consumed and accumulates in the kidney where it can cause irreversible damage (Waldbott, 1978). Berry (1977) determined the percentage of the population exposed to cadmium (at a level greater than 18 ng/m^3) in four income groups in Chicago, Illinois – the poorest group (incomes between

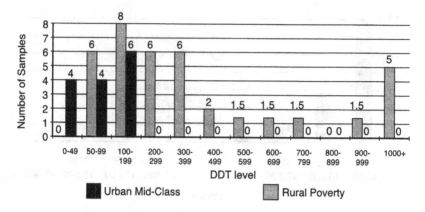

Figure 7.2. Comparison of DDT in poor and mid-class women. (Adapted from Woodard *et al.*, 1976.)

US$ 7500 and $10,000) had the largest percentage (11%) of people exposed to cadmium. Of those people exposed to cadmium 8.5% had incomes from US$ 10,000 to $17,500; fewer than 1% had incomes US$ 17,500 to $20,000; and even fewer still had earnings greater than US$ 20,000.

Levels of DDT (dichlorodiphenyltrichloroethane) in breast milk samples from the mid-1970s were obtained from urban middle class white women in Nashville, Tennessee, and black women in rural poverty from counties in Mississippi (Bolivar County) and Arkansas (Lee County) (Woodard *et al.*, 1976). Figure 7.2 shows that most samples from urban middle class white women are at the very low end of the range, while the samples from poor rural black women occur across the entire range of concentrations and many more samples have relatively high concentrations. In fact, there are no samples from middle class urban white women in the moderate or high range. Figure 7.2 illustrates that pollutant burdens while greater among the poor, are not necessarily greater in urban populations.

The problem of pollutant exposure and income or social class as defined by a variety of characteristics, is not restricted to the USA. Kim and Ferguson (1993) presented an especially thorough analysis of factors related to heavy metal exposure in Christchurch, New Zealand. They examined the relationship of cadmium and lead levels in house dust with several household characteristics. When households were grouped by the degree of carpet wear (Figure 7.3) the homes with new carpets had the lowest mean values for cadmium. As carpet wear increased to 'threadbare', the cadmium level in house dust increased regularly. The same relationship is seen with carpet wear and lead. When households were grouped by

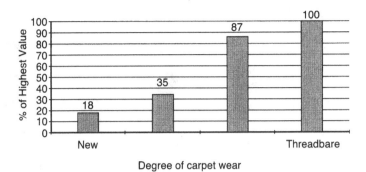

Figure 7.3. Cadmium levels in house dust by degree of carpet wear. (Adapted from Kim and Ferguson, 1993.)

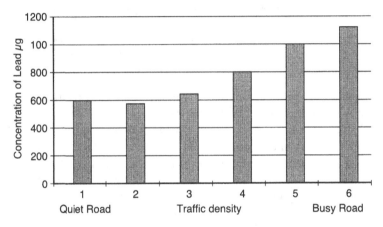

Figure 7.4. House dust lead concentrations in relation to traffic density on adjacent roads. (Adapted from Kim and Ferguson, 1993.)

proximity to roadways with varying traffic levels, those next to the most heavily trafficked roads had the highest levels of lead (Figure 7.4). Painted homes compared to brick homes also had higher levels. Among the painted ones, older homes had higher levels as well. In these instances, residence plays a key role in determining exposure whether the source of the pollution is agricultural chemicals, industry or transportation.

Pollutant disposal practices also affect social classes differentially. In the USA and perhaps elsewhere, ethnicity and race are important risk factors for exposure to pollution emanating from hazardous waste sites. In the USA there is a disproportionate percentage of black Americans and Hispanic Americans in communities near to operating hazardous waste landfills (US General Accounting Office, 1983; Commission for Racial Justice, 1987; Bullard, 1990; Mott, 1995). In 1983, the US General Accounting Office conducted a study of the racial and economic status of

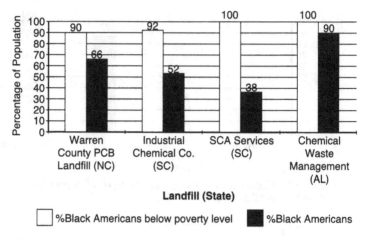

Figure 7.5. Race and income in census areas of EPA region IV hazardous-waste landfills. (Adapted from the US General Accounting Office, 1983.)

populations near off site hazardous waste landfills operating in the US EPA's (Environmental Protection Agency) Region IV (Alabama, Florida, Georgia, Kentucky, Mississippi, North Carolina, South Carolina, Tennessee). One-fifth of the population in these states is black. Of the four sites in these states, three of them are located in predominantly black communities (Figure 7.5). At all four sites, nearly all black Americans live below the federal poverty line, i.e. US$ 11,611 for a family of four in 1987). In the USA, where income and ethnicity or race are interrelated, it is difficult to separate their individual effects on pollutant exposure.

In 1987, the Commission for Racial Justice examined the distribution of hazardous waste sites in the USA in relation to the ethnicity and economic characteristics of the communities surrounding the sites. There were 27 hazardous waste sites operating in the 48 contiguous states of the USA. These had a total capacity of 127,897 acre feet ($\approx$ 52.55 million m^3). One-third of these sites were located in just a few southern states – Alabama, Louisiana, Oklahoma, South Carolina and Texas – and these sites provided 60% of the nation's total hazardous waste landfill capacity (Bullard, 1990). Figure 7.6 shows the percentage of black or Hispanic Americans living in the surrounding postal codes of the nine sites in southern states. In the areas surrounding the three largest landfills, 65% of the population, on average, is black whereas only 28% of the population of these five southern states is black. The three largest landfills in minority areas account for 58.6% of the regions hazardous waste capacity and 35% of the nation's. Statistical analysis of these and other data show that sites are more likely to occur in areas where minority group membership is high

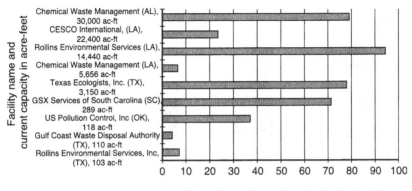

Figure 7.6. Ethnicity of communities around operating hazardous-waste landfills in the southern USA. (Adapted from the Commission for Racial Justice, 1987.)

rather than just where income is low (Commission on Racial Justice, 1987).

Hispanic ethnicity is an often overlooked risk factor for pollution exposure in the USA. The National Coalition of Hispanic Health and Human Services Organizations recently examined data from the EPA and compared the percentages of Hispanic, non-Hispanic black and non-Hispanic white persons living where US EPA air quality standards are unmet (National Coalition of Hispanic Health and Human Services Organizations, 1992; Metzger *et al.*, 1995). Hispanic Americans are consistently more likely to live where one, two or three air quality standards are unmet (Figure 7.7). Indeed, Hispanic Americans are three times more likely to live in areas where four or more standards are unmet compared to white Americans. Hispanic Americans are also more than twice as likely as whites or non-Hispanic blacks to live in areas where particulate matter in the air is elevated above government standards. They are far more likely to live where both carbon monoxide levels and lead levels are elevated. These differences in exposure are defined by residence. A reasonable conclusion from this review of the distributions of several pollutants is that residential patterns or more specifically, residential segregation by ethnicity, income and race are a major allocator of pollutant exposure. This exposure in turn has great consequences for human biology and health differentials.

A second important allocator of risk for pollutant exposure is occupation. Occupation and ethnicity frequently coincide. Seventy-one percent of all seasonal agricultural workers in the USA are Hispanic (National Coalition of Hispanic Health and Human Services Organizations (COSSMHO), 1992). As such, they are exposed to higher levels of agricultural chemicals, particularly organophosphates and organochlorides, than other segments of society. These materials have been related to important

120   *L.M. Schell and S.A. Czerwinski*

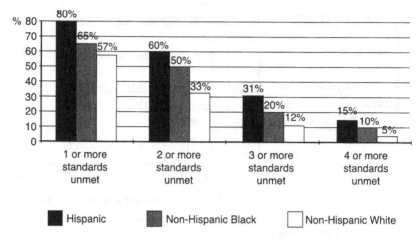

Figure 7.7. Percentage of persons living in areas failing to meet the US Environmental Protection Agency ambient air quality standards, by race/ethinicity. (Adapted from Wernette & Nieves, 1992.)

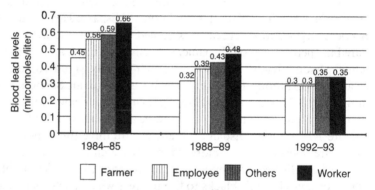

Figure 7.8. Blood lead levels in a Swiss population: men. (Adapted from Wietlisbach *et al.*, 1995.)

health outcomes such as reproductive impairment and cancer. Occupation may be an important marker of pollutant exposure outside the USA as well. A comprehensive study of lead levels in a Swiss sample demonstrated the effect of occupation (Wietlisbach *et al.*, 1995). Figure 7.8 shows four occupational groups – farmers, white collar employees, blue-collar workers and others including executives, retirees and the self-employed. Blood lead levels are greatest among men employed in blue collar occupations and lowest amongst farmers. More recent surveys indicate that differences in occupational exposures are still significant though they have become less pronounced in recent years (Wietlisbach *et al.*, 1995).

To summarize, pollutant exposure and burdens vary significantly by

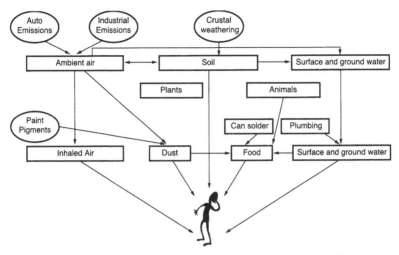

Figure 7.9. Pathways of lead exposure. (Adapted from EPA, 1986.)

social factors particularly by residence and occupation. These factors are related to income, and in the USA certainly, to race and ethnicity. This differential exposure may be termed risk allocation – the allocation of risk to certain groups based on culturally or socially defined attributes.

The interaction of pollution and social inequality: lead as an example

To elucidate the consequences of unequal pollution exposure, it is useful to focus on one pollutant. Lead pollution is an appropriate example because it is one of the most widespread and longstanding pollutant problems we face, and it represents a class of pollutants that affects human development. Lead is ubiquitous in the human environment. Most people's lead exposure is from mundane sources rather than occupational ones (Figure 7.9) (EPA, 1986). Everyday lead exposure is from crustal weathering, and from two anthropogenic sources – industrial emissions and automobile emissions. These sources add lead to air, water, soil and dust. Paint pigments, plumbing and lead based solder also contribute. From these sources, lead enters air, dust, food and water where it is accessible to plants and animals. As a person matures, a lead burden is acquired primarily by ingestion of lead-containing dust, water and food. Respiration also adds a lesser fraction. Children are at greater risk for lead toxicity compared to adults because children are exposed to more lead dust through play activities, and once exposed, they absorb a far greater percentage of ingested lead from the gastrointestinal tract than do adults due to increased metabolism during growth (Moore, 1986). Children also absorb more lead through respiration since both their respiratory rate and their exchange rate per unit body mass is higher compared with adults. The foetus may be at greatest risk because of the sensitivity of the developing organism to toxic insult.

The effects of lead differ depending on whether it is the result of acute poisoning or a chronic, mundane exposure. When an adult is exposed to a heavy lead dose at work, or when a child consumes lead paint chips, he or she receives an acute exposure that can alter haem synthesis, cause brain damage, encephalopathy, seizures and even death (Damstra, 1977; Mushak *et al.*, 1989). Brain damage in children can be permanent. Lead poisoning, where lead in the blood rises to 80 μg/dL or more is rare today. Now, the more common situation is one of a chronic exposure to ambient lead in air, food and water that produces elevations in blood lead levels that are small to moderate in comparison to elevations from acute exposure.

The first population survey of blood lead levels in the USA was conducted from 1976 to 1980 as part of the National Health and Nutrition Examination Survey (NHANES II) (Annest *et al.*, 1982). Figure 7.10 shows the percentage of children aged 6 months to 5 years with lead levels at or above 30 μg/dL (the action level at the time) stratified by ethnicity and income. Eleven percent of low income children of all races exceed the action level. However, nearly 19% of poor black American children exceed the action level. The results are similar when reorganized by urban residential status, again reinforcing the fact that residential segregation and race are related in the USA. These results alarmed the public health community and produced a surge of research in the USA and elsewhere on lead.

Much of this research focused on the relationship between lead levels in children's blood and their performance on tests of cognitive and neurological development. Several longitudinal studies have followed pregnant women and their children until school age and most have observed significant negative effects of lead on children's cognitive and behavioural development, speech and language function, and attention, even after taking into account the appropriate social and biological factors that should be controlled (Bellinger et al, 1986, 1991, 1992; Dietrich *et al.*, 1987). A remarkable study by Herbert Needleman and colleagues (1990) linked lead burden in childhood as measured in deciduous teeth to outcomes in early adulthood. The investigators determined that young adults' whose deciduous teeth had high lead levels when they were children, were far more likely to have a serious reading disability. They were also far less likely to complete secondary school (Needleman, 1990). Certainly these results

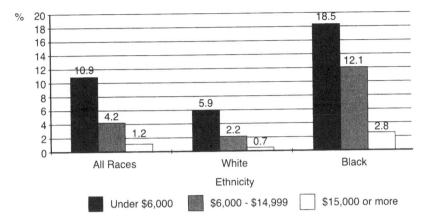

Figure 7.10. Percentage of children aged 6 months to 5 years with blood levels of 30 μg/dL or greater, US 1976–80. (Adapted from Annest *et al.* 1982.)

must be replicated in other studies before they can be accepted. A conservative conclusion, however, can be drawn from the large number of studies whose results show with consistency that children with higher lead levels reach developmental milestones more slowly and perform worse on tests of cognitive development. Based on these studies (Lansdown and Yule, 1986; Mushak *et al.*, 1989) public health officials in the USA now have set 10 μg/dL as the acceptable level of lead in blood (Centers for Disease Control, 1991). This new level is down from the 60 μg/dL that was acceptable in the 1960s, and the 25 μg/dL that was acceptable as recently as 1985.

Between 1988 and 1991 the Third National Health and Nutrition Examination Survey (NHANES III) was conducted and determined that lead levels have decreased substantially in children in the USA. While lead levels have decreased, so too has the cut-off level for lead which is considered safe. Research conducted in the 1980s and 1990s shows that lead produces detrimental effects at levels far below 30 μg/dL which was considered safe 10 years earlier when NHANES II was conducted. Using the current Centers for Disease Control action level of 10 μg/dL as the cut-off, the data from NHANES III are not encouraging. Figure 7.11 shows the relationship of income to the blood lead level of children from 1 to 5 years of age (Brody *et al.*, 1994). As in the earlier survey poor children are at greater risk than those from well-to-do homes – the risk is about four times greater. Income in this case was measured by a Poverty Income Ratio (PIR) which is total family income divided by a family-size specific poverty threshold (low: < 1.3; mid: 1.3–2.999; and high: > 3.0). When the children are grouped by ethnicity a more striking picture is apparent (Figure 7.12). Poor black children have a far higher rate of elevated lead in

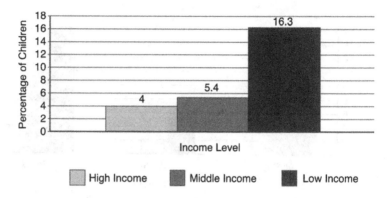

Figure 7.11. Percentage of children aged 1 to 5 years of age with blood lead levels 10 μg/dL or greater by income level. (Adapted from Brody *et al.*, 1994.)

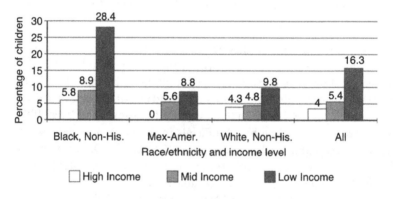

Figure 7.12. Percentage of children aged 1 to 5 years of age with blood lead levels 10 μg/dL or greater by race/ethnicity and income level. (Adapted from Brody *et al.*, 1994.)

the blood than any other group by a great margin and are six to seven times more likely to have an elevated lead level than their white peers.

Urban residence continues to be a risk factor for elevated blood lead levels. When the same ethnic groups are stratified by degree of urbanization, 37% of the young non-Hispanic black American children who live in the central areas of large cities have lead levels above the new action level. In contrast, only 6.1% of non-Hispanic white children in the same environment have elevated lead levels and only 5.2% of non-Hispanic white children in non-city environments have elevated levels. There is thus a seven-fold difference in rates. This is an elevated relative risk with tremendous importance since such lead levels are associated with differences in cognitive development. The effect extends beyond these children with elevated lead levels and into the classroom. In a typical first grade (i.e. age

6–7) classroom of 30 black urban children, 11 may have a lead-associated deficit in learning ability and attention. This will affect their education and probably the education of their classmates since their teacher may be forced to devote extra time to the impaired children (Needleman *et al.*, 1996).

Recreating the environment of impairment

The distribution of elevated lead burdens among children in the USA is not the result of a new influx of lead into the children's environment or of recent environmental exposure. Indeed, lead levels in the environment and in people have decreased most recently. The lead burden of young children today is the result of a multi-generational experience with lead. Because lead passes through the placenta, infants are born already carrying lead, acquired transplacentally and attributable to their mother's own personal life history of mundane exposure. Whenever a pollutant crosses the placenta, past environmental burdens are recreated to some degree in the next generation. In addition, there are social forces that not only recreate environmental burdens but even concentrate or focus them on some groups, raising pollutant burdens further.

A model depicting the interrelation of poverty, lead exposure and lead health effects in a multi-generational system is displayed in Figure 7.13. It begins with residence during childhood in a neighbourhood where ambient lead levels in the environment are high. These neighbourhoods or 'lead belts' exist throughout the USA, but are usually concentrated in the central part of many older cities. Here lead dust from neglected housing and exhaust emissions is present in the air and concentrated in the soil. Numerous studies have shown that childhood residence in the lead belt neighbourhood results in high lead exposure and absorption (ATSDR, 1988). A high blood lead level has many effects on the individual, most importantly cognitive and behavioural impairments. These impairments have been described in many studies and form the basis for the tolerance limits of lead burden set by the Centers for Disease Control in the USA and elsewhere (Centers for Disease Control, 1991). These impairments may increase the chances of doing poorly in school and of leaving school prior to completing secondary school (Lyngbye, 1988; Raab *et al.*, 1989; Needleman *et al.*, 1990), as suggested by Needleman's work referenced earlier. Poor school performance may contribute to poor preparation for employment, early fertility and other outcomes. These relationships are fairly well known from the education literature.

Another pathway that is not well characterized is a nutritional one. Poor school performance could promote poor knowledge of dietary require-

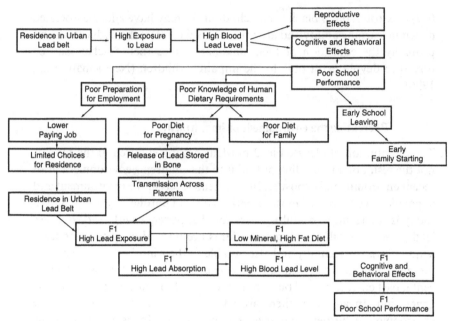

Figure 7.13. Risk focusing model for transgenerational effects of lead exposure. (*Source:* Schell, 1992.)

ments. While this does not result directly from not completing school, it would be one of several outcomes related to sub-optimal education that would be more likely. Poor knowledge of dietary requirements may result in a sub-optimal diet during pregnancy and sub-optimal household nutrition once children are born. A poor diet during pregnancy may increase the release of lead from the mother's skeleton (since in adults, 90% of lead is stored in bone) and into the blood. Lead passes easily across the placenta (Barltrop, 1969) and into the foetus. Laboratory research and case-studies suggest that lead can be released from the skeleton during pregnancy (Thompson *et al.*, 1985; Silbergeld, 1986) but the relationship of diet and nutritional status to neonatal lead levels has not been carefully studied. Preliminary results from the first phase of The Albany Pregacy Infancy Lead Study (APILS) suggest that there are links between mother's nutritional status and her newborn's lead level. Anthropometric data obtained from pregnant women of low socio-economic status in Albany, New York show a strong relationship between larger maternal size and lower lead levels in their newborns (Czerwinski *et al.*, 1993) In children, a diet that is poor in minerals and high in fats may increase the risk of lead absorption as has been shown in many animal studies and in a recent human study (Lucas, 1996). Among people already at risk for lead exposure, such diets may be more common at least in the USA as we can see from the

NHANES data. Dietary data obtained from pregnant women in the Albany Pregnancy Infancy Lead Study show a relatively poor quality of diet, high in fat and low in mineral intake (Czerwinski *et al.*, 1995). This type of diet may contribute to increased lead absorption, retention and transfer during pregnancy in this population.

The parents' poor preparation for employment increases the risk of lead exposure for the next generation because poor job preparation increases the risk of obtaining a lower-paying job or being unemployed and both can lead to limited choices of residence and an increased chance of residence in a lead belt neighbourhood. A young pregnant female from such a neighbourhood is less likely to seek early prenatal care and thus would be less likely to receive adequate nutritional counselling. With a high fat, mineral poor diet, she may be more likely to transfer lead across the placenta to the foetus. She may consequently deliver a child with a considerable lead burden. If the postnatal diet is also inadequate the child will absorb greater amounts of lead from its lead-laden environment. This next generation will ultimately begin school with a substantial lead burden destined to continue the cycle of cognitive/behavioural effects, poor school performance and early fertility.

This negative cycle of exposure, deficit and downward mobility, leading to more exposure and even more deficit, can be termed risk focusing (Schell, 1992, 1997). It encompasses the socio-cultural process of allocating exposure to toxic materials to groups whose members enter the group partly because of previous exposure to toxic materials. A path diagram (Figure 7.14) developed specifically to model the interaction of lead-child development and social inequality may be abstracted and its basic structure applied to interactions between social factors and other pollutants that impact upon cognitive development. There are two paths depicted, one taken by the wealthy and another taken by poor. Low socio-economic status is associated with occupations and residences that lead to greater risk of toxic exposure, which leads to toxic insult and disability. This in turn results in poorer qualifications for employment leading to lower occupational status which returns to lower socio-economic status. However, individuals of higher socio-economic status have occupations and residences with less exposure to toxic materials thereby preserving them from insult and disability. This in turn results in better qualifications for employment and higher paying occupations with less toxic exposure. Choices of residence are greater and more likely to be away from environmental sources of toxic exposure and insult. This system increases the distance between rich and poor, between exposed and unexposed. It contributes to and strengthens social stratification.

Lead is only an example of many pollutants that may exist in a similar

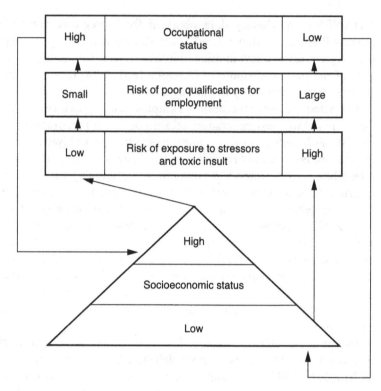

Figure 7.14. Model of risk focusing. (*Source:* Schell, 1992.)

relationship to human development and social class. Other pollutants and substances which affect child development including mercury, PCBs (polychlorinated biphenyls), and other factors such as heavy alcohol or cigarette use may be modelled with this structure as well (see Table 7.1). Exposure to these substances is also mediated through social factors, and that exposure eventually contributes to biological difference. For human biologists this is an example of true bio-cultural interaction. Culture affects biological variation and biological difference affects socio-cultural patterns.

Social stratification, or more conservatively, social distinction, acts as an allocator of risk of pollutant exposure and its consequences. Risk focusing is a bio-cultural process that embeds social stratification in biological differences, making social differences appear to have a biological basis even though they originate in the social order and arrangement of pollutant exposure. This may occur largely before birth, be invisible and appear to be a biological characteristic of some groups when it is actually a product of the social structure.

Table 7.1. *Neurobehavioural toxins*

Heavy Metals:	
	lead
	mercury
	cadmium
Solvents:	
	dry cleaning chemicals
	household cleaners
	paint thinner
Pesticides:	
	ddt
	organophosphates
	carbamates
Drugs:	
	cigarette smoking
	marijuana
	alcohol
	cocaine
	methadone
Organic Compounds:	
	PCBs
	PBBs

(Adapted from Riley & Vorhees, 1986.)

References

ATSDR (Agency for Toxic substances and Disease Registry). (1988). The nature and extent of lead poisoning in children in the United States: a report to congress. Atlanta: US Department of Health and Human Services.

Annest, J. L., Mahaffey, K. R., Cox, D. H, & Roberts, J. (1982). Blood lead levels for persons 6 months–74 years of age: United States, 1976–80. *NCHS Advance Data* **79**, 1–23.

Barltrop, D. (1969). Transfer of lead to the human foetus. In *Mineral Metabolism in Paediatrics*, ed. D. Barltrop, pp. 135–51. Philadelpia: F. A. Davis Co.

Bellinger, D., Leviton, A. & Needleman, H. (1986). Low level lead exposure and infant development in the first year. *Neurobehavioral Toxicology and Teratology* **8**, 151–61.

Bellinger, D., Sloman, J., Leviton, A. & Rabinowitz, M. (1991). Low-level lead exposure and children's cognitive function in the preschool years. *Pediatrics* **87**, 219–27.

Bellinger, D., Stiles, K. & Needleman, H. (1992). Low-level lead exposure, intelligence and academic achievement: a long-term follow-up study. *Pediatrics* **90**, 855–61.

Berry, B. J. (1977). *The Social Burdens of Environmental Pollution: A Comparative Metropolitan Data Source*. Cambridge: Ballinger Publishing Company.

Brody, D., Pirkle, J., Kramer, R., Flegal, K., Matte, T. & Gunter, E. (1994). Blood lead levels in the US population. Phase 1 of the Third National Health and Nutrition Examination Survey (NHANES III, 1988–1991). *Journal of the*

American Medical Association **272**, 277–83.

Bullard, R. D. (1990). *Dumping in Dixie: Race, class, and environmental quality.* San Francisco: Westview Press.

Centers for Disease Control (1991). *Preventing Lead Poisoning in Young Children.* Atlanta: U.S. Department of Health and Human Services, Public Health Service.

Commission for Racial Justice (1987). *Harzardous Wastes and Race in the United States: A National Report on the Racial and Socio-Economic Characteristics of Communities with Hazardous Waste Sites.* New York: Commission for Racial Justice.

Czerwinski, S. A., Schell, L. M., Stark, A. D. & Grattan, W. (1993). Maternal morphology and neonatal blood lead levels. [Abstract] *American Journal of Physical Anthropology* (Supplement 16), 77–8.

Czerwinski, S. A., Schell, L. M. & Stark, A. D. (1995). Dietary assessment of socioeconomically disadvantaged pregnant women. [Abstract] *American Journal of Human Biology* **7**, 121.

Damstra T. (1977). Toxicological properties of lead. *Environmental Health Perspectives* **19**, 297–307.

Dietrich, K. N., Krafft, K., Bornschein, R. L., Hammond, P., Berger, O., Succop, P. A. & Bier, M. (1987). Low-level lead exposure: effect on neurobehavioral development in early infancy. *Pediatrics* **80**, 721–30.

EPA (U.S. Environmental Protection Agency) (1986). *Air quality critieria for lead.* Research Triangle Park: U.S. Environmental Protection Agency.

Kim, N. & Ferguson, J. (1993). Concentrations and sources of Cd, Cu, Pb, and Zn in house dust in Christchurch, New Zealand. *Science of the Total Environment* **138**, 1–21.

Lansdown, R. & Yule, W. (Ed) (1986). *Lead toxicity: history and environmental impact.* Baltimore: Johns Hopkins University Press.

Lucas, S. R. (1996). Relationship between blood lead and nutritional factors in preschool children: a cross-sectional study. *Pediatrics* **97**, 74–8.

Lyngbye, T., Hansen, O. N. & Grandjean, P. (1988). Neurological deficits in children: medical risk factors and lead exposure. *Neurobehavioral Toxicology and Teratology* **10**, 531–7.

Metzger, R., Delgado, J. L. & Herrell, R. (1995). Environmental health and Hispanic children. *Environmental Health Perspectives* **103** (Supplement 6), 25–32.

Moore, M. R. (1986). Lead in humans. In *Lead toxicity: history and environmental impact*, ed. R. Lansdown & W. Yule, pp. 54–95. Baltimore: The Johns Hopkins University Press.

Mott, L. (1995). The disproportionate impact of environmental health threats on children of color. *Environmental Health Perspectives* **103** (Supplement 6), 33–5.

Mushak, P., Davis, J., Crocetti, A. & Grant, L. (1989). Prenatal and postnatal effects of low-level lead exposure: integrated summary of a report to the U.S. congress on childhood lead poisoning. *Environmental Research* **50**, 11–36.

National Coalition of Hispanic Health and Human Services Organizations (COSSMHO). (1992). *The state of Hispanic health.* Washington: National

Coalition of Hispanic Health and Human Services Organizations.

Needleman, H. L., Schell, A., Bellinger, D., Leviton, A. & Allred, E. (1990). The long-term effects of exposure to low doses of lead in childhood, an 11-year follow-up report. *New England Journal of Medicine* **322**, 83–8.

Needleman, H. L., Riess, J. A., Tobin, M. J., Biesecker, G. E. & Greenhouse, J. B. (1996). Bone lead levels and delinquent behavior. *Journal of the American Medical Association* **275**, 363–9.

Raab, G., Fulton, M., Thomson, G., Laxen, D., Hunter, R. & Hepburn, W. (1989). Blood lead and other influences on mental abilities – results from the Edinburgh lead study. In *Lead exposure and child development: an international assessment*, ed. M. Smith, L. Grant & A. Sors, pp. 183–200. Boston: Kluwer Academic Publishers.

Schell, L. (1992). Risk focusing: an example of biocultural interaction. In *Health and lifestyle change*, ed. R. Huss-Ashmore, J. Schall & M. Hediger, pp. 137–44. Philadelphia: Masca, University of Pennsylvania.

Schell, L. M. (1997). Culture as a stressor: a revised model of biocultural interaction. *American Journal of Physical Anthropology* **102**, 67–77.

Silbergeld, E. K. (1986). Maternally mediated exposure of the fetus: in utero exposure to lead and other toxins. *Neurotoxicology* **7**, 557–68.

Thompson, G., Robertson, E. & Fitzgerald, S. (1985). Lead mobilization during pregnancy. *Medical Journal of Australia* **143**, 131.

US General Accounting Office. (1983). *Siting of hazardous waste landfills and their correlation with racial and economic status of surrounding communities.* Washington, D.C. General Accounting Office.

Waldbott, G. L. (1978). *Health effects of environmental pollutants*, 2nd edn. St. Louis: The C.V. Mosby Company.

Wernette, D. & Nieves, L. (1992). Breathing polluted air: minorities are disproportionately exposed. *EPA Journal* **18**, 16–17.

Wietlisbach, V., Rickenbach, M., Berode, M. & Guillemin, M. (1995). Time trend and determinants of blood lead levels in a Swiss population over a transition period (1984–1993) from leaded to unleaded gasoline use. *Environmental Research* **68**, 82–90.

Woodard, B. T., Gerguson, B. B. & Wilson, D. J. (1976). DDT levels in milk of rural indigent blacks. *American Journal of Diseases of Childhood* **130**, 400–3.

8 Educational potential and attainment: long-term implications of childhood undernutrition

SARA STINSON

Introduction

Social inequality has numerous negative consequences for children's nutritional status. Among the many potentially deleterious effects of undernutrition is its impact on cognitive functioning. This chapter briefly reviews the vast literature dealing with whether undernutrition early in life has biological effects that have lasting negative consequences for mental functioning. The effects of undernutrition on learning have been studied in disadvantaged populations in both rich and poor countries (for reviews see Gorman, 1995; Grantham-McGregor, 1995; Wachs, 1995). This chapter will deal almost exclusively with poor children living in poor countries. These are the children who experience the most disadvantaged environments and in whom we might therefore expect to see the greatest effects of social inequality.

The main focus of this chapter will be on whether there are biological effects of early undernutrition that influence functioning in the 'real world' rather than on how early undernutrition affects scores on tests of mental ability. Although we might wish for a large variety of measures of functioning, in practice, information related to school performances is most frequently collected. While school achievement measures just one aspect of function, it is an aspect that can have important consequences for poor children's lives. Long-term education may provide the skills that allow them to move out of extreme poverty; even a few years of school will give them the basic literacy to be better able to succeed with the dominant culture; for minority ethnic groups these first few years of school can teach them the national language.

A variety of research designs has been used to address the question of whether undernutrition has long-term biological effects on cognitive functioning. Studies on humans can be divided into three main types: investigations in which children who were severely undernourished are restudied years after the nutritional insult; studies of the effects of food supplementation in mild to moderately undernourished children; and research in which anthropometric measurements are used to indicate nutritional status and comparisons are made between children who differ anthropometrically (Gorman, 1995; Grantham-McGregor, 1995; Wachs, 1995). There has also been a great deal of research on the effects of experimentally produced undernutrition in laboratory animals. Animal studies have the advantage of being able to use a truly controlled experimental design and have shown that undernutrition early in life has lasting negative cognitive effects, even in rehabilitated animals (Strupp and Levitsky, 1995). However, because animal studies cannot provide information on the practical consequences of these impairments for children's functioning and achievement, they will not be discussed in this chapter.

Follow-up of severely undernourished children

In terms of providing information about the long-term effects of early undernutrition, there are several advantages and limitations of studies that follow-up children who had been hospitalized for severe undernutrition early in life. These studies can provide information on the long-term consequences of early undernutrition because children are restudied at least several years after the episode of clinical malnutrition, and in some cases over a decade later. Since these studies generally consider older children, measures of school performance are frequently collected in addition to tests of cognitive functioning. There are also limitations to these studies. The children studied have been hospitalized for severe undernutrition, so one could question the extent to which they are representative of the much more common mild to moderate undernutrition. Probably the major problem with this research design is determining to what extent any cognitive problems in the previously malnourished children are the biological consequence of undernutrition rather than the result of a disadvantaged environment. To try to disentangle these influences, the previously malnourished child generally is compared to a sibling or a child from a similar environment, but neither of these is truly a controlled comparison. Even in the same neighbourhood, the households of severely undernourished children are likely to compare negatively to other households in characteristics such as economic status, parental education, family cohesion, and amount of intellectual stimulation provided in the home (Richardson, 1980; Grantham-McGregor, 1995). Since all of these factors can affect mental development, it is probable that comparisons between severely malnourished children and children from similar surroundings measure not just the biological effects of undernutrition, but the effects of differing environments as well. Comparisons with siblings suffer from the opposite sort of constraint. Siblings of severely undernourished children are likely to be more undernourished than other children in the same locale (Birch *et al.*, 1971; Hertzig *et al.*, 1972; Grantham-McGregor, 1995). Thus sibling comparisons are apt to underestimate any biological consequences of undernutrition. Neither sibling comparisons nor comparisons with children from the same neighbourhood control for the effects of hospitalization (Grantham-McGregor, 1995).

A number of studies have examined the school performance or classroom behaviour of children who suffered from severe undernutrition early in life (Richardson *et al.*, 1972, 1973; Graham and Adriazen, 1979; Pereira *et al.*, 1979; Moodie *et al.*, 1980; Galler *et al.*, 1983, 1984, 1990). The design of the Jamaican study of Richardson and colleagues is particularly interesting because it involved not just a previously malnourished (index) boy and his brother, but also a comparison boy from the same school classroom for both the sibling and the index child. These children were compared in terms of their scores on the Wechsler Intelligence Scale for Children, their scores on the Wide Range Achievement Test (WRAT), and teachers' ratings of their performance and behaviour.

Figure 8.1 shows the scores of the four groups of boys on the WRAT for reading, spelling and arithmetic. Although the IQs of the index child and his sibling were significantly different (Hertzig *et al.*, 1972), in those achievement tests which measured knowledge of specific subjects, there were no significant differences between the two groups (Richardson *et al.*, 1973). Both the previously malnourished boy and his sibling had lower scores than their comparison children from the same classroom. These differences were significant for all tests for the index boys, but only for reading for the siblings. However, because the comparison children for the previously malnourished boys scored higher than the comparison boys for the siblings, the siblings do differ significantly from the index comparison children on all three tests. These results illustrate the problem of choosing an appropriate comparison for a previously severely undernourished child; the results are different depending on whether the comparison is with a sibling or a classmate.

The teachers of the four groups of boys were asked to rate the children on their school behaviour (Richardson *et al.*, 1972). The ratings included behaviours such as memory, distractibility, ability to pay attention in class, frequency of class participation, co-operation with the teacher and how well the child got along with his classmates. The previously malnourished

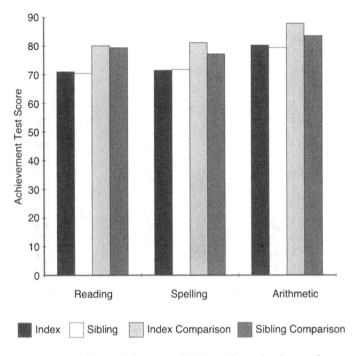

Figure 8.1. Wide Range Achievement Test scores in previously severely malnourished Jamaican boys (index), siblings of the previously malnourished boys (sibling), boys from the same classroom as the previously malnourished boy (index comparison), and boys from the same classroom as the sibling (sibling comparison). There were no significant differences in test performance between index boys and their siblings. The index boys scored significantly lower than their classroom comparisons on all tests. The siblings scored significantly lower than their classroom comparisons in reading. (Data from Richardson *et al.*, 1973).

boys had significantly lower teacher ratings than their comparison classmates for all of these types of behaviour, while for the siblings and their classmate comparisons only the difference in distractibility was statistically significant (see Figure 8.2). Galler *et al.* (1983) report similar behavioural differences between previously malnourished children and comparison children in Barbados.

The teachers also rated the school work of each of the four groups of Jamaican boys and the boys' school grades were recorded (Richardson *et al.*, 1973). The results of these comparisons were similar to the results for behaviour. Previously malnourished children received significantly lower grades and lower ratings of the quality of their school work than their classroom comparisons. While the siblings had somewhat lower grades and teacher ratings than their classmate comparisons, none of these differences reached statistical significance. Galler *et al.* (1984, 1990) also

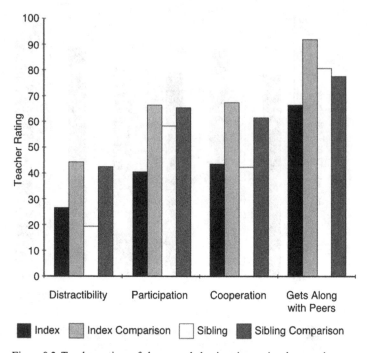

Figure 8.2. Teacher ratings of classroom behaviour in previously severely malnourished Jamaican boys (index), boys from the same classroom (index comparison), siblings of the previously malnourished boys (sibling), and boys from the same classroom as the siblings (sibling comparison). Higher values indicate better teacher ratings. Previously malnourished boys had significantly lower teacher ratings than their classroom comparison for all types of behaviour; siblings of the previously malnourished boys had lower ratings than their classroom comparisons only for distractibility. The previously malnourished boys and their siblings were rated by different teachers because they were in different classes, so their ratings are not comparable. (Data from Richardson *et al.*, 1972.)

found worse school performance in previously malnourished children as compared to healthy classmates in Barbados.

The results of the Jamaica study indicate that previously malnourished children had lower IQs than their siblings but were similar in terms of tests of knowledge in specific subjects. In contrast to the achievement test scores, teachers' ratings of behaviour and school performance were lower for the previously malnourished boys in comparison to their classmates, while siblings differed much less in comparison to their classmates. This implies that teachers' ratings of school work and the grades they assigned to children were influenced by both behaviour and achievement. Since teachers' evaluations of their students' performance can have a major impact on school success, these results suggest that severe undernutrition early in life does have an important effect on future functioning.

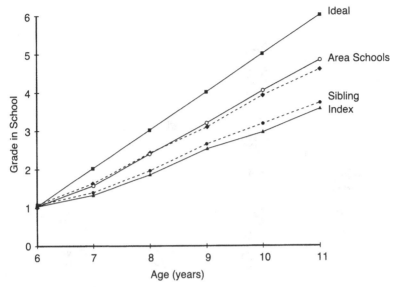

Figure 8.3. School progress among children in poor areas of Lima, Peru. The ideal pattern is that a child is in 1st grade at age 6, 2nd grade at age 7, 3rd grade at age 8, etc. Children in two elementary schools in poor areas of the city (area schools) were about one year behind in terms of school progress. Previously malnourished children (index) and their siblings averaged about two years behind, and there were no significant differences between previously malnourished children and their siblings. (Adapted from Graham and Adriazen, 1979.)

Another way to look at how previously malnourished children do in school is to look at how they progress from grade to grade in school. Graham and Adriazen (1979) found that Peruvian children who had been hospitalized for severe malnutrition did not differ in school progress from their siblings, although both groups were behind in school compared to children from the same neighbourhoods (see Figure 8.3). Similar findings were reported by Moodie *et al.* (1980) for South African children followed-up 15 years after being hospitalized for kwashiorkor. However, very different results were found in a similar study in Vellore, India (Pereira *et al.*, 1979).

In this case, the previously malnourished children were much more likely than their siblings to be behind in school, although the two groups were similar in terms of the percentage of children who had dropped out of school or never attended school (see Figure 8.4). These studies do not provide clear-cut evidence that children who suffered from an early episode of severe malnutrition necessarily progress less well in school than do their siblings, although they do suggest that previously malnourished

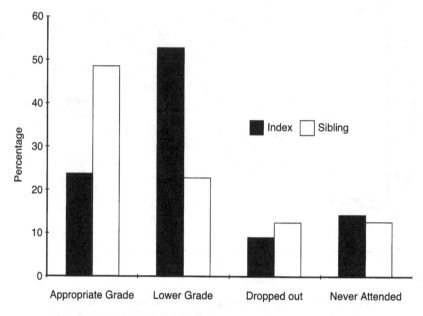

Figure 8.4. School progress in previously malnourished children (index) and their siblings in Vellore, India. Previously malnourished children were more likely to be behind in school. (Data from Pereira *et al.*, 1979.)

children do less well than children experiencing the same disadvantaged environment.

Supplementation studies

A second research design which has been used to study the effects of undernutrition on mental development is that in which the diets of moderately undernourished infants and young children are supplemented with additional food. Compared to follow-up studies of severely undernourished children, supplementation studies offer some benefits and some limitations in determining the enduring effects of undernutrition. Since these studies involve mild to moderately undernourished children, they are more representative of the vast majority of undernutrition in the world. A major advantage over follow-up studies is that supplementation studies are more experimental in nature. Children are frequently randomly assigned to groups receiving supplement or no supplement, so there is not the problem of selecting an appropriate comparison group that so confounds follow-up studies of severely malnourished children. In terms of the focus of this chapter there are also limitations to supplementation studies. For most of these studies, long-term follow-up results have not been published. This means that in most cases there is only information about the effects of supplementation up to the end of the period of supplementation, generally in the preschool years. Because older children generally are not studied, indicators of school achievement are usually not available – so measures of the practical effects of supplementation are frequently lacking.

Studies of the effects of general nutritional supplementation of mild to moderately undernourished children have used somewhat different research designs. Investigations in Colombia (Herrera *et al.*, 1980; Waber *et al.*, 1981), Jamaica (Grantham-McGregor *et al.*, 1991), and Indonesia (Husaini *et al.*, 1991) involved comparisons between children who received a supplement and children who received no supplement. The Guatemalan INCAP study (Engle *et al.*, 1993; Pollitt *et al.*, 1993) was a comparison of children who received a more nutritious supplement with children who received a less nutritious supplement. In the Taiwanese Bacon-Chow study, children were not supplemented directly; their mothers were supplemented before pregnancy, and during pregnancy and lactation (Joos *et al.*, 1983).

The results of Pollitt and Oh's (1994) analysis of these five supplementation studies are shown in Figure 8.5. Based on a meta-analysis of the combined data they found that supplemented/more supplemented children differed significantly from unsupplemented/less supplemented children in motor performance for tests administered between 8 and 15 months of age and between 18 and 24 months of age. The difference between the two groups in scores on mental tests was not significant in the younger age group and barely reached statistical significance in the older age group. The finding that motor development was more affected than mental development is in keeping with the idea that one way in which undernutrition influences mental ability is by changing the way the child interacts with the environment (Husaini *et al.*, 1991; Pollitt *et al.*, 1993). Supplemented children are more advanced in motor development so they are likely to be more active in interacting with their environment and may also elicit different kinds of responses from the people in their environment.

One supplementation study that provides results for older children is the Guatemalan INCAP study. The design of this study (Martorell *et al.*, 1995) involved two supplements: atole, a high energy and high protein gruel and fresco, a low energy, no protein drink; both supplements contained vitamins and minerals. Each supplement was distributed at feeding stations in two rural Guatemalan villages between 1969–77, with attendance and intake being recorded for pregnant and lactating women and children under 7 years of age. The INCAP study is unique among supple-

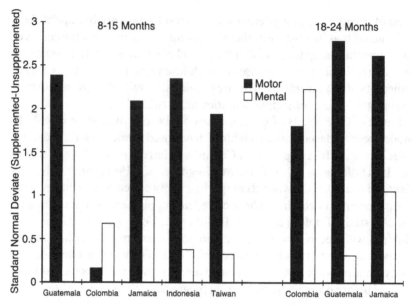

Figure 8.5. The difference in tests of mental and motor performance between unsupplemented/less supplemented children and supplemented/more supplemented children. The differences between the two groups are expressed in terms of the significance level of a one tailed t-test converted to the z value from a normal distribution. The differences between the supplemented/more supplemented and unsupplemented/less supplemented children were greater for motor performance than for mental performance in both age groups. (Data from Pollitt and Oh, 1994.)

mentation studies in that it has now included a long-term follow-up. The children who participated in the supplementation study were restudied in 1988–89 when they were between 11 and 24 years of age.

Figure 8.6 shows the results of a reanalysis of the data from the original INCAP study (Pollitt *et al.*, 1993). Children in atole and fresco villages were compared at the ages of 3, 4, 5 and 6 years of age in terms of their performance on a number of cognitive tests. After adjusting for sex, attendance at the feeding centres, and socio-economic status, there was a significant difference between the two groups of villages at 4 and 5 years of age. Scores on the cognitive tests were higher in children in the atole villages, but the difference between the two groups of villages was dependent on socio-economic status. As socio-economic status increased, the scores of children in the fresco villages increased, but the scores of children in the atole villages did not. Thus the greatest difference between children receiving the two supplements was in the lowest socio-economic group.

Similar results were found in the INCAP follow-up, as shown in Figure 8.7. After adjusting for socio-economic status, sex, attendance at feeding

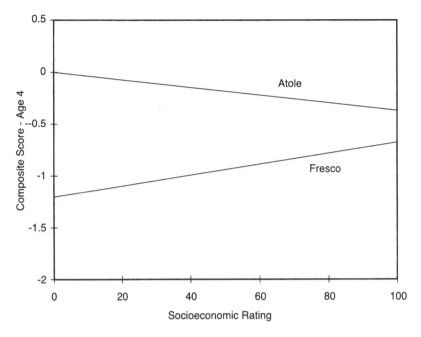

Figure 8.6. Regression of the first factor score from factor analysis of a number of cognitive tests on socio-economic rating among 4-year-old children in the INCAP study. Scores on the cognitive tests were higher in atole than in fresco villages. Socio-economic status was positively associated with performance in fresco subjects, but not in atole subjects. (Adapted from Pollitt *et al.*, 1993.)

centres, age at school entry and maximum grade attained, the scores on tests of knowledge, numeracy, reading and vocabulary were higher in children who had received atole. Again, there was a significant interaction between supplement and socio-economic status, with scores in fresco villages increasing as socio-economic status increased. These results indicate that supplementation had its greatest effect in the most disadvantaged and that the effects of supplementation were still evident over 10 years after supplementation ended.

Although there were significant effects of supplementation on test performance at follow-up, supplementation explained only a small percentage of the total variance in test scores, ranging from 1% to 4% (Pollitt *et al.*, 1993). It should be remembered that since children in both atole and fresco villages were supplemented in some form, this is a conservative test of the effects of supplementation. In addition, attendance at the feeding stations and the amount of supplement consumed varied among individuals (Schroeder *et al.*, 1993), so there were individuals in the atole villages who did not consume much supplement.

Some earlier results from the INCAP study suggest that any positive

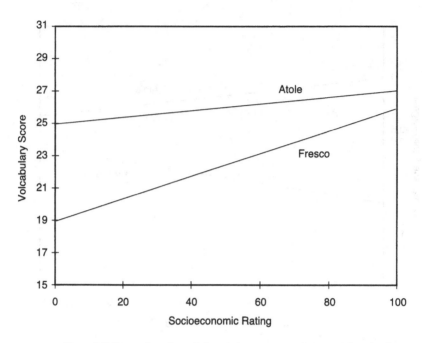

Figure 8.7. Regression of vocabulary test scores on socio-economic rating for participants in the INCAP follow-up. Vocabulary test scores were higher in children from atole villages than in children from fresco villages. As socio-economic status increased, the scores of children who received fresco increased, but this was not true for children who received atole. (Adapted from Pollitt *et al.*, 1993.)

effects of atole on cognitive functioning might not have a large effect on measures of school performance because of differences between atole and fresco villages (Pollitt *et al.*, 1993). As shown in Figure 8.8, school progress through the first grade between 1975 and 1982 was more satisfactory in fresco villages than in atole villages. In fresco villages, a greater percentage of children enrolled in school and a greater percentage were promoted, while a smaller percentage repeated the first grade and a smaller percentage dropped out of school. The differences in school performance between fresco and atole villages are likely to reflect dissimilarities in the schools and may also be related to the fact that parents in fresco villages were more likely to be literate and had attended school longer (Pollitt *et al.*, 1993). Such educational differences could influence parental attitudes toward education. Whatever the cause of the differences between atole and fresco villages, these dissimilarities could overwhelm any biological effects of supplementation.

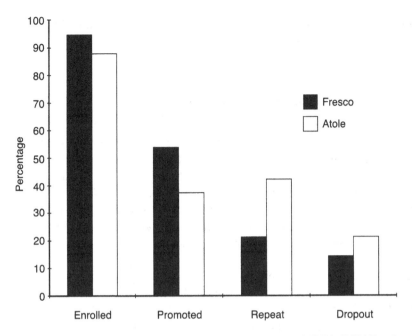

Figure 8.8. Progress in the first grade between 1975 and 1982 in INCAP atole and fresco villages. A greater percentage of children in fresco villages showed satisfactory progress. (Data from Pollitt *et al.*, 1993.)

Correlation studies

A large number of studies have used anthropometric measurements as indicators of nutritional status and correlated these measures with school performance and tests of cognitive function. These have primarily been investigations of moderately undernourished children, so they are representative of this most common type of undernutrition. They also give indicators of the practical effects of undernutrition since they frequently include measures of school achievement. Most of these studies have found significant correlations between some anthropometric indicator of nutritional status and measures of school performance or mental function, in many cases after controlling for potential confounding factors such as socio-economic status (Wachs, 1995). This lends further support to the idea that undernutrition has a negative impact on cognitive function, although the limitations of correlation studies do not make these results a very strong test. The environments of children who differ in body size are likely to differ in other ways that influence mental function. For this reason, comparisons of children who differ anthropometrically are not just measuring the biological effects of nutrition. In addition, anthropometric differences among children may not reflect different nutritional histories since body size is influenced by genetic factors.

Factors influencing school attendance

As indicated by some of the results discussed above, school attendance is hardly universal among poor children in poor countries. If children with poorer nutritional status are less likely to attend school, this would compound any negative biological effects of undernutrition on cognition.

Figure 8.9 shows differences between 8- to 13-year-old children who had attended at least one year of school and children who had never attended school in three of the four rural Guatemalan villages that participated in the INCAP study (Irwin *et al.*, 1978). Children who had never attended school had significantly lower scores on a series of cognitive tests administered at preschool age, were of lower socio-economic status, and had mothers who were less likely to report that they attempted to teach their preschool children in the home. Further analysis showed that after controlling for socio-economic status and family values toward education, scores on the preschool test were significantly correlated with the age at which a child first entered school for boys and girls, and with the number of years the child had attended school for girls only. After controlling for preschool cognitive performance and family values toward education, house quality (one of the measures of socio-economic status) was significantly correlated with age at first enrolment for girls and with length of attendance for boys (see Figure 8.10).

These results suggest that there is definite inequality in terms of which children attend school and how long they attend school, which is related to characteristics of the child and the family. Children who attend school are likely to have greater cognitive abilities, to be from higher socio-economic status homes, and to have parents who take a more active role in their education.

Nutritional status has also been found to influence school attendance. Moock and Leslie (1986) found that both height-for-age and weight-for-height among elementary school age children in southern Nepal were significantly related to the probability of attending school. Height-for-age was also significantly associated with the grade attained, as was also found in a study in China (Jamison, 1986).

The relationship between nutritional status, as indicated by anthropometric measurements, and school attendance can also be explored among Amerindian Chachi children in northwest Ecuador. The data for the analysis come from a cross-sectional growth study (Stinson, 1989) in which

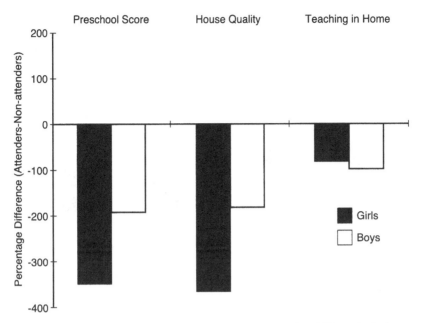

Figure 8.9. Comparison of 8- to 13-year-old Guatemalan children who had attended at least one year of school with children who had not attended school. The zero line is the value for children who had attended school; the bars indicate the percentage difference for children who had not attended school. Children who had not attended school scored significantly lower on cognitive tests given before the children entered school (preschool score), house quality (a measure of the size and construction of the house), and mothers' reports of trying to teach their preschool children in the home (teaching in home). (Data from Irwin *et al.*, 1978.)

most measurements were taken in schools, but some were collected by going from house to house. Although the study was not designed to examine factors influencing school attendance, children can be divided into those measured in school and those measured in their homes. This division results in a conservative test of differences between school attenders and non-attenders because children who attended school may have been classified as non-attenders.

It is certainly possible that children who were measured in their homes actually were attending school but had been absent on the day measurements were conducted or had attended school at some point in the past. As shown in Figure 8.11, there is a major difference in height-for-age between Chachi children who were measured in school and those who were measured at home – i.e. not in school when the measurements were conducted. This is especially true in the younger ages so that for children under 14, those attending school are significantly taller than those not attending school.

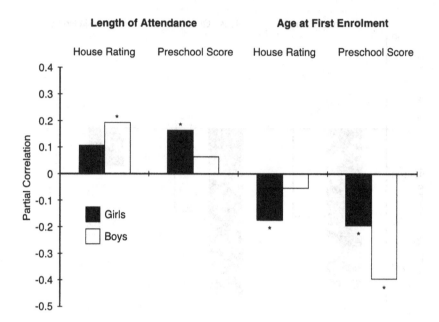

Figure 8.10. Partial correlations of socio-economic status (house rating) and scores on preschool cognitive tests with age of school enrolment and length of school attendance for 8- to 13-year-old Guatemalan children. Partial correlations for house rating controls for preschool scores and family values toward education; partials for preschool scores control for socio-economic status and family values toward education. Asterisks indicate statistically significant correlations. (Data from Irwin *et al.*, 1978.)

There are several possible explanations for the difference in body size between school attenders and non-attenders at younger ages. Although most Chachi children have birth certificates, parents are not usually aware of the exact ages of their children. Thus, decisions about when a child is ready to start school may be based on parents' perceptions of a child's age. Children who are large for their age may be more likely be sent to school early, while the entrance of small children might be delayed. This may be the explanation for the very large difference in height-for-age between attenders and non-attenders in the 6- to 8-year-age range. Another factor that probably plays a role in explaining the body size differences at the younger ages is socio-economic status. Height is positively correlated with socio-economic status among the Chachi (Stinson, 1996), and given the results of Irwin *et al.*'s (1978) Guatemalan study it would be expected that higher socio-economic status children would be more likely to attend school.

It is more difficult to explain the lack of difference between children attending and not attending school at the older ages. One possibility is that

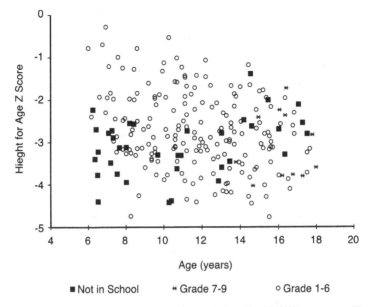

Figure 8.11. Height-for-age Z scores of Ecuadorian Chachi children measured in school or measured in their homes. Below the age of 14 years, children measured in school were significantly taller than children measured at home.

children who were not attending school at the time of the study had attended earlier but had since dropped out of school. A second possibility is more directly related to body size. As can be seen from Figure 8.11, most of the older children were in elementary school (there is an elementary school in almost every village, but there are only two secondary schools in the region, so only a small percentage of children continue beyond elementary school). It is possible that adolescents who are large for their age feel uncomfortable being in school with 8- and 9-year-olds. Therefore there may be some tendency for the tallest teenagers to leave elementary school at earlier ages. Finally, given the small number of older children measured in their homes, the possibility of sampling error cannot be discounted.

Although the approach used in this analysis of Chachi data is basically correlational and could be criticized on those grounds, the results do suggest that children's nutritional status influences whether they attend school, particularly at younger ages.

Conclusions

A variety of research designs has been used to examine whether there are lasting effects of undernutrition on mental functioning. In studies that have employed standardized tests, most but not all (e.g. Richardson *et al.*, 1973) have found that undernutrition is negatively associated with performance. These results suggest that undernutrition early in life has enduring effects on cognitive function. The practical consequences of any impairment are less clear, with only some data indicating that early undernutrition also has a negative impact on school performance. Most undernourished children endure many deprivations in addition to poor nutrition. In many cases these inequalities may overwhelm and obscure any biological consequences of undernutrition.

There is evidence that in disadvantaged environments, school attendance is influenced by children's nutritional status and mental abilities. If this is the case, any biological disadvantage for children who have suffered from early undernutrition would be magnified by the fact that they are less likely to receive an adequate education. This would further compound the serious inequalities in the quality of education available to children.

References

Birch, H. G., Piñeiro L., Alcalde, E., Toca, T. & Cravioto, J. (1971). Relation of kwashiorkor in early childhood and intelligence at school age. *Pediatric Research* 5, 579–85.

Engle P. L., Gorman K., Martorell R. & Pollitt E. (1993). Infant and preschool psychological development. *Food and Nutrition Bulletin* 14, 201–14.

Galler, J. R., Ramsey, F., Solimano, G. & Lowell, W. E. (1983). The influence of early malnutrition on subsequent behavioral development. II. Classroom behavior. *Journal of the American Academy of Child Psychiatry* 22, 16–22.

Galler, J. R., Ramsey, F. & Solimano, G. (1984). The influence of early malnutrition on subsequent behavioral development. III. Learning disabilities as a sequel to malnutrition. *Pediatric Research* 18, 309–13.

Galler, J. R., Ramsey, F. C., Morley, D. S., Archer, E. & Salt, P. (1990). The long-term effects of early kwashiorkor compared with marasmus. IV. Performance on the national high school entrance examination. *Pediatric Research* 28, 235–9.

Gorman, K. S. (1995). Malnutrition and cognitive development: evidence from experimental/quasi-experimental studies among the mild-to-moderately malnourished. *Journal of Nutrition* 125, 2239S–2244S.

Graham, G. G. & Adriazen, B. T. (1979). Status in school of Peruvian children severely malnourished in infancy. In *Behavioral Effects of Energy and Protein Deficits*, ed. J. Brozek, pp. 185–94. Washington: NIH.

Grantham-McGregor, S. (1995). A review of studies of the effect of severe malnutrition on mental development. *Journal of Nutrition* 125, 2233S–2238S.

Grantham-McGregor, S. M., Powell, C. A., Walker, S. P. & Himes, J. H. (1991). Nutritional supplementation, psychosocial stimulation, and mental development of stunted children: the Jamaican Study. *Lancet* 338, 1–5.

Herrera, M. G., Mora, J. O., Christiansen, N., Ortiz, N., Clement, J., Vuori, L., Waber, D., De Paredes, B. & Wagner, M. (1980). Effects of nutritional supplementation and early education on physical and cognitive development. In *Life-Span Developmental Psychology. Intervention*, ed. R. R. Turner & H. W. Reese, pp. 149–84. New York: Academic Press.

Hertzig, M. E., Birch, H. G., Richardson, S. A. & Tizard, J. (1972). Intellectual levels of school children severely malnourished during the first two year of life. *Pediatrics* **49**, 814–24.

Husaini, M. A., Karyadi, L., Husaini, Y. K., Sandjaja, Karyadi, D. & Pollitt, E. (1991). Developmental effects of short-term supplementary feeding in nutritionally-at-risk Indonesian infants. *American Journal of Clinical Nutrition* **54**, 799–804.

Irwin, M., Engle, P. L., Yarbrough, C., Klein, R. E. & Townsend, J. (1978). The relationship of prior ability and family characteristics to school attendance and school achievement in rural Guatemala. *Child Development* **49**, 415–27.

Jamison, D. T. (1986). Child malnutrition and school performance in China. *Journal of Development Economics* **20**, 299–309.

Joos, S. K., Pollitt, E., Mueller, W. H. & Albright, D. L. (1983). The Bacon Chow Study: maternal nutritional supplementation and infant behavioral development. *Child Development* **54**, 669–76.

Martorell, R., Habicht, J.-P. & Rivera, J. A. (1995). History and design of the INCAP longitudinal study (1969–77) and its follow-up (1988–89). *Journal of Nutrition* **125**, 1027S–1041S.

Moock, P. R. & Leslie, J. (1986). Childhood malnutrition and schooling in the Terai region of Nepal. *Journal of Development Economics* **20**, 33–52.

Moodie, A. D., Bowie, M. D., Mann, M. D. & Hansen, J. D. L. (1980). A prospective 15-year follow-up study of kwashiorkor patients. Part II. Social circumstances, educational attainment and social adjustment. *South African Medical Journal* **58**, 677–81.

Pereira, S. M., Sundararaj, R. & Begum, A. (1979). Physical growth and neurointegrative performance of survivors of protein-energy malnutrition. *British Journal of Nutrition* **42**, 165–71.

Pollitt, E. & Oh, S-Y. (1994). Early supplementary feeding, child development, and health policy. *Food and Nutrition Bulletin* **15**, 208–14.

Pollitt, E., Gorman, K. S., Engle, P. L., Martorell, R. & Rivera, J. (1993). Early supplementary feeding and cognition: effects over two decades. *Monographs of the Society for Research in Child Development* **58**(7), 1–99.

Richardson, S. A. (1980). The long range consequences of malnutrition in infancy: a study of children in Jamaica, West Indies. In *Topics in Paediatrics 2, Nutrition in Childhood*, ed. B. A. Warton, pp. 163–76. Tunbridge Wells: Pitman.

Richardson, S. A., Birch, H. G., Grabie, E. & Yoder, K. (1972). The behavior of children in school who were severely malnourished in the first two years of life. *Journal of Health and Social Behavior* **13**, 276–84.

Richardson, S. A., Birch, H. G. & Hertzig, M. E. (1973). School performance of children who were severely malnourished in infancy. *American Journal of Mental Deficiency* **77**, 623–32.

Schroeder, D. G., Kaplowitz, H. & Martorell, R. (1993). Patterns and predictors of participation and consumption of supplement in an intervention study in rural Guatemala. *Food and Nutrition Bulletin* **14**, 191–200.

Stinson, S. (1989). Physical growth of Ecuadorian Chachi Amerindians. *American*

Journal of Human Biology **1**, 697–707.

Stinson, S. (1996). Early childhood growth of Chachi Amerindians and Afro-Ecuadorians in northwest Ecuador. *American Journal of Human Biology* **8**, 43–53.

Strupp, B. J. & Levitsky, D. A. (1995). Enduring cognitive effects of early malnutrition: a theoretical reappraisal. *Journal of Nutrition* **125**, 2221S–2232S.

Waber, D. P., Vuori-Christiansen, L., Ortiz, N., Clement, J. R., Christiansen, N. E., Mora, J. O., Reed, R. B. & Herrera, M. G. (1981). Nutritional supplementation, maternal education, and cognitive development of infants at risk of malnutrition. *American Journal of Clinical Nutrition* **34**, 807–13.

Wachs, T. D. (1995). Relation of mild-to-moderate malnutrition to human development: correlational studies. *Journal of Nutrition* **125**, 2245S–2254S.

9 *Affluence in developing countries and natural selection in humans*

KERIN O'DEA

Introduction

Patterns of chronic degenerative diseases (such as cardiovascular diseases, many cancers, diabetes, obesity and related conditions and non-infective bowel diseases) vary widely both within and between populations. These 'diseases of affluence' are strongly linked with a westernized lifestyle. Non-insulin-dependent diabetes (NIDDM) is a prime example. It is a serious and increasing health problem in populations worldwide, the only exceptions being the few indigenous and agrarian populations continuing to live traditional lifestyles. Examples of groups with no diabetes mellitus include the Mapuche Indians of Chile, rural non-Austronesian Melanesians of Papua New Guinea and small groups of Australian Aborigines living a traditionally oriented lifestyle (O'Dea *et al.*, 1988a; King and Rewers, 1993). Without exception these populations are lean and lead physically active lives, and although their diets vary, in no instance are they energy-dense or high in fat.

Although diabetes has been recognized for hundreds and in some cases thousands of years, it is only in the last 20–30 years that it has affected large numbers of people. Until the 1960s and 1970s it occurred primarily in affluent western societies, and only among the wealthy in the non-industrialized world. However, the epidemiology of NIDDM has undergone a dramatic change in the 1980s and 1990s, and while the situation has certainly not improved in the industrialized world, it has deteriorated markedly in populations previously relatively unaffected. Thus, the greatest increase in prevalence of NIDDM is among populations who have been exposed to a rapid westernization of diet and lifestyle in the latter part of the twentieth century. Indigenous minorities in affluent societies, such as the USA, Canada and Australia, are among those populations worst affected, with prevalence rates of 25–50% in the 30–64 year age group (King and Rewers, 1993; O'Dea *et al.*, 1993).

151

Lifestyle change unmasks susceptibility to NIDDM

The strikingly different prevalences of diabetes, which have been reported for populations of the same ethnicity under different lifestyle conditions, highlight the critical role played by lifestyle factors in precipitating diabetes by unmasking genetic susceptibility. Marked increases in diabetes prevalence have been documented as populations urbanize, and when they migrate from a low risk environment (traditional lifestyle) to a more affluent one. Westernization is accompanied by a very predictable and consistent pattern of changes in energy expenditure and composition and availability of the diet. Traditional lifestyles of both hunter–gatherers and subsistence agriculturalists were characterized by high levels of physical activity and diets with low energy density. Hunter–gatherers derive their diet from wild animals (much leaner than domesticated meat animals) and a wide variety of uncultivated plant foods (high nutrient density, but low energy density). Subsistence agriculturalists derive their diet from one or more starchy staples supplemented by small quantities of other plant foods and meat (very low in fat and bulky). With westernization, the energy expenditure needed for survival falls markedly (due to multiple forms of labour-saving mechanization), while at the same time energy availability is maintained or increased. The resulting imbalance between energy demand and supply explains the increases in population mean body mass index, and in obesity and associated conditions such as NIDDM which appear to inevitably accompany modernization.

Impact of westernization on risk of NIDDM

There are many dramatic illustrations of the impact of migration on the prevalence of lifetyle-related chronic diseases. For example, the prevalence of NIDDM in mainland China is generally low, although it increases with increasing socio-economic status. However, when the Chinese migrate and adopt a more affluent lifestyle the prevalence of NIDDM rises to moderate (Singapore) or high (Mauritius) levels. Similarly, Asian Indians show a strong tendency to diabetes when they urbanize in India (Ramachandran *et al.*, 1988; 1992) or elsewhere (UK, Fiji, South Africa, Mauritius) (e.g. Maher and Keen, 1985; Dowse *et al.*, 1990; Motala *et al.*, 1993). The public health implications of increasing rates of diabetes in heavily populated areas of the developing world, such as China and India, are enormous.

Most indigenous populations throughout the world have been subjected to rapid and profound lifestyle changes in the twentieth century, and many are now experiencing epidemics of obesity and diabetes. Among the most well-documented are the Pima Indians in the USA, where more than half of the adult population has diabetes, and Micronesian Nauruans in the Pacific where ~40% of adults are diabetic. Less well-documented are populations from Australia and Papua New Guinea (PNG) in which very high rates of diabetes have been reported only relatively recently.

Papua New Guinea

The Wanigela people of PNG appear to be one of the most diabetes-prone populations in the world (Dowse *et al.*, 1994). In the urban population of Koki (people originating predominantly from the rural village of Wanigela) the prevalence of diabetes almost doubled in the 14 years between 1977 and 1991 (from 16.7% to 27.5%). At the same time there was an almost doubling of impaired glucose tolerance (IGT), so that by 1991 more than 60% of those 25 years and older had abnormal glucose tolerance. Even the rural population of Wanigela showed an increasing prevalence of diabetes and IGT over the eight-year period between the baseline survey in 1983 and the follow-up in 1991, with diabetes increasing from 9.8% to 11.1% and IGT from 6.7% to 14.1%. In contrast, the rural village of Kalo had low diabetes prevalence (< 2%) both at baseline and at follow-up, supporting the argument that the Wanigela people are particularly susceptible to the diabetogenic effects of westernization (Dowse *et al.*, 1994).

Australian Aborigines

Australian Aborigines and Torres Strait Islanders are highly susceptible to diabetes when they westernize (Duffy *et al.*, 1981; O'Dea *et al.*, 1988b, 1993; O'Dea, 1992). Most Aborigines in Australia today live a westernized lifestyle, deriving their diet completely or in large part from western foods and leading sedentary, inactive lives. Relative to the rest of the Australian population, Aborigines have a 20-year shorter life expectancy (AIHW, 1994), up to four times higher prevalence of coronary heart disease (CHD) and, in the 20–50 year age group, more than 10 times the prevalence of diabetes (O'Dea *et al.*, 1993). Like other high prevalence populations, the age of onset is much earlier than in the lower risk European populations. Cases of NIDDM have been reported in Aboriginal teenagers, and the prevalence of diabetes reaches its maximum at the age of about 40 years, which is more than 30 years earlier than in Europeans (Glatthaar *et al.*, 1985). However, the prevalence of diabetes varies widely between Aboriginal communities even when they are located in the same region. The differences in prevalence appear to be accounted for in large part by differences in the prevalence of overweight and obesity (O'Dea *et al.*, 1993; Gault *et al.*, 1996). Increasing BMI is associated with increased risk of NIDDM at all ages, although the slope of the relationship rises with age to plateau at about age 40. Increasing BMI is also associated with increases in the classical CHD risk factors of hypertension, hypertriglyeridemia and hypercholesterolemia (O'Dea, 1992; O'Dea *et al.*, 1993; Gault *et al.*, 1996).

Recent prospective data from two central Australian Aboriginal communities which were followed for four to eight years indicated that BMI at baseline was a very strong predictor of diabetes, with no new cases of diabetes occurring in anyone with a BMI of 22 kg/m^2 or less even in the older age groups. Given the strength of this association, it is particularly disturbing to note the sharp rise in prevalence of obesity in young Aboriginal people in these same communities (O'Dea unpub. data).

Impact of temporary reversion to hunter–gatherer lifestyle on NIDDM in Australian Aborigines

Australian Aborigines experience very high rates of NIDDM now that they no longer live traditionally. For older people in some communities in remote parts of the country the transition to a westernized lifestyle has occurred within their own lifetime. Consequently, these people have retained the knowledge and ability to survive as hunter–gatherers. We have taken advantage of this unusual opportunity to examine the impact of temporary reversion to the hunter–gatherer lifestyle on the metabolic abnormalities of NIDDM in these people (O'Dea, 1984). In 1982, a group of middle-aged, overweight, diabetic Aborigines returned to their traditional country in northwestern Australia to live off the land on bush foods for seven weeks. This 'reverse lifestyle change' had a remarkable effect on their health: they lost weight (although most were still overweight at the end); their fasting blood glucose levels fell to near the healthy non-diabetic range; their blood lipids fell to normal levels; blood pressure fell; and bleeding times increased substantially (interpreted as indicating a reduced thrombosis tendency). This study indicated the very powerful effect of lifestyle on health and that the changes can occur very rapidly. Three factors that are integral to the hunter–gatherer lifestyle were responsible for these health benefits: weight loss; regular physical activity; and a low fat high fibre diet derived from very lean meat, fish, shellfish and a wide range of wild vegetables and fruits. The public health challenge is to develop strategies to apply these lifestyle principles to everyday life at the community level in an attempt to reduce the severity of diabetes and its complications. Excessive body fat, physical inactivity and diet have all been shown to increase the risk of NIDDM in populations. Some of the evidence for this will be summarized below.

The link between excessive weight gain and NIDDM

The strong linkage between increasing prevalence of overweight and obesity and an increase in the prevalence of NIDDM has been observed in all high risk populations, although the weight gain needed to trigger diabetes on a population basis can vary substantially (as it does on an individual basis). Whether this can be fully accounted for by genetic differences in susceptibility ('thrifty genotype') remains an open question in the absence of convincing molecular evidence. Indeed, as westernization becomes more pervasive and more and more populations develop high rates of diabetes, it appears that populations of European origin may be the exception in being relatively resistant to the diabetogenic effects of the western lifestyle (Stern, 1991). What is clear in numerous populations is that susceptibility to diabetes (whether inherited or acquired) is unmasked by excessive weight gain, frequently occurring in adolescence or early adult life. Even in populations not previously considered to be high risk, such as those of African origin, a clear gradient of risk of NIDDM highly correlated with obesity has been reported (Cooper, 1994).

The relationship between body weight and risk of NIDDM in populations of European origin has been addressed convincingly in two large prospective studies among female and male health care professionals in the USA – the Nurses' Health Study and the Health Professionals Follow-up Study (Colditz *et al.*, 1990; Chan *et al.*, 1994). In a cohort of 113,861 women aged 30–55 years and initially free of known diabetes, Colditz *et al.* (1990) reported that the most striking predictor of diabetes eight years later was BMI. In comparison with women of BMI less than 22 kg/m^2 (Relative Risk (RR) = 1.0), women at the upper end of the normal weight range (BMI = 25) had a RR of 3.1. For women at the upper end of the overweight range (BMI = 30) the RR was 20, and for those who were clearly obese (BMI > 35) there was a 60-fold increase in risk. Obesity was a much stronger risk factor than family history of diabetes, which conferred only a two-fold additional risk. Unfortunately, no data were available in this study on body fat distribution. In the Health Professionals Follow-up Study, Chan *et al.* (1994) reported a similarly strong relationship between baseline BMI and future risk of NIDDM in men, and showed that both body weight and a centralized distribution of body fat were important components of this risk.

Prevention of NIDDM through weight loss in the obese

An important conclusion from both of these studies is that the single most important intervention to reduce diabetes prevalence in a population may be to reduce the population mean BMI. The results of a study of a group of severely obese subjects with IGT who underwent gastric bypass surgery appear to support this as yet unproven possibility (Long *et al.*, 1994). Of the 136 subjects who participated in the study, 109 had bariatric surgery for weight loss, with the remaining 27, who did not have surgery, acting as controls. They were followed for an average of 5.8 years. The control group did not lose weight during the follow-up period, whereas the gastric bypass group lost 60% of their excess body weight. Six of the 27 control subjects developed diabetes, equivalent to 4.72 cases per 100 person years. In contrast, only one of the 109 bypass patients developed diabetes, giving a conversion rate of only 0.15 cases per 100 person years. (The one patient who developed diabetes in this group failed to maintain weight loss post surgery.) This study indicated a > 30-fold reduced risk of developing NIDDM after weight loss in severely obese subjects with IGT.

Physical activity and risk of NIDDM

The two major lifestyle changes which accompany westernization are access to a readily available energy dense diet (rich in fat and refined carbohydrate) and reduced physical activity. Numerous prospective studies have demonstrated the beneficial effects of regular physical activity in protecting against the development of NIDDM (Helmrich *et al.*, 1991). It appears that the greatest benefit is for the most sedentary – that is, a moderate level of regular exercise in very inactive people will confer a greater health benefit than regular vigorous exercise in moderately active people (Jennings *et al.*, 1986). There is also evidence accumulating that the type of exercise which is most beneficial metabolically is low intensity, long duration exercise such as walking (one hour per day at least 5 times per week) (Lemieux and Després, 1994). These data can form the basis of very positive public health messages – with the focus on facilitating an increasing proportion of the population incorporating regular low intensity exercise into their daily routines.

Diet and risk of NIDDM

The role of diet in the pathogenesis and treatment of NIDDM continues to be the subject of lively debate. Given the strong association between excess body fat and risk of NIDDM, it is possible that the role of diet is mediated primarily through effects on body weight and/or energy intake. Over the long-term the two are highly correlated and difficult to disentangle.

Some of the best data suggesting that high fat diets are associated with the development of NIDDM comes from the San Luis Valley study which reported that high fat intake is associated with the diagnosis of diabetes in people with previously undiagnosed NIDDM and with progression from IGT to NIDDM, independent of energy intake, age, sex, ethnicity and obesity (Marshall *et al.*, 1991, 1994). Diet has certainly been shown to be important in the treatment of NIDDM. Diets high in fibre and complex carbohydrate (in particular, soluble fibre and slowly digested carbohydrate) and low in fat have been shown to produce improvements in glycemic control and lipid profile in NIDDM. These effects appear to be due to a combination of high fibre and low fat, since improved metabolic control was not observed in the absence of fibre, but was demonstrated with a low fat low carbohydrate diet (O'Dea *et al.*, 1989).

More recently, due to concerns about long-term compliance with the high fibre, low fat diet, an alternative approach has been investigated in which saturated fat is replaced by oleic acid-rich oils rather than fibre-rich carbohydrate. Such 'modified fat' diets have been associated with even better improvements in metabolic control in NIDDM than the high fibre, low fat diets and appear to have better long-term compliance (Campbell *et al.*, 1994; Garg, 1994). Furthermore, despite their relatively high fat content, the modified fat diet is not associated with weight gain and may result in a more favourable body fat distribution (Walker *et al.*, 1995; 1996).

Trends in prevalence of overweight and obesity: differential effects of socio-economic status

Despite the recognition by public health authorities worldwide of the deleterious health consequences of overweight and obesity, particularly in relation to the rising prevalence of NIDDM, the trends in prevalence of these conditions are upward in both industrialized and developing countries. The patterns vary according to level of economic development of a country and socio-economic status of its people (Sobal and Stunkard, 1989). In all countries for which data are available, population mean BMI is rising and there are increasing proportions of obesity. However, in affluent societies such as Australia, the USA and Europe, obesity is more prevalent among those in the lower socio-economic strata (Pietinen *et al.*, 1996), while in developing countries, overweight and obesity are more common among the affluent. Diabetes prevalence rates mirror these overweight/obesity patterns, with diabetes being more common in the underprivileged in affluent societies (e.g. black women in the USA), but more common among the affluent in poorer countries such as China and India (Ramachandran *et al.*, 1988). Similar socio-economic gradients of obesity and diabetes have been observed among high risk populations such as Hispanics in the USA, where those of higher socio-economic status (better educated, higher income) have lower rates of obesity and diabetes (Monterrosa *et al.*, 1995). However, differences in prevalence of overweight and obesity do not fully account for these socio-economic differentials in diabetes prevalence within populations (Stern *et al.*, 1991)

Differentiating between genetic and lifestyle influences in the pathogenesis of NIDDM

Thrifty genotype?

The thrifty genotype was proposed by Neel in 1962 to explain the increasing incidence of diabetes in the western world (Neel, 1962). Since then it has been invoked frequently to explain the epidemics of obesity and NIDDM in populations all over the world as they have made the rapid transition to a westernized lifestyle in the twentieth century. The 'feast and famine' conditions prevailing throughout most of human history could have selected strongly for a 'thrifty' metabolism, facilitating efficient fat deposition in times of food abundance and providing an energy buffer in times of scarcity. Food shortages were not confined to hunter–gatherers. Indeed, they may have been even more significant after the advent of agriculture (Brown and Konner, 1987). Archaeological evidence of shorter stature and other skeletal evidence of nutritional stress are consistent with severe and regular food shortages in agricultural societies. It has been argued that humans could not have adapted to conditions of continual surplus, because they have simply never existed in the past (Brown and Konner, 1987). Dowse *et al.* (1991) invoked the thrifty genotype hypothesis to explain the decline in incidence of glucose intolerance in Nauruans, and argued that the increased mortality and lower fertility of diabetic Nauruans may be early indicators of a fall in population frequency of the diabetic genotype. However, although the thrifty genotype is an attractive and plausible hypothesis and there is considerable indirect evidence for the concept, its molecular basis is yet to be established.

Thrifty phenotype?

While the role of lifestyle factors (obesity, physical inactivity, excess energy intake) in adulthood in precipitating NIDDM is widely recognized, there is increasing interest in factors operating in early life to heighten risk of diabetes and a range of other chronic conditions. Barker (1992) has observed that poor nutrition in foetal and early postnatal life is associated with higher risk of NIDDM, hypertension and cardiovascular disease in later life. It was originally proposed that the link with NIDDM was mediated via effects on beta-cell development, which predisposed to beta-cell dysfunction and diabetes in later life – the 'thrifty phenotype' hypothesis (Hales and Barker, 1992). More recent data from the same group have indicated that the effects are mediated by effects on insulin action rather than on insulin secretion (at least in the English cohorts examined). Phillips *et al.* (1994) have evidence that thinness at birth and obesity in adulthood are each independently associated with insulin resistance, and that individuals who were both thin at birth and obese as adults were the most insulin resistant. The mechanism by which early malnutrition increases insulin resistance has not yet been elucidated, but may be related to changes in sensitivity/responsiveness to glucocorticoids.

It is also possible that effects of undernutrition during foetal life may be expressed differently in different populations in response to variations in the type and/or severity of the insult during pregnancy. The different natural history of NIDDM in the black population of southern Africa is consistent with such a possibility – Joffe *et al.* (1992) have speculated that their characteristic pattern of accelerated beta-cell decompensation leading to insulinopaenic diabetes in NIDDM patients may be related to decreased beta-cell mass secondary to severe malnutrition in early life. Such a scenario would be consistent with data from experimental animals (Dhari *et al.*, 1993), which is discussed briefly below.

Not only is undernutrition during foetal life a risk factor for NIDDM in later life, but so is overnutrition. Pettit *et al.* (1988; 1993) have reported that offspring, aged 19–24 years, of Pima Indian women with NIDDM were more obese and had much higher prevalence of diabetes (45%) than offspring of non-diabetic women (1.4%) or women who developed diabetes after the pregnancy (8.6%). They concluded that metabolic disturbances affecting the intra-uterine environment can have significant effects on body composition and metabolism of the offspring that favour the development of obesity and NIDDM many years later.

Foetal programming of adult carbohydrate metabolism

There are critical periods for organ development during foetal life. Growth retardation at these critical times appears to have long lasting consequences. For example, Hoet and co-workers have demonstrated that the offspring of rats fed low protein diets during pregnancy have reduced beta-cell proliferation, and reduced islet size and vascularization (Dahri *et al.*, 1993). When tested at 10 weeks of age the offspring had impaired glucose tolerance and reduced insulin secretion. In similar studies, Hales and co-workers have demonstrated dramatic effects of protein deprivation in pregnancy on the activities of key enzymes in the regulation of glucose metabolism in the offspring (Desai *et al.*, 1995). They reported that the activity of the enzyme phosphoenolpyruvate carboxykinase doubled, while that of glucokinase halved, and that these effects were evident throughout adult life. They concluded that the 'enzymic "setting" of the liver is biased by a factor of 400% towards the production rather than the utilization of glucose.'

The role of events in very early life in 'programming' organ development and/or metabolism in ways which greatly increases susceptibility to 'diseases of affluence' such as NIDDM in later life could be an extremely important component of the current epidemic in popupations undergoing a rapid transition to a western lifestyle.

Possible consequences of rapid transition from poverty to affluence

In the Australian Aboriginal population, low birth weight is still common (Kliewer and Stanley, 1989), and frequently co-exists with obesity and diabetes in the parents (O'Dea, 1992). The potential multiplier effect of low birth weight on the risk of NIDDM and related conditions in adulthood needs to be acknowledged in the development of comprehensive long-term approaches to reducing the prevalence of these conditions in high risk populations. All of the populations which are now experiencing extremely high prevalences of NIDDM, or sharply increasing prevalences, have been exposed to rapid lifestyle changes over the period in which diabetes rates have risen. They have made (or are in the process of making) the transition from a traditional lifestyle (with low fat, bulky diets, high levels of physical activity and lean body weights) to the western lifestyle (readily available energy-dense food supply, sedentary lifestyle and increasing levels of body fat). The traditional lifestyle of many populations was frequently associated with high prevalence of low birth weight. Thus, many of the high risk populations experienced low birth weight followed by excessive weight gain in adulthood as they underwent the lifestyle transition. There is evidence that it is this particular combination of thinness at birth and obesity in adulthood which is associated with the highest risk of the insulin resistance syndrome and related conditions such as NIDDM, even in countries such as the UK where the intensity of the lifestyle transition is considerably less extreme than that in most of the high risk populations (Phillips *et al.*, 1994).

It appears that many populations around the world that are currently undergoing 'development' are experiencing the double jeopardy of poor nutrition in early life and westernization of diet and lifestyle in later life – a combination which may greatly increase their risk of NIDDM and its complications, and thereby impose impossible burdens on what are often already inadequate healthcare systems. The epidemic of NIDDM and its complications represent, therefore, an extraordinarily serious and rapidly escalating public health challenge.

References

AIHW (1994). *Australia's health 1994: the fourth biennial health report of the Australian Institute of Health and Welfare.* Canberra: AGPS.

Barker, D. J. P. (1992). *Fetal and Infant origins of Adult Disease.* London: BMJ Publications.

Brown, P. J. & Konner, M. (1987). An anthropological perspective on obesity. *Annals of New York Academy of Sciences* **499**, 29–46.

Campbell, L. V., Borkman, M., Marmot, P. E., Storlien, L. H. & Dyer, J. A. (1994). The high-monounsaturated fat diet as a practical alternative for NIDDM. *Diabetes Care* **17**, 177–82.

Chan, J. M., Rimm, E.B., Colditz, G.A., Stampfer, M. J. & Willett, W. C. (1994). Obesity, fat distribution, and weight gain as risk factors for clinical diabetes in men. *Diabetes Care* **17**, 961–9.

Colditz, G. A., Willett, W. C., Stampfer, M. J., Manson, J. E., Hennekens, C. H., Arky, R. A. & Speizer, F. E. (1990). Weight as a risk factor for clinical diabetes in women. *American Journal of Epidemiology* **132**, 501–4.

Cooper, R. (1994). Diabetes and the thrifty gene. *Lancet* **344**, 1648.

Dahri, S., Snoeck, A., Reusens-Billen, B., Remacle, C. & Hoet, J. J. (1993). Islet function in offspring of mothers on low protein diet during gestation. *Journal of Physiology (London)* **467**, 292.

Desai, M., Crowther, N. J., Ozanne, S. E., Lucas, A. & Hales, C. N. (1995). Adult glucose and lipid metabolism may be programmed during fetal life. *Biochemical Society Transactions* **23**, 331–5.

Dowse, G. K., Gareeboo, H., Zimmet, P. Z., Alberti, G. M. M., Tuomilehto, J., Fareed, D., Brissonnette, L. G. & Finch, C. F. (1990). High prevalence of NIDDM and impaired glucose tolerance in Indian, Creole, and Chinese Mauritians. *Diabetes* **39**, 390–6.

Dowse, G. K., Zimmet, P.Z., Finch, C.F. & Collins, V. R. (1991). Decline in incidence of epidemic glucose intolerance in Nauruans: implications for the 'thrifty genotype'. *American Journal of Epidemiology* **133**, 1093–104.

Dowse, G. K., Spark, R.A., Mavo, B., Hodge, A.M., Erasmus, R. T., Gwalimu, M., Knight, L. T., Koki, G. & Zimmet, P. Z. (1994). Extraordinary prevalence of non-insulin-dependent diabetes mellitus and bimodal plasma glucose distribution in the Wanigela people of Papua New Guinea. *Medical Journal of*

Australia **160**, 767–74.

Duffy, P., Morris, H. & Neilson, G. (1981). Diabetes mellitus in the Torres Strait region. *Medical Journal of Australia* 1(Suppl.), 8–11.

Garg, A. (1994). High-monounsaturated fat diet for diabetic patients. *Diabetes Care* **17**, 242–6.

Gault, A., O'Dea, K., Rowley, K., McLeay, T. & Traianedes, K. (1996). Abnormal glucose tolerance and other coronary heart disease risk factors in an isolated Aboriginal community in central Australia. *Diabetes Care* **19**, 1269–73.

Glatthaar, C., Welborn, T. A., Stenhouse, N. S. & Garcia-Webb, P. (1985). Diabetes and impaired glucose tolerance. A prevalence estimate based on the Busselton 1981 survey. *Medical Journal of Australia* **143**, 436–40.

Hales, C. N. & Barker, D. J. P. (1992). Type 2 (non-insulin dependent) diabetes mellitus: the thrifty phenotype hypothesis. *Diabetologia* **35**, 595–601.

Helmrich, S. P., Ragland, D. R., Leung, R.W. & Paffenbarger, R. S. (1991). Physical activity and reduced occurrence of non-insulin-dependent diabetes mellitus. *New England Journal of Medicine* **325**, 147–52.

Jennings, G. L., Nelson, L., Nestel, P. *et al.* (1986). The effects of changes in physical activity on major cardiovascular risk factors, haemodynamics, sympathetic function, and glucose utilization in man: a controlled study of four levels of activity. *Circulation* **73**, 30–9.

Joffe, B. I., Panz, V.R., Wing, J. R., Raal, F. J. & Seftel, H. C. (1992). Pathogenesis of non-insulin-dependent diabetes mellitus in the black population of southern Africa. *Lancet* **340**, 460–2.

King, H. & Rewers, M. (WHO Ad Hoc Diabetes Reporting Group). (1993). Global estimates for prevalence of diabetes mellitus and impaired glucose tolerance in adults. *Diabetes Care* **16**, 157–77.

Kliewer, E. V., Stanley, F. J. (1989). Aboriginal and white births in Western Australia, 1980–86. Part I: Birthweight and gestational age. *Medical Journal of Australia* **151**, 493–502.

Lemieux, S. & Després, J. P. (1994). Metabolic complications of visceral obesity: contribution to the Aetiology of type 2 diabetes and implications for prevention and treatment. *Diabète et Métabolisme (Paris)* **20**, 375–93.

Long, D. S., Swanson, M. S., O'Brien, K., Pories, W. J., MacDonald, Jr. K. G., Caro, J. F. & Leggett Frazier, N. (1994). Weight loss in severley obese subjects prevents the progression of impaired glucose tolerance in type II diabetes. *Diabetes Care* **17**, 372–5.

Maher, H. M. & Keen, H. (1985). The Southall Diabetes Survey: prevalence of known diabetes in Asians and Europeans. *British Medical Journal* **291**, 1081–4.

Marshall, J. A., Hamman, R. F. & Baxter, J. (1991). High-fat, low-carbohydrate diet and the etiology of non-insulin-dependent diabetes mellitus; the San Luis Valley diabetes study. *American Journal of Epidemiology* **134**, 590–603.

Marshall, J. A., Hoag, S., Shetterly, S. & Hamman, R. F. (1994). Dietary fat predicts conversion from impaired glucose tolerence to NIDDM – the San Luis Valley diabetes study. *Diabetes Care* **17**, 50–6.

Monterrosa, A. E., Haffner, S. M., Stern, M. P. & Hazuda, H. P. (1995). Sex difference in lifestyle factors predictive of diabetes in Mexican-Americans. *Diabetes Care* **18**, 448–56.

Motala, A., Mahomed, A. K., Omar, A.K. & Gouws, E. (1993). High risk of progression to NIDDM in South-African Indians with impaired glucose tolerance. *Diabetes* **42**, 556–63.

Neel, J. V. (1962). Diabetes mellitus: a 'thrifty' genotype rendered detrimental by 'progress'? *American Journal of Human Genetics* **14**, 353–62.

O'Dea, K. (1984). Marked improvement in carbohydrate and lipid metabolism in diabetic Australian aborigines after temporary reversion to traditional lifestyle. *Diabetes* **33**(6), 596–603.

O'Dea, K. (1992). Diabetes in Australian Aborigines: impact of the western diet and lifestyle. *Journal of Internal Medicine* **232**, 103–17.

O'Dea, K., White, N.G. & Sinclair, A. J. (1988a). An investigation of nutrition-related risk factors in an isolated Aboriginal community in Northern Australia: advantages of a traditionally-orientated life-style. *Medical Journal of Australia* **148**, 177–80.

O'Dea, K., Traianedes, K., Hopper, J. & Larkins, R. G. (1988b). Impaired glucose tolerance, hyperinsulinemia and hypertriglyceridemia in Australian Aborigines from the desert. *Diabetes Care* **11**, 23–9.

O'Dea, K., Traianedes, K., Ireland, P., Niall, M., Sadler, J., Hopper, J. & De Luise, M. (1989). The effects of diet differing in fat, carbohydrate, and fibre on carbohydrate and lipid metabolism in Type II diabetes. *Journal of the American Dietetic Association* **89**, 1076–86.

O'Dea, K., Patel, M., Kubisch, D., Hopper, J. & Traianedes, K. (1993). Obesity, diabetes and hyperlipidemia in a central Australian Aboriginal community with a long history of acculturation. *Diabetes Care* **16**, 1004–10.

Pettit, D. J., Aleck, K. A., Baird, H. R., Carraher, M. J., Bennett, P. H. & Knowler, W. C. (1988). Congenital susceptibility to NIDDM: role of intrauterine environment. *Diabetes* **37**, 622–8.

Pettit, D. J., Nelson, R. G., Saad, M. F., Bennett, P. H. & Knowler, W. C. (1993). Diabetes and obesity in the offspring of Pima Indian women with diabetes during pregnancy. *Diabetes Care* **16**, 310–14.

Phillips, D. I. W., Barker, D. J. P., Hales, C. N. & Osmond, C. (1994). Thinness at birth and insulin resistance in adult life. *Diabetologia* **37**, 150–4.

Pietinen, P., Vartiainen, E. & Männistö, S. (1996). Trends in body mass index and obesity among adults in Finland from 1972 to 1992. *International Journal of Obesity* **20**, 114–20.

Ramachandran, A., Jali, M. V., Mohan, V., Snehalatha, C. & Viswanathan, M. (1988). High prevalence of diabetes in an urban population in south India. *British Medical Journal* **297**, 587–90.

Ramachandran, A., Dharmaraj, D., Snelhalatha, C. & Viswanathan, M. (1992). Prevalence of glucose intolerance in Asian Indians: urban–rural difference and significance of upper body adiposity. *Diabetes Care* **15**, 1348–55.

Sobal, J. & Stunkard, A. J. (1989). Socioeconomic status and obesity: a review of the literature. *Psychological Bulletin* **105**, 260–75.

Stern, M. P. (1991). Primary prevention of type II diabetes mellitus: Kelly West Lecture. *Diabetes Care* **14**, 399–410.

Stern, M. P., Knapp, J. A., Hazuda, H. P. *et al.* (1991). Genetic and environmental determinants of type-II diabetes in Mexican-Americans. Is there a descending limb to the modernization diabetes relationship? *Diabetes Care* **14**(7), 649–54.

Walker, K. Z., O'Dea, K., Nicholson, G. C. & Muir, J. G. (1995). Dietary composition, body weight and non-insulin dependent diabetes: comparison of high-fibre, high-carbohydrate, and modified-fat diets. *Diabetes Care* **18**, 401–3.

Walker, K. Z., O'Dea, K., Johnson, L., Sinclair, A., Nicholson, G. C. & Muir, J. G. (1996). Body fat distribution and non-insulin dependent diabetes: comparison of a fiber-rich high carbohydrate low (23%) fat diet and a 35% fat diet high in monounsaturated fat. *American Journal of Clinical Nutrition* **63**, 254–60.

10 *Physical activity, sport, social status and Darwinian fitness*

ROBERT M. MALINA

Calls for increasing daily physical activity in all segments of the population from early childhood through to old age are common in discussions of public health policy. The increased energy expenditure that accompanies regular physical activity contributes to more efficient function of various systems, weight maintenance, reduced risk of several degenerative diseases, reduced risk of mortality and overall improvement of quality of life (Bouchard *et al.*, 1994; US Department of Health and Human Services, 1996). Physical activity comprises a range of types of behaviour which occur in a variety of forms and contexts, e.g. leisure time activity, occupational activity, play, dance, sport, exercise, etc. Sport is the primary medium of physical activity for many children and youth, and high performance, elite sport is perhaps the most visible form of physical activity.

Social conditions can influence choice of and access to physical activity. These include, among others, social class, socio-economic status, parental education, quality of school physical education programmes, local recreation and sport programmes, area of residence, ethnicity and others. These factors, among others, are not mutually exclusive. Opportunity for regular participation in physical activity and sport thus may not be equitable, and, as sport becomes more elite, it is very selective and exclusive.

Physical activity may influence reproductive function, and through its interaction with physical fitness, may influence risk for several diseases. Thus, physical activity and physical fitness have the potential to influence Darwinian fitness. This chapter explores several interrelationships among physical activity, sport, social status and Darwinian fitness. Some of these issues have been discussed elsewhere (Malina, 1991a). Where appropriate, these will be expanded in the context of the theme, human biology and social inequality. Note that the term fitness has totally different meanings in the Darwinian and physical contexts, although the terms are occasionally confused.

165

Darwinian fitness

Darwinian fitness refers to reproductive efficiency and fertility. It is generally defined in terms of the average reproductive success of an individual. According to Dobzhansky (1962, p. 125):

> the genetic fitness of a genotype, and by extension of an individual, is measured by the contribution it makes, relative to other genotypes or individuals, to the gene pool of the succeeding generations. The fittest parent is the parent of the greatest number of surviving children.

Characteristics of an individual, including cultural context, which influence survival to reproductive age and subsequent reproduction affect their Darwinian or genetic fitness. Components of Darwinian fitness include mate selection, mating success, social status, age at marriage, fecundity, longevity, rate of reproduction and progeny survival to maturity.

Physical activity and physical fitness may be components of mate selection, and are also significantly influenced by cohabitation and common lifestyle. Spouse correlations for indicators of physical activity and physical fitness are positive but low, approximately 0.2 to 0.3 (Bouchard et al., 1997). They are similar in magnitude to those for measures of body size and body composition. Assortative mating, specifically the genetic component of related phenotypic characteristics, may be of significance because it will not only lead to an increase in the frequency of homozygotes for genetic loci that may be relevant, but will also inflate the additive genetic variance for a given characteristic.

The potential influence of a physically active lifestyle and high physical fitness on Darwinian fitness may operate through improved physiological function, reduced risk for several diseases, and less so on increased longevity, while a sedentary lifestyle and low physical fitness may be associated with enhanced risk of mortality at relatively young ages. Some extreme forms of physical activity, specifically prolonged rigorous training for some sports, may interfere with reproductive function and thus potentially influence Darwinian fitness. Alterations in the menstrual cycle associated with training for sport are well documented. In addition, the decision by some female athletes to postpone childbearing for the sake of continuation in sport may directly influence Darwinian fitness.

Physical activity and physical fitness

Physical activity refers to 'any body movement produced by the skeletal muscles and resulting in a substantial increase over the resting energy expenditure' (Bouchard and Shephard, 1994, p. 77). The measurement and quantification of physical activity and its major correlate, energy expenditure, are difficult tasks and need further attention. Most discussions refer to an estimated level of habitual physical activity, e.g., hours/week or an activity score, derived from questionnaires, interviews, diaries and heart rate integrators. Physical activity is behaviour that occurs in a variety of forms and contexts. It is a bio-cultural process – energy is expended (biologically) in active behaviour that occurs within a cultural context. Physical activity thus cannot be approached from an exclusively biological, or from an exclusively cultural, perspective.

A related construct of physical activity is physical fitness, which is an adaptive state. It is, to a large extent, the outcome of responses to a variety of environments. The concept of physical fitness has evolved from a primary performance focus to a health-related focus. The latter is operationalized in three components: cardio-vascular fitness; musculo-skeletal function of the lower trunk, i.e., abdominal strength and lower back flexibility; and body composition, specifically subcutaneous fatness (Malina, 1991b, 1995). The concept of fitness has been expanded to include morphological, muscular, motor, cardio-vascular and metabolic components (Bouchard and Shephard, 1994). The morphological component refers largely to fatness and relative fat distribution, while the metabolic component refers to a variety of risk factors, including serum lipids, blood pressures and glucose metabolism.

Levels of habitual activity and specific measures of physical fitness vary with growth, maturation and ageing independent of physical activity. Specific indicators or components of physical fitness have variable heritabilities (Bouchard *et al.*, 1997). However, regular physical activity and lifestyle have the potential to influence physical fitness through the lifespan (Malina, 1990, 1994a, 1995; Spirduso, 1995). It is also generally assumed that habits of and attitudes towards physical activity developed during childhood continue through adolescence into adulthood. However, indicators of physical activity track at only low to moderate levels during childhood and adolescence, from adolescence into adulthood, and across various ages in adulthood. Measures of performance- and health-related physical fitness track significantly across childhood and adolescence, but correlations are low to moderate. Limited data that span adolescence into adulthood indicate somewhat higher inter-age correlations for flexibility, static strength and power (Malina, 1996a). Data for different periods in adulthood are limited.

Physical activity and sport

Sport is perhaps the most visible form of physical activity. It is the major source of physical activity for most children. Mass participation in youth sports at the community level is a major feature of daily living in many parts of the world. The majority of North American children, for example, participate in organized sports by 5–7 years of age. Participation increases during childhood, but subsequently declines after about 12–13 years of age, which parallels declining rates of participation in physical activities across adolescence (Malina, 1991b, 1992, 1995). This age period also includes adolescence and puberty with their multiple demands on youth.

A major limiting factor to organized sport participation is the availability of local resources, in particular human resources in the form of adults to coach and supervise programmes. The vast majority of coaches at this level are volunteers. Club sports, e.g. swimming, gymnastics, figure skating and ballet, and in some areas ice hockey and soccer (European football), require parental financial support, so that economic considerations limit access to these sports for many children and youth. An additional concern is safety, specifically for children and youth in urban 'inner city' areas. Increased selectivity and time demands, as sports become more rigorous, and increased levels of competition are other constraints on organized sport participation with increasing age during adolescence.

At the more specialized, elite levels, sport is extremely selective and exclusive, i.e. inequitable. Selection criteria and practices for specific sports vary with programmes. Community-based programmes emphasize mass participation; age and willingness to participate are the primary criteria. However, some programmes emphasize the elite and have as their objective the identification and subsequent training of young athletes with potential for success in regional, national and/or international competition – high performance sports (Malina, 1993, 1996b). The selection/exclusion process begins early and is rather systematic. It begins at about five years of age in some sports (e.g. gymnastics); criteria and timing of evaluation for other sports, including ballet, vary. Identifying and selecting the potentially talented athlete at an early age is the first step in a relatively long-term process. The perfection of talent is another matter which requires long hours of systematic and often repetitive training under the scrutiny of demanding coaches, dietary regulation and perhaps manipulation, other forms of manipulation (e.g. chemical), and often separation from family.

The term 'natural selection' is sometimes used in discussing the exclusion or 'weeding out' process in ballet (Hamilton, 1986) and sport (Bompa, 1985). Needless to say, the use of the term 'natural selection' in this context has no relevance to natural selection in the Darwinian sense.

On the one hand, economic resources are often a limiting factor in securing access to sport programmes, facilities, expert coaching and related requisites for success. On the other hand, potential exploitation of youngsters, especially those from minority and impoverished backgrounds, is a factor in some sports, e.g. interscholastic basketball and American football in many communities. High school athletes in the USA are, in turn, often the objects of rigorous and competitive selective recruiting practices by collegiate and university level athletic coaches. Thus, by the time an adolescent becomes an intercollegiate athlete, he/she has been the product of selective practices beginning in late childhood or early adolescence.

Ethnic and social class variation in habitual physical activity

Ethnicity and socio-economic status are to a large extent coterminous in the USA. Blacks and Hispanics accounted for about 12% and 9%, respectively, of the total United States population in 1990, while about 33% of blacks and 29% of Hispanics were living below the poverty level in 1991 compared to only 9% of whites (US Bureau of the Census, 1992). The inequality among ethnic minorities in the USA is also evident in the Human Development Index (HDI), 'a measure of people's ability to live a long and healthy life, to communicate and to participate in the life of the community and to have sufficient resources to obtain a decent living' (United Nations Development Programme, 1993, p. 104). The index is based on longevity (life expectancy at birth), knowledge (adult literacy and years of schooling) and income (utility of income), and is considered a minimal measure. Within the USA, whites have an HDI similar to that for Japan (.983), while blacks and Hispanics have HDI's that are, respectively, similar to those for Trinidad (.877) and Estonia (.872). There is also a sex difference within ethnic groups; the HDI is highest in white females (.990), followed by white males (.974), black females (.895) and Black males (.858). Sex-specific indices for Hispanics were not reported (United Nations Development Programme, 1993).

Youth

American black and Hispanic youth, on average, spend relatively less time in moderate to vigorous physical activity. In the 1990 Youth Risk Behavior Survey of a nationally representative sample of American high school youth, only about 50% of boys and 24% of girls in grades 9–12 (i.e. ages 14–18) reported participation in vigorous physical activity three or more days per week (Heath *et al.*, 1994). The percentage of Black boys (43%) meeting the criterion of vigorous physical activity three or more days per week was lower than those for white (51%) and Hispanic (50%) boys, while the percentages of black and Hispanic girls meeting the criterion (17% and

16%, respectively) were lower than that for white (27%) girls. Similar trends were apparent in several regional studies using different activity measures and criteria (Aaron *et al.*, 1993; Wolf *et al.*, 1993; Myers *et al.*, 1996). In contrast, patterns of participation in vigorous exercise among high school youth in San Diego, California, did not differ by ethnicity and socio-economic status (Sallis *et al.*, 1996), which perhaps reflects geographic and seasonal variation, and perhaps method of estimating physical activity. However, socio-economic status did influence access to lessons and classes in specific activities.

Although determinants of physical activity in youth of different ethnic backgrounds have not received detailed attention (Sallis *et al.*, 1992), limited availability of activity programmes, neighbourhood safety and convenience of facilities are probably important concerns, as is also the case for organized sports.

Adults

Ethnic differences in physical activity are also evident in adults. In the National Health Interview Survey of 1985, black men and women were more sedentary than white men and women, and were somewhat less likely to be considered moderately active or very active (Schoengorn, 1986). The data, however, were not stratified by education, an indicator of socio-economic status. Results from the Minnesota Heart Study also indicated, overall, significantly less estimated energy expenditure in leisure time physical activity among blacks than among whites within each sex (Folsom *et al.*, 1991). When stratified by level of education, the ethnic difference in estimated energy expended in total leisure time physical activity was apparent only among those with a high school education or less, while the differences between blacks and whites with college education were small and not significant. With vigorous leisure time physical activity as the criterion, white males within each educational category expended more energy in 'heavy' leisure time physical activity than black males, while white females with a high school education or less expended more energy in 'heavy' leisure time physical activity than black females with the same level of education. Among those with college education, however, energy expenditure in 'heavy' leisure time physical activity did not differ between black and white women.

Corresponding data for Hispanic samples are apparently not available. Nevertheless, generally similar trends were evident in a survey of working women. American women of colour (African-, Asian-, and Hispanic-American) were less likely to participate regularly in sports and fitness activities; only 46% participated in a sport or fitness activity over the past year compared to 61% of white women (Women's Sports Foundation, 1993).

Ethnic variation in habitual physical activity among adults is consistent with the prevailing observation in studies of socio-economic status as a determinant of physical activity. Adults from better-off groups spend more time in physical activity than those from the lowest income groups (Stephens *et al.*, 1985). The ethnic variation represents, to some extent, inequalities in income and access to participation in sports and fitness activities, and perhaps discrimination. However, the number of American black athletes in some sports, e.g. track, basketball and American football, is far more than the proportion in the general population. The over-representation of black males in track and basketball, more so than black females, as well as the lack of representation of black athletes in sports such as swimming, tennis and gymnastics, begs issues related to equality of access (or to 'equity') and to potential population variability in physical performance.

Given the health benefits of physical activity, it is possible that increased activity may reduce the risk for several diseases which show ethnic differentials, e.g. higher prevalence of hypertension in American blacks and of diabetes in Mexican Americans. Ethnic and socio-economic variation in estimated physical activity or sedentariness may thus be confounding factors. For example, sedentary behaviour was independently associated with an increased prevalence of hypertension in black women, but not in black men (Ainsworth *et al.*, 1991). Some diseases and risk factors for disease may be prevented or their progress slowed by the increased levels of regular physical activity, among other aspects of lifestyle, such as diet and smoking.

Training, sport and sexual maturation of girls

Later mean ages at menarche are commonly reported for adolescent and adult athletes in a variety of sports (Malina, 1983; Beunen and Malina, 1996). It is suggested by some that the later mean ages at menarche in athletes are a consequence of regular training before menarche, i.e. training 'delays' menarche (Malina, 1983,1994b). The term 'delay' in association with increased sport activity is also used in epidemiological (Merzenich *et al.*, 1993) and auxological (Eveleth and Tanner, 1990) studies of menarche, and in public health statements (American Medical Association/American Dietetic Association, 1991).

'Delayed' menarche, defined as an age at menarche of 14 years or older, is also suggested as a reproductive risk of athletic training (Constantini and Warren, 1994; Lee, 1994) or as a reproductive disorder (Warren, 1995), although the specific risk is not indicated. Use of the term 'delay' is misleading and perhaps naïve; it implies that training or physical activity is causing menarche to be later than normal or expected.

The confusion associated with later ages at menarche in athletes is related in part to the methods of estimating age at menarche – prospective, *status quo* or retrospective. Prospective data for young girls regularly training in a sport are generally short-term and limited to small, select samples in several sports. *Status quo* data for young athletes actively involved in systematic training provide sample or population estimates. They are quite limited and are not entirely consistent with later ages at menarche reported in retrospective studies of late adolescent and adult athletes, which comprise the vast majority of data for athletes (Malina, 1983, 1994b, 1996b; Beunen and Malina, 1996).

The later mean ages of athletes in a variety of sports and correlations with training before menarche are often used to infer that training prior to menarche 'delays' this maturational event. Training is not ordinarily quantified, and the distinction between initial training in a sport and systematic, formal training is not made. The data are associational, and association does not imply a cause–effect sequence between training and sexual maturation. There may in fact be other confounding and/or modifier variables. In some of the analyses, those who take up systematic, serious training after menarche, e.g. some early maturers, are excluded in discussions of the assumed training effect. It is also important to note that not all athletes experience menarche late. Ranges of reported ages at menarche in elite university athletes in seven sports (swimming, diving, tennis, golf, track and field, basketball and volleyball), are 9.2–17.7 years in 292 white athletes and 9.3–17.3 years in 80 black athletes, and in 314 white non-athlete students attending the same university, 9.1–17.4 years, completely overlap, and are well within the range of normal variability (Malina, unpub. data).

How can the later mean ages at menarche in athletes be interpreted without exclusive reliance on training as the aetiological factor? Menarche is a biological event. In adequately nourished individuals, age at menarche is a highly heritable characteristic (Malina and Bouchard, 1991). There is a familial tendency for later maturation in athletes. Mother–daughter and sister–sister correlations in families of athletes (Brooks-Gunn and Warren, 1988; Stager and Hatler, 1988; Baxter-Jones *et al.*, 1994; Malina *et al.*, 1994) are similar to those in the general population (Malina *et al.*, 1994).

Menarche is also influenced by a number of socially or bio-culturally mediated variables, e.g. socio-economic differentials in some countries

(Bielicki *et al.*, 1986) and positive secular changes in age at menarche in association with improved health and nutritional circumstances over time (Tanner, 1962; Malina, 1979). Sport-specific selective factors must be considered as a part of this bio-cultural matrix in athletes. Gymnastics selects for short stature, perhaps a somewhat more muscular physique, and late maturation. It is not uncommon for gymnastics coaches to ask for photographs of the potential gymnast's mother when she was an adolescent (Press, 1992). Female divers have many characteristics in common with gymnasts, and many divers have their initial sport experience in gymnastics (Malina and Geithner, 1993). Ballet has rigid selection criteria which place an emphasis on thinness and linearity of physique (Hamilton, 1986), both of which are associated with later maturation.

The number of children in the family is also associated with menarche. Girls from larger families tend to attain menarche later than those from smaller families, and the estimated magnitude of the sibling number effect (controlling for birth order) in athletes, range from 0.15 to 0.22 years per additional sibling, overlaps that in non-athletes, 0.08 to 0.19 years per additional sibling; estimated sibling number effects in studies of non-athletes not controlling for birth order range from 0.11 to 0.18 years per additional sibling (Malina *et al.*, 1997). Although data are not extensive, athletes tend to be from larger families than non-athletes (Malina *et al.*, 1982, 1997).

The sibling number effect is often attributed to marginal nutrition or increased frequency of diseases, generally in conjunction with lower socio-economic status. It is difficult, however, to implicate such conditions in samples of athletes, who have been reared, on average, under better-off economic, health and nutritional circumstances. Black athletes who are largely from lower socio-economic circumstances than white athletes, are an exception, but attain menarche at an earlier age, on average (Malina, unpub. data). Gymnasts, figure skaters and ballet dancers may also be exceptions. They are generally from better-off socio-economic circumstances, but are a highly select sample. The diets of elite young gymnasts and figure skaters are routinely closely monitored (Press, 1992; Ryan, 1995) and perhaps manipulated. Young East German female gymnasts were kept on a dietary regime 'intended to maintain the optimal body weight [for performance], i.e., a slightly negative energy balance, and thus (had) a limited energy depot over a long period' (Jahreis *et al.*, 1991, p. 98). This may in fact be chronic mild undernutrition. And, significant numbers of young ballerinas have persistent problems with disordered eating (Hamilton *et al.*, 1988).

Other factors which may interact with marginal caloric status and altered eating habits merit closer attention. These may include the psychological and emotional stress associated with maintaining body weight when the natural course of growth is to gain, year-long training (often before school in the morning and after school in the late afternoon), frequent competitions, altered social relationships with peers and perhaps overbearing and demanding coaches.

It is possible that the number of children in the family provides an indirect estimate of birth spacing or birth crowding, and perhaps foetal programming. Some evidence suggests that age at menarche is influenced by birth interval, the shorter the interval, the later the age at menarche of the next child (Douglas, 1966; Leistol, 1982). Heavier birth weight was associated with a later age at menarche; the difference between the highest and lowest quintiles was 2.2 months, leading the authors to hypothesize that: 'menarcheal age is linked to programmed patterns of gonadotrophin release established *in utero*, when the fetal hypothalamus is imprinted, and is subsequently modified by weight gain in childhood' (Cooper *et al.*, 1996, p. 814). Other data, however, indicate no effect of birth weight on age at menarche after controlling for family size and birth order (Roberts *et al.*, 1986).

Conditions during foetal development may influence the hypothalamic–pituitary–gonadal axis independently of birth weight. The axis is active in foetal life and early infancy, but is then suppressed for almost a decade until the onset of puberty (Grumbach and Kaplan, 1990). Environmental conditions near the time of birth have also been related to menarche. In a study of secular changes in age at menarche in Norway, trends in the Gross Domestic Product (an index of socio-economic development) in the year of birth and age at menarche parallelled each other closely, more so than the Gross Domestic Product in the year of or years near menarche, leading to the following hypothesis:

> During the period around or after birth, the processes leading to menarche are sensitive to a set of environmental stimuli. After this period, the reproductive system becomes clearly less susceptible. During adolescence, the maturation process may be influenced somewhat, but probably not to any great extent as long as the conditions are not adverse.
>
> *(Leistol, 1982, p. 534.)*

Early influences may also interact with other environmental influences during childhood and adolescence, including aspects of the home environment. Results of several studies suggest a role for household composition and stress as potential factors associated with an earlier age at menarche (Jones *et al.*, 1972; Surbey, 1990; Graber *et al.*, 1995). The correlation between stressful life events and age at menarche, though low, was similar in small-scale ($n = 75$, $r = -0.12$, Graber *et al.*, 1995) and large-scale ($n = 1104$, $r = -0.14$, Surbey, 1990) studies.

The mechanisms that link a variety of factors to menarche, e.g. number of children in the family, household composition, family relationships, birth interval and nutritional status, among others, are not known. The involvement of families and coaches in the training of child and adolescent athletes presents additional dimensions to family-athlete and coach-athlete relationships and interactions. Psychological and emotional stresses associated with training and competition *per se* are additional concerns.

The interactions are likely to operate along the hypothalamic–pituitary–ovarian axis which regulates the onset and progress of puberty, including first menstruation. Puberty is a brain-driven event (Grumbach and Kaplan, 1990); and of course, the brain and central nervous system are the filter through which environmental factors and stresses are processed. Given the complexity of factors related to menarche, it is essential that they be considered before inferring causality for regular training before and during puberty as a factor influencing the timing of this maturational event in presumably healthy adolescent athletes. If training for sport is related to later menarche, it most likely interacts with or is confounded by other factors so that the specific effect of training *per se* may be impossible to extract. Nevertheless, two comprehensive discussions of exercise and female reproductive health offer the following conclusions on training and menarche:

> although menarche occurs later in athletes than in non-athletes, it has yet to be shown that exercise delays menarche in anyone.
>
> *(Loucks* et al., *1992, p. S288)*

and,

> the general consensus is that while menarche occurs later in athletes than in non-athletes, the relationship is not causal and is confounded by other factors.
>
> *(Clapp and Little, 1995, pp. 2–3.)*

Significance of later menarche

An age at menarche of 14 years is defined as 'delayed' (Warren, 1995), while lack of spontaneous menstruation by 16 or 17 years of age is considered primary amenorrhoea (Behrman, 1967; Shangold, 1994). Using the criteria of Warren (1995), 41% of the university athletes and 21% of the non-athlete university students discussed earlier would have 'delayed' menarche. Assuming 16 means the 16th birthday, 8% of the athletes and 1% of the non-athletes would be classified as primary amenorrhoeic. Although training is often implicated as a cause of primary amenorrhoea, 'the most common cause ... particularly among athletes, is constitutional delay' (Shangold, 1994, p. 169).

A later age at menarche potentially implies later entry into reproductive life (although many other factors are involved) and thus a decrease in fertility potential. An association between later menarche with a relative decrease in fertility and an increase in sterility has been suggested (Behrman, 1967), but a prospective study of about 2000 women indicated no relationship between age at menarche and total fertility and risk of still-birth (Sandler *et al.*, 1984). Age at menarche has been related to spontaneous abortions, but results are equivocal. Risk and rates of spontaneous abortion declined with an increasing age at menarche in prospective (Sandler *et al.*, 1984) and retrospective (Leistol, 1980) studies, while another retrospective study noted a 'U-shaped' relationship between age at menarche and miscarriage rates in the first pregnancy (Martin *et al.*, 1983).

Later menarche is associated with a prolonged hypo-oestrogenic state, which is associated with reduced accretion of bone mass (Loucks *et al.*, 1992). The association between reduced bone mass and reproductive fitness, if any, has not been established.

Training, sport and menstrual dysfunction

Menstrual dysfunction is more common in athletes than in the general population, and the greater prevalence is often attributed to training in athletes or otherwise active women (Loucks *et al.*, 1992; Constantini and Warren, 1994; Shangold, 1994). The data have methodological limitations, use variable definitions of dysfunction, and rely quite often on runners and ballet dancers to the exclusion of other athletes. Further, observations on adolescent athletes, including some athletes in the late teenages, may be confounded by normal variation in the establishment of mature menstrual function. Menstrual cycles immediately following menarche tend to be anovulatory and quite irregular. The development of more or less regular menstrual cycles takes place over several years after menarche so that some of the menstrual irregularity observed in adolescent athletes may in fact reflect the normal process of sexual maturation involved in the establishment of regular menstrual cycles (Malina and Bouchard, 1991). Nevertheless, moderate weekly energy expenditure (601–930 kcal; 2515–3891 kJ) in physical activity is associated with anovulatory cycles in regularly menstruating adolescent girls, after controlling for age at menarche and gynaecological age (Bernstein *et al.*, 1987).

The two most reported forms of menstrual dysfunction are oligomenorrhoea – occasional menstrual periods or periods that occur at intervals between 35 and 90 days – and secondary amenorrhoea – the cessation of periods after menarche has occurred. Absence of menstruation for three or more months, or for five or six months are commonly used as criteria for secondary amenorrhoea (Rebar, 1990; Constantini and Warren, 1994). Other aspects of menstrual dysfunction infrequently studied in association with physical activity and training are luteal phase deficiency, anovulatory cycles in eumenorrhoeic women and hypo-oestrogenic amenorrhoea (Constantini and Warren, 1994; Shangold, 1994).

Estimated prevalences of menstrual dysfunction (oligomenorrhoea and secondary amenorrhoea) in women athletes are variable. It is reported more often in runners (6% to 43%) and ballet dancers (58% to 79%) than in cyclists and swimmers (12% to 16%), which contrasts to estimates between 2% and 5% in the general population (Constantini and Warren, 1994). Estimates for athletes in other sports are limited. Among 96 university athletes (20.4 ± 1.3 years) followed longitudinally over an academic year (nine months), menstrual dysfunction was recorded in athletes in all seven sports (see p. 172). If golfers were excluded (1 of 10), 58% of the athletes experienced menstrual dysfunction. Within three sports – track, basketball, volleyball – the prevalence of menstrual dysfunction was more than twice as common in white (71%) than in black (31%) athletes (Ryan, 1996). The prevalence of menstrual dysfunction in a retrospective study of Norwegian athletes in 35 sports ($n = 339$, 19.4 ± 4.3 years) and age- and home community-matched control subjects (20.6 ± 4.4 years) was 42% and 28%, respectively (Sundgot-Borgen and Larsen, 1993). The estimate for non-athlete controls is among the highest reported. Further, the wide range of variation in age would imply that some of the dysfunction may be associated with normal variation in the establishment of mature menstrual function. Menstrual dysfunction varied with type of sport (although some of the classifications are arbitrary): endurance – 62%; aesthetic – 60%; weight-dependent – 50%; technical – 37%; ball game – 28%; and power – 22% (Sundgot-Borgen and Larsen, 1993).

The aetiology of menstrual dysfunction in athletes is not known with certainty. Evidence from prospective studies does not implicate exercise alone as a causal factor in menstrual disturbances or amenorrhoea (Barr and Prior, 1994), while a causal link between exercise alone and menstrual disorders has not yet been established (Loucks *et al.*, 1992; Loucks, 1994). Hyper-androgenism, hyper-prolactinemia, low levels of absolute or relative fatness and psychological stress are not causative factors in athletic amenorrhoea (Loucks *et al.*, 1992). However, Constantini and Warren (1995) have suggested that menstrual dysfunction in swimmers is associated with mild hyper-androgenism, in contrast to hypo-oestrogenism observed in runners and ballet dancers with menstrual dysfunction. These observations would seem to suggest variation in the mechanism for menstrual dysfunction in athletes in different sports.

Nevertheless, strenuous physical activity is associated with menstrual dysfunction:

> data from unrandomized, prospective experiments strongly indicate that abruptly imposed, prolonged, and intense physical activity can induce reproductive morbidities in at least some women ... There does appear to be a dose-response gradient in the effect, because more gradually imposed, brief, and less strenuous physical activity has been ineffective at inducing ... amenorrhoea.
>
> *(Loucks, 1994, p. 947)*

The mechanism(s) for menstrual dysfunction associated with strenuous physical activity is (are) not presently known, but is (are) being investigated primarily in the context of two hypotheses (Loucks *et al.*, 1992; Loucks,1994). The first proposes that activation of the adrenal axis during strenuous and prolonged exercise inhibits the gonadotrophic hormone releasing hormone (GnRH) pulse generator in the hypothalamus. Slightly elevated cortisol levels have been observed in amenorrhoeic athletes, which is consistent with the hypothesis. The second hypothesis proposes energy deficit as the causative factor, i.e. energy intake does not meet the energy expended in regular training. Evidence suggests that luteinizing hormone (LH) pulsatility depends on energy availability (Loucks and Heath, 1994), and it has been suggested that reduced pulsatile release of LH is related to menstrual dysfunction in distance runners (Velduis *et al.*, 1985).

In the prospective study of 96 university athletes, food intake and training logs were kept for one week during the first month of heavy training. Athletes with menstrual irregularities had lower absolute and relative estimated energy intakes than eumenorrhoeic athletes. The estimated daily energy deficit in athletes with menstrual dysfunction was 981 ± 1117 kcal (4105 ± 4674 kJ) compared to an estimated daily energy deficit of 295 ± 844 kcal (1234 ± 3531 kJ) in eumenorrhoeic athletes (Ryan, 1996). An issue related to the energy deficit hypothesis is the relatively high prevalence of eating disorders in athletes. Only one of 45 eumenorrhoeic university athletes had a diagnosed eating disorder, while 33 of 51 athletes with menstrual disorders had a diagnosed eating disorder (Ryan, 1996). Eating disorders are difficult to document in the athletic population due to secrecy and denial associated with the disorder, thus, athletes who experienced menstrual dysfunction without a diagnosis could have had an undiagnosed eating disorder. Only one of 23 black athletes had a diagnosed eating disorder, while 13 of 24 white athletes in three sports (track, basketball, volleyball) had diagnosed eating disorders. The ethnic variation in the prevalence of eating disorders among athletes may reflect differences in perceptions of body weight and in social pressures to stay thin.

Later menarche is also associated with menstrual dysfunction. Mean ages at menarche increase from eumenorrhoeic to oligomenorrhoeic to amenorrhoeic subjects both among athletes (Sundgot-Borgen and Larsen, 1993; Ryan, 1996) and non-athletes (Sundgot-Borgen and Larsen, 1993). There is, however, considerable overlap in ages at menarche among athletes in the three menstrual function groups, so that it is difficult to implicate later menarche as a specific factor.

Implications of menstrual dysfunction

The different manifestations of menstrual dysfunction can be considered reproductive morbidities (Louck, 1994). From the perspective of Darwinian fitness, one can inquire as to the potential long-term effects of oligomenorrhoea, secondary amenorrhoea, luteal phase deficiency and anovulatory cycles in eumenorrhoeic women on the reproductive system. Luteal phase deficiency is associated with infertility (Shangold, 1994). Among infertile women with ovulatory failure, those with a history of vigorous exercise for one hour or more per day were more often represented among nulligravid cases (Greene *et al.*, 1986).

It is generally assumed that menstrual dysfunctions associated with regular training are reversible, although 'this reversibility has not been completely proven' (Baker, 1981, p. 694). The long-term implications of menstrual dysfunctions for reproductive fitness and perhaps reproductive health need further study. The current generation of female athletes has had more opportunities in sport, more systematic and rigorous training and perhaps greater competitive stresses than earlier generations of female athletes. They represent an ideal sample for the study of the long-term implications of reproductive morbidities. Needless to say, a bio-cultural approach is essential, recognizing social and cultural factors associated with reproductive behaviour. The suggestion of Shangold (1984, p. 71) merits evaluation:

> Although infertility is reported to be no higher for runners than for the general population, this may be true because many women runners who are actually infertile have not yet tested their fertility, and therefore are unaware that they are infertile.

Training and sexual maturation in boys

The potential effects of training for sport have not generally been considered as a significant factor affecting the sexual maturation of boys. This may not be surprising since early and average maturation are characteristic of the majority of young male athletes (Malina, 1994b). With the exception of constant emphasis on weight regulation in some sports, environmental stresses related to sport, such as anxiety and sleep problems, undoubtedly affect boys as well as girls. Wrestling is the primary sport among males that has an emphasis on weight regulation. The emphasis on weight control, however, is short-term, and longitudinal observations during a season indicate no significant effects on maturation and hormonal profiles (Roemmich, 1994). Under conditions of experimental food restriction in mice and rats, however, reproductive development continues in males but is blocked in females (Loucks, 1994).

It has been proposed that males are 'better prepared physically for metabolic demands during the development of reproductive maturity' (Warren, 1983, p. 370). This hypothesis presumably includes the demands associated with rigorous physical training. However, longitudinal studies of males indicate no effect of regular training for sport on the timing and tempo of indicators of sexual, skeletal and somatic maturation (Malina, 1994b). In the context of Warren's (1983) hypothesis, it has also been proposed that 'the significant gains in strength and muscle mass which are possible in prepubertal boys undergoing resistance training could accelerate pubertal onset' (Cumming *et al.*, 1994, pp. 56–7). However, several studies of resistance training in prepubertal boys, on the one hand, indicate gains in muscular strength without muscular hypertrophy; pubertal boys, on the other hand, increase in both strength and muscle mass in response to resistance training (Sale, 1989).

Males, of course, do not have an outcome variable of pubertal maturation equivalent to menarche. Responses of gonadotrophic hormones and testosterone to acute or chronic exercise in adolescent males vary with pubertal status and are equivocal. The limited data include observations over a swim season, after a marathon, in response to maximal cycle ergometry and in response to an incremental treadmill exercise to maximum (see Malina, 1991a).

Training and reproductive function in adult males

Regular endurance running is associated with possible alterations in the hypothalamic–pituitary–gonadal axis similar to those observed in trained women with menstrual dysfunction. The data indicate reduced levels of testosterone and alterations in LH pulse frequency, amplitude and/or area under the LH curve (see Malina, 1991a). In addition to possibly altered hypothalamic–pituitary–gonadal axis function in endurance trained males, some evidence suggests potential disruption of sperm production. A cross-sectional study of endurance runners and resistance-trained weightlifters indicated lower testosterone levels compared to matched controls, and subclinical alterations in sperm density, motility and morphology in only the endurance runners (Arce *et al.*, 1993). A one-year longitudinal study of men training for a marathon indicated changes in several hormones that were within the normal physiological range; reductions in the volume of semen and in the motility and morphology of sperm, but not in sperm count were also noted (Jensen *et al.*, 1995).

The implications of these observations for male reproductive function and fertility are not known. Presumably the observed changes associated with endurance training are reversible as suggested for menstrual function in females. Nevertheless, reduction in the percentage of morphologically normal sperm may influence fertility rates (Kruger *et al.*, 1988). Disordered eating behaviour also occurs in males. Anorectic behaviour, i.e. obsession with leanness, marked weight loss and elevated stress, have been suggested in some marathon runners (Ayers *et al.*, 1985). Unfortunately, the behavioural correlates of endurance training in males have not received the same attention as those in females, e.g. current emphasis on the female athlete triad – disordered eating, secondary amenorrhoea and osteoporosis (Smith, 1996).

Physical activity, sport, mortality and morbidity

Mortality and morbidity prior to sexual maturity and during the reproductive years are components of Darwinian fitness. Several aspects relating sport and physical activity to mortality and morbidity are subsequently considered.

Sudden death in sport

Deaths at relatively young ages associated with physical activity, specifically in the context of sport, are rare. However, sudden deaths do occur in young individuals across a wide age spectrum in the context of sport, excluding, of course, motor sports, mountain climbing and other sport activities with an inherent risk for accidental death. Maron *et al.* (1995) presented the clinical profile of 25 children and adolescents 3–19 years of age who had died from cardiac arrest while participating in organized or recreational sport from 1977–95. The 25 cases – 24 males and one female – collapsed with cardiac arrest after receiving an unexpected blow to the chest. Death ensued from *commotio cordis* or 'cardiac concussion'. Of the 25 deaths, 16 occurred in baseball, two in softball, four in ice hockey and one each in American football, karate and lacrosse. The blow was most often inflicted by a pitched, thrown or batted baseball or softball (18), a hockey puck (2) or a lacrosse ball (1). Other blows to the chest came from a football helmet, the heel of a hockey stick, a body check and a karate kick. Of these cases approximately 60% (16) occurred in organized sports, while approximately 40% (9) occurred in recreational settings, including informal play at home. Interestingly, the blow that struck the youngster in each case was described as not extraordinary. The specific cause of the deaths is not known with certainty, but may be related to the thinness of the chest wall in children and adolescents, which yields to the force of the projectile or blow, thus facilitating the transmission of the force to the heart. The sex difference in prevalence relates most likely to the greater number of young males involved in sport, although other unidentified factors may be involved.

Two other recent reports have considered cases of sudden death in adolescents and young adults associated with vigorous exercise (Van Camp et al., 1995; Maron et al., 1996). The former described 160 cases of non-traumatic death in high school and college athletes, 13–23 years of age, in a variety of sports between 1983 and 1993 (Van Camp et al., 1995). The ratio of males to females was about 10 to 1, 146 males (16.9 ± 2.0 years) to 14 females (16.2 ± 2.4 years). The latter described the clinical profile of 134 athletes, 12–40 years of age, who suddenly died of cardio-vascular-related complications in sport between 1985 and 1995 (Maron et al., 1996). Ninety percent of the athletes died during or immediately after a training session or an athletic competition. The ratio of males to females was again about 10 to 1 – 120 males to 14 females – and the median age at death was 17 years.

In both studies, the major contributor to sudden death from cardio-vascular causes during sport was hypertrophic cardio-myopathy, a pathological thickening of the walls of the left ventricle that obstructs blood flow from the left ventricle to the aorta. The majority of cases had no symptoms. In the study of high school and college athletes (Van Camp et al., 1995), 136 of the 160 cases had adequate information to identify cause of death. Among athletes with cardio-vascular conditions, hypertrophic cardio-myopathy was the cause of death in 50 of 92 males and in 1 of 8 females. In the other study across a broader age spectrum (Maron et al., 1996), hypertrophic cardio-myopathy was cited in 48 of 134 deaths. The second most common cardio-vascular cause of death in both studies was congenital anomalies of the coronary arteries, 16% and 13% in each study, respectively. A variety of rare cardiovascular conditions as well as several apparently 'normal hearts' were represented in the remainder of cases of sudden death in young athletes.

Basketball and American football were the two sports most represented among the athletes described in both studies. Overall, 12 sports for males and six for females were represented among the cases of high school and college athletes (Van Camp *et al.*, 1995). In the study of Maron *et al.* (1996), 48% of the cases of hypertrophic cardio-myopathy in these two sports were of African-American (black) ancestry compared to 26% in athletes of European-American (white) ancestry. This is in part due to the relatively larger numbers of youth and young adults who participate in these sports in general, and in particular to the larger number of black youth and young adults who participate in these two sports compared to their proportion in the general population. An additional factor is the lack of baseline data on hypertrophic cardio-myopathy in the American black population (Maron *et al.*, 1996).

Deaths resulting from sporting activities are rare in children, adolescents and young adults. Estimated rates for non-traumatic sports death in high school and college athletes are 7.5 and 1.3 per million athletes participating per year in males and females, respectively. The estimated rate is higher among college male athletes than among high school athletes, 14.5 and 6.6 million athletes participating per year, respectively (Van Camp *et al.*, 1995). Corresponding estimates for traumatic deaths are not available.

Several of the sudden deaths in sport were associated with the sickle cell trait (HB AS), which has a prevalence of about 8% in the American black population (Kark *et al.*, 1987). Sudden unexplained deaths during rigorous exercise associated with basic military training have also been reported among American black recruits with HB AS. The estimated death rate was 0.31/1000 in black recruits with Hb AS, 17–34 years of age; the rate increased with age from 0.12/1000 at 17–18 years to 1.36/1000 at 26–30 years (Kark *et al.*, 1987). The deaths were attributed to cardiac arrest, exertional heat stroke, heat stress or rhabdomyolosis. Thus, a genotype which apparently has a selective advantage in malarial environments, may present a risk for cardio-vascular mortality under conditions of physical exertion.

Physical inactivity and low physical fitness as risk factors for mortality

In contrast to sudden death, physical inactivity and low levels of cardio-respiratory physical fitness are independent risk factors in all-cause mortality and specifically in cardio-vascular disease mortality (Powell *et al.*, 1987; Blair, 1994). More recently (Blair *et al.*, 1996), follow-up of a large cohort of well-educated men and women from 1970 through to 1989 indicated low physical fitness (treadmill run time, lowest quintile) as an independent risk for all-cause mortality in men (relative risk 1.52, 95% CI

1.28–1.82) and women (relative risk 2.10, 95% CI 1.36–3.21). Mean ages of the decedents were 52.1 ± 11.4 years in men and 53.3 ± 11.2 years in women, so that a significant number of those who died were still in the reproductive years. The relative risks for mortality from cardio-vascular disease associated with low fitness were higher, 1.70 (95% CI 1.28–2.25) in males and 2.42 (95% CI 0.99–5.92) in females (Blair *et al.*, 1996). The mortality gradients by physical fitness category (low, moderate, high) suggest a dose-response relationship, but the largest differences were between the moderate and low fit groups. In addition, moderate and high levels of physical fitness appeared to afford some degree of protection against other independent risk factors for all-cause and cardio-vascular mortality – smoking, increased cholesterol, elevated systolic blood pressure and abnormal electrocardiogram in men and smoking in women (Blair *et al.*, 1996).

Physical activity and cancer risk

Although the evidence is less substantial and there are many confounding factors, physical inactivity may be a risk factor for mortality from several cancers (Lee, 1994). For example, physically active individuals tend to be lighter and leaner (lower body mass index) and tend to follow a healthier diet, factors which are independently associated with reduced risk for colon cancer. Data from former college athletes suggest a lower prevalence of breast cancer compared to non-athletes after controlling for age and other related variables, e.g. estimated leanness–fatness, smoking, age at menarche, number of pregnancies, use of oral contraceptives and post-menopausal hormones, and family history (Frisch *et al.*, 1985). Although menarche occurs, on average, later in athletes than in non-athletes (Malina, 1983, 1994a,b), mean ages at menarche of the sample of former college athletes and non-athletes differed only slightly but significantly, 13.1 compared to 12.9 years (Frisch *et al.*, 1985). The former college athletes and non-athletes were college alumnae in classes between 1925 and 1981. Mean age at menarche in American girls declined from about 14.0 years in 1900 to 12.8 years in the 1960s (Malina, 1979), so that this sample of former college students were likely to be somewhat earlier maturing and perhaps from better-off socio-economic circumstances (seeing as they attended college) than the general population.

In a case-control design of women diagnosed with breast cancer ≈ 40 years of age, weekly time spent in physical activity, including sport teams, after menarche was significantly associated with reduced risk. The odds ratio of breast cancer among women who spent ≈ 3.8 hours per week in physical activity during their reproductive lifetime was 0.42 compared to inactive women (Bernstein *et al.*, 1994). The apparent protective effect of regular physical activity was stronger in women who had a full-term pregnancy compared to nulliparous women, while age at menarche did not alter risk.

Later menarche is associated with reduced risk for breast cancer (Carter and Micozzi, 1986). A related factor may be altered menstrual function and in turn hormonal milieu associated with training (Lee, 1994). Issues related to later menarche and menstrual dysfunction in athletes have already been discussed. Physique, independent of leanness or fatness, may be an additional factor in the lower risk for breast cancer in women athletes. Women athletes in a variety of sports tend to be more mesomorphic and less endomorphic (Carter and Heath, 1990), and to have more androgynous physiques (androgyny index = [3 × biacromial breadth] − bicristal breadth) than non-athletes (Malina and Merrett, 1995).

Overview

Studies considered in the preceding discussion do not include significant numbers from minority populations. The risk of mortality from cardio-vascular disease are greater in American blacks than in American whites, and deaths from heart disease occur, on average, at earlier ages in blacks than in whites. Hypertension, obesity and adult-onset diabetes are also more prevalent among American blacks, and all are independent risk factors for cardio-vascular disease (Giel, 1988). The incidence of breast cancer does not differ between younger black and white women (< 40 years), but is greater among older white women (Polednak, 1989). American black women, including athletes, attain menarche, on average, earlier than white women (Malina and Bouchard, 1991). Black adolescents and adults also have lower levels of habitual physical activity (see above). Data dealing with physical fitness of black and white adults are not available, but black adolescent females have lower levels of aerobic fitness (Pivarnik *et al.*, 1995).

Physical activity, sport and longevity

Evidence for favourable longevity of male former athletes is not conclusive, but a habitually active lifestyle among college graduates is associated with a modest gain in longevity (Malina, 1991a). Samples in most of the currently available data are probably white. Given the altered composition of athletic teams at the intercollegiate and professional levels in some sports over the past two or three decades, e.g. basketball, American

football and track, future studies will need to control for ethnic background, in addition to potential confounding factors such as use of performance enhancing substances, extended schedules and frequent travel.
Future studies of college alumni will also need to control for ethnicity as
the number of ethnic minorities in higher education increases.

Opportunities for girls and women in sport have historically been unequal ('inequitable') in the USA. However, enhanced opportunities for
girls and women in sport since the early 1970s in the USA, and continued
participation of many women in sport through the collegiate years and
beyond, beg the issue of longevity of female athletes. Such analyses are
perhaps a generation into the future.

Conclusion

Physical activity may influence reproductive function, and through its
interaction with physical fitness, may influence risk for several diseases and
mortality. Thus, physical activity and physical fitness have the potential to
influence Darwinian fitness. Several potential inter-relationships among
physical activity, sport, social status and Darwinian fitness have been
considered in several contexts within the theme, human biology and social
inequality. It is difficult to implicate training for sport *per se* as the
determinant of later mean ages at menarche observed in athletes. Training
for sport does not influence the timing and tempo of sexual maturation of
boys. However, training for sport, especially endurance sports, is associated with reproductive morbities in adult women and men. The implications of these morbidities for fertility are not known. A habitually active
lifestyle is associated with reduced cardio-vascular morbidity and mortality, and reduced risk for several cancers. Low physical fitness is an independent risk factor for all-cause and cardio-vascular mortality, while
sudden death at young ages, though rare, is associated with strenuous
physical exertion and sport. The sudden deaths associated with sport, and
a significant percentage of the mortality associated with low physical
fitness occurred in individuals who were, respectively, at pre-reproductive
ages or still in the reproductive years; hence, they have implications for
Darwinian fitness.

References

Aaron, D. J., Kriska, A. M., Dearwater, S. T., Anderson, R. L., Olsen, T. L.,
    Cauley, J. A. & Laporte, R. E. (1993). The epidemiology of leisure time
    physical activity in an adolescent population. *Medicine and Science in Sports
    and Exercise* **25**, 847–53.

Ainsworth, B. E., Keenan, N. L., Strogatz, D. S., Garrett, J. M. & James, S. A. (1991). Physical activity and hypertension in Black adults: The Pitt County Study. *American Journal of Public Health* **81**, 1477–9.

American Medical Association/American Dietetic Association (1991). *Targets for Adolescent Health: Nutrition and Physical Fitness*. Chicago: American Medical Association.

Arce, J. C., De Souza, M. J., Pescatello, L. S. & Luciano, A. A. (1993). Subclinical alterations in hormone and semen profile in athletes. *Fertility and Sterility* **59**, 398–404.

Ayers, J. W. T., Komesu, Y., Romain, T. & Ansbacher, R. (1985). Anthropometric, hormonal and psychologic correlates of semen quality in endurance trained male athletes. *Fertility and Sterility* **43**, 917–21.

Baker, E. R. (1981). Menstrual dysfunction and hormonal status in athletic women: a review. *Fertility and Sterility* **36**, 691–6.

Barr, S. I. & Prior, J. C. (1994). The menstrual cycle: effects on bone in premenopausal women. In *Advances in Nutritional Research*, ed. by H. C. Draper, vol. 9, pp. 287–307. New York: Plenum.

Baxter-Jones, A. D. G., Helms, P., Baines-Preece, J. & Preece, M. (1994). Menarche in intensively trained gymnasts, swimmers and tennis players. *Annals of Human Biology* **21**, 407–15.

Behrman, S. J. (1967). Adolescent amenorrhea. *Annals of the New York Academy of Sciences* **142**, 807–12.

Bernstein, L., Ross, R. K., Lobo, R. A., Hanisch, R., Krailo, M. D. & Henderson, B.E. (1987). The effects of moderate physical activity on menstrual cycle patterns in adolecence: implications for breast cancer prevension. *British Journal of Cancer* **55**, 681–5.

Bernstein, L., Henderson, B. E., Hanisch, R., Sullivan-Halley, J. & Ross, R. K. (1994). Physical exercise and reduced risk of breast cancer in young women. *Journal of the National Cancer Institute* **86**, 1403–8.

Beunen, G. & Malina, R. M. (1996). Growth and biological maturation: relevance to athletic performance. In *The Child and Adolescent Athlete*, ed. O. Bar-Or. Oxford: Blackwell Science.

Bielicki, T., Waliszko, A., Hulanicka, B. & Kotlarz, K. (1986). Social class gradients in menarcheal age in Poland. *Annals of Human Biology* **13**, 1–11.

Blair, S. N. (1994). Physical activity, fitness, and coronary heart disease. In *Physical Activity, Fitness, and Health*, ed. C. Bouchard, R. J. Shephard & T. Stephens, pp. 579–90. Champaign: Human Kinetics.

Blair, S. N., Kampert, J. B., Kohl, H. W., Barlow, C. E., Macera, C. A., Paffenbarger, R. S. & Gibbons, L. W. (1996). Influences of cardiorespiratory fitness and other precursors on cardiovascular disease and all-cause mortality in men and women. *Journal of the American Medical Association* **276**, 205–10.

Bompa, T. O. (1985). *Talent Identification*. Ottawa: Coaching Association of Canada.

Bouchard, C. & Shephard, R. J. (1994). Physical activity, fitness, and health. In *Physical Activity, Fitness, and Health*, ed. by C. Bouchard, R. J. Shephard & T. Stephens, pp. 77–88. Champaign: Human Kinetics.

Bouchard, C., Shephard, R. J. & Stephens, T. (Eds.) (1994). *Physical Activity, Fitness, and Health*. Champaign: Human Kinetics.

Bouchard, C., Malina, R. M. & Perusse, L. (1997). *Genetics of Fitness and Physical*

Performance. Champaign: Human Kinetics.

Brooks-Gunn, J. & Warren, M. P. (1988). Mother-daughter differences in menarcheal age in adolescent girls attending national dance company schools and non-dancers. *Annals of Human Biology* **15**, 35–43.

Carter, C. L. & Micozzi, M. S. (1986). Genetic factors in human breast cancer. *Yearbook of Physical Anthropology* **29**, 161–80.

Carter, J. E. L. & Heath, B.H. (1990). *Somatotyping – Development and Applications*. Cambridge: Cambridge University Press.

Clapp, J. F. & Little, K. D. (1995). The interaction between regular exercise and selected aspects of women's health. *American Journal of Obstetrics and Gynecology* **173**, 2–9.

Constantini, N. W. & Warren, M. P. (1994). Physical activity, fitness, and reproductive health in women: Clinical observations. In *Physical Activity, Fitness, and Health*, ed. C. Bouchard, R. J. Shephard & T. Stephens, pp. 955–66. Champaign: Human Kinetics.

Constantini, N. W. & Warren, M. P. (1995). Menstrual dysfunction in swimmers: A distinct entity. *Journal of Clinical Endocrinology and Metabolism* **80**, 2740–4.

Cooper, C., Kuh, D., Egger, P., Wadsworth, M. & Barker, D. (1996). Childhood growth and age at menarche. *British Journal of Obstetrics and Gynaecology* **103**, 814–17.

Cumming, D. C., Wheeler, G. D. & Harber, V. J. (1994). Physical activity, nutrition, and reproduction. *Annals of the New York Academy of Sciences* **709**, 55–74.

Dobzhansky, Th. (1962). *Mankind Evolving: The Evolution of the Human Species*. New Haven: Yale University Press.

Douglas, J. W. B. (1966). The age of reaching puberty: some associated factors and some educational implications. *Scientific Basis of Medicine Annual Reviews*, pp. 91–105 (as cited by Leistol, 1982).

Eveleth, P. B. & Tanner, J. M. (1990). *Worldwide Variation in Human Growth*, 2nd edn. Cambridge: Cambridge University Press.

Folsom, A. R., Cook, T. C., Sprafka, J. M., Burke, G. L., Norsted, S. W. & Jacobs, D. R. (1991). Differences in leisure-time physical activity levels between Blacks and Whites in population-based samples: The Minnesota Heart Survey. *Journal of Behavioral Medicine* **14**, 1–9.

Frisch, R. E., Wyshak, G., Albright, N. L., Albright, T. E., Schiff, I., Jones, K. P., Witschi, J., Shiang, E., Koff, E. & Marguglio, M. (1985). Lower prevalence of breast cancer and cancers of the reproductive system among former college athletes compared to non-athletes. *British Journal of Cancer* **52**, 885–91.

Giel, D. (1988). Fitness and exercise issues for Black Americans. *Physician and Sportsmedicine* **16**, 162–3.

Graber, J. A., Brooks-Gunn, J. & Warren, M. P. (1995). The antecedents of menarcheal age: heredity, family environment, and stressful life events. *Child Development* **66**, 346–59.

Greene, B. B., Daling, J. R., Weiss, N. S., Liff, J. M. & Koepsell, T. (1986). Exercise as a risk for infertility with ovulatory dysfunction. *American Journal of Public Health* **76**, 1432–6.

Grumbach, M. M. & Kaplan, S. L. (1990). The neuroendocrinology of human puberty: an ontogenetic perspective. In *Control of the Onset of Puberty*, ed. M.

M. Grumbach, P. C. Sizonenko & M. L. Aubert, pp. 1–62. Baltimore: Williams and Wilkins.

Hamilton, L. J., Brooks-Gunn, J., Warren, M. P. & Hamilton, G. W. (1988). The role of selectivity in the pathogenesis of eating problems in ballet dancers. *Medicine and Science in Sports and Exercise* **20**, 560–5.

Hamilton, W. G. (1986). Physical prerequisites for ballet dancers: selectivity that can enhance (or nullify) a career. *Journal of Musculoskeletal Medicine* **3**, 61–6.

Heath, G. W., Pratt, M., Warren, C. W. & Kahn, L. (1994). Physical activity patterns in American high school students. *Archives of Pediatric and Adolescent Medicine* **148**, 1131–6.

Jahreis, G., Kauf, E., Frohner, G. & Schmidt, H. E. (1991). Influence of intensive exercise on insulin-like growth factor I, thyroid and steroid hormones in female gymnasts. *Growth Regulation* **1**, 95–9.

Jensen, C. E., Wiswedel, K., McLoughlin, J. & van der Spuy, Z. (1995). Prospective study of hormonal and semen profiles in marathon runners. *Fertility and Sterility* **64**, 1189–96.

Jones, B., Leeton, J., McLeod, I. & Wood, C. (1972). Factors influencing the age at menarche in a lower socioeconomic group in Melbourne. *Medical Journal of Australia* **2**, 533–5.

Kark, J. A., Posey, D. M., Schumacher, H. R. & Ruehle, C. J. (1987). Sickle cell trait as a risk factor for sudden death in physical training. *New England Journal of Medicine* **317**, 781–7.

Kruger, T. F., Acosta, A. A., Simmons, K. F., Swanson, R. J., Matta, J. F. & Oehninger, S. (1988). Predictive value of abnormal sperm morphology in in vitro fertilization. *Fertility and Sterility* **49**, 112–17.

Lee, I.-M. (1994). Physical activity, fitness, and cancer. In *Physical Activity, Fitness, and Health*, ed. C. Bouchard, R. J. Shephard & T. Stephens. Champaign: Human Kinetics.

Leistol, K. (1980). Menarcheal age and spontaneous abortion: a causal connection. *American Journal of Epidemiology* **111**, 753–8.

Leistol, K. (1982). Social conditions and menarcheal age: the importance of early years of life. *Annals of Human Biology* **9**, 521–37.

Loucks, A. B. (1994). Physical activity, fitness, and female reproductive morbidity. In *Physical Activity, Fitness, and Health*, ed. C. Bouchard, R. J. Shephard & T. Stephens, pp. 943–54. Champaign: Human Kinetics.

Loucks, A. B. & Heath, E. M. (1994). Dietary restriction reduces luteinizing hormone (LH) pulse frequency during waking hours and increases LH pulse amplitude during sleep in young menstruating women. *Journal of Clinical Endocrinology and Metabolism* **78**, 910–15.

Loucks, A. B., Vaitukaitis, J., Cameron, J. L., Rogol, A. D., Skrinar, G., Warren, M. P., Kendrick, J. & Limacher, M. C. (1992). The reproductive system and exercise in women. *Medicine and Science in Sports and Exercise* **24**, S288–93.

Malina, R. M. (1979). Secular changes in size and maturity: causes and effects. *Monographs of the Society for Research in Child Development* **44** (179), 59–102.

Malina, R. M. (1983). Menarche in athletes: A synthesis and hypothesis. *Annals of Human Biology* **10**, 1–24.

Malina, R. M. (1990). Growth, exercise, fitness, and later outcomes. In *Exercise, Fitness, and Health*, ed. C. Bouchard, R. J.Shephard, T. Stephens, J. R. Sutton & B. D. McPherson, pp. 637–53. Champaign: Human Kinetics.

Malina, R. M. (1991a). Darwinian fitness, physical fitness and physical activity. In *Applications of Biological Anthropology to Human Affairs*, ed. C. G. N. Mascie-Taylor & G. W. Lasker, pp. 143–84. Cambridge: Cambridge University Press.

Malina, R. M. (1991b). Fitness and performance: Adult health and the culture of youth. *American Academy of Physical Education Papers* No. 24, pp. 30–8. Champaign: Human Kinetics.

Malina, R. M. (1992). Physical activity and behavioural development during childhood and youth. In *Physical Activity and Health*, ed. N. G. Norgan, pp. 101–20. Cambridge: Cambridge University Press.

Malina, R. M. (1993). Youth sports: readiness, selection and trainability. In *Kinanthropometry IV*, ed. W. Duquet & J. A. P. Day, pp. 285–301. London: Spon.

Malina, R. M. (1994a). Physical activity: Relationship to growth, maturation, and physical fitness. In *Physical Activity, Fitness, and Health*, ed. C. Bouchard, R. J. Shephard & T. Stephens, pp. 918–30. Champaign: Human Kinetics.

Malina, R. M. (1994b). Physical growth and biological maturation of young athletes. *Exercise and Sports Science Reviews* **22**, 389–433.

Malina, R. M. (1995). Physical activity and fitness of children and youth: questions and implications. *Medicine, Exercise, Nutrition, and Health* **4**, 123–35.

Malina, R. M. (1996a). Tracking of physical activity and physical fitness across the lifespan. *Research Quarterly for Exercise and Sport* **67**, S48–57.

Malina, R. M. (1996b). The young athlete: biological growth and maturation in a biocultural context. In *Children and Youth in Sport: A Biopsychosocial Perspective*, ed. F. L. Smoll & R. E. Smith, pp. 161–86. Dubuque: Brown and Benchmark.

Malina, R. M. & Bouchard, C. (1991). *Growth, Maturation, and Physical Activity*. Champaign: Human Kinetics.

Malina, R. M. & Geithner, C.A. (1993). Background in sport, growth status, and growth rate of Junior Olympic divers. In *U.S. Diving Sport Science Seminar Proceedings*, ed. R. M. Malina & J. L. Gabriel, pp. 26–35. Indianapolis: United States Diving.

Malina, R. M. & Merrett, D. M. S. (1995). Androgyny of physique of women athletes: Comparisons by sport and over time. In *Essays on Auxology*, ed. R. Hauspie, G. Lindgren & F. Falkner, pp. 355–63. Welwyn Garden City: Castlemead Publications.

Malina, R. M., Bouchard, C., Shoup, R. F. & Lariviere, G. (1982). Age, family size and birth order in Montreal Olympic athletes. In *Physical Structure of Olympic Athletes: Part I. The Montreal Olympic Games Anthropological Project*, ed. J. E. L. Carter, pp. 13–24. Basel: S. Karger.

Malina, R. M., Ryan, R. C. & Bonci, C.M. (1994). Age at menarche in athletes and their mothers and sisters. *Annals of Human Biology* **21**, 417–22.

Malina, R. M., Katzmarzyk, P. T., Bonci, C. M., Ryan, R. C. & Wellens, R. E. (1997). Family size and age at menarche in athletes. *Medicine and Science in Sports and Exercise* **29**, 99–106.

Maron, B. J., Poliac, L. C., Kaplan, J. A. & Mueller, F. O. (1995). Blunt impact to the chest leading to sudden death from cardiac arrest during sports activities. *New England Journal of Medicine* **333**, 337–42.

Maron, B. J., Shirani, J., Poliac, L. C., Mathenge, R., Roberts, W. C. & Mueller, F.

O. (1996). Sudden death in young competitive athletes: Clinical, demographic, and pathological profiles. *Journal of the American Medical Association* **276**, 199–204.

Martin, E. J., Brinton, L. A. & Hoover, R. (1983). Menarcheal age and miscarriage. *American Journal of Epidemiology* **117**, 634–6.

Merzenich, H., Boeing, H. & Wahrendorf, J. (1993). Dietary fat and sports activity as determinants for age at menarche. *American Journal of Epidemiology* **138**, 217–24.

Myers, L., Strikmiller, P. K., Webber, L. S. & Berenson, G. S. (1996). Physical and sedentary activity in school children grades 5–8: The Bogalusa Heart Study. *Medicine and Science in Sports and Exercise* **28**, 852–9.

Pivarnik, J. M., Bray, M. S., Hergenroeder, A. C., Hill, R. B. & Wong, W. W. (1995). Ethnicity affects aerobic fitness in U.S. adolescent girls. *Medicine and Science in Sports and Exercise* **27**, 1635–8.

Polednak, A.P. (1989). *Racial and Ethnic Differences in Disease.* New York: Oxford University Press.

Powell, K. E., Thompson, P. D., Caspersen, C. J. & Kendrick, J. S. (1987). Physical activity and the incidence of coronary heart disease. *Annual Review of Public Health* **8**, 253–87.

Press, A. (1992). Old too soon, wise too late? *Newsweek*, 10 August, pp. 22–4.

Roberts, D. F., Wood, W. & Chinn, S. (1986). Menarcheal age in Cumbria. *Annals of Human Biology* **13**, 161–70.

Rebar, R. (1990). Disorders of menstruation, ovulation, and sexual response. In *Principles and Practice of Endocrinology and Metabolism*, ed. K. Becker, pp. 798–814. Philadelphia: Lippincott.

Roemmich, J. N. (1994). *Weight loss effects on growth, maturation, growth related hormones, protein nutrition markers, and body composition of adolescent wrestlers.* Doctoral dissertation, Kent State University, Kent, OH.

Ryan, J. (1995). *Little Girls in Pretty Boxes: The Making and Breaking of Elite Gymnasts and Figure Skaters.* New York: Warner Books.

Ryan, R. C. (1996). *Menstrual status in elite female athletes: An evaluation of multiple sports.* Doctoral dissertation, University of Texas, Austin.

Sale, D. G. (1989). Strength training in children. In *Youth, Exercise, and Sport*, ed. C. Gisolfi & D. R. Lamb, pp. 165–216. Carmel: Benchmark Press.

Sallis, J. F., Simons-Morton, B. G., Stone, E. J., Corbin, C. B., Epstein, L. H., Faucette, N., Iannotti, R. J., Killen, J. D., Klesges, R. C., Petray, C. K., Rowland, T. W. & Taylor, W. C. (1992). Determinants of physical activity and interventions in youth. *Medicine and Science in Sports and Exercise* **24**, S248–57.

Sallis, J. F., Zakarian, J. M., Hovel, M. F. & Hofstetter, R. (1996). Ethnic, socioeconomic, and sex differences in physical activity among adolescents. *Journal of Clinical Epidemiology* **49**, 125–34.

Sandler, D. P., Wilcox, A. J. & Horney, L. F. (1984). Age at menarche and subsequent reproductive events. *American Journal of Epidemiology* **119**, 765–74.

Schoenborn, C. A. (1986). Health habits of U.S. adults, 1985: The 'Almeda 7' revisited. *Public Health Reports* **101**, 571–80.

Shangold, M. M. (1984). Exercise and the adult female: hormonal and endocrine effects. *Exercise and Sports Science Reviews* **12**, 53–79.

Shangold, M. M. (1994). Menstruation and menstrual disorders. In *Women and Exercise: Physiology and Sports Medicine*, ed. M. M. Shangold & G. Mirkin, 2nd ed., pp. 152–71. Philadelphia: Lea & Febiger.

Smith, A. D. (1996). The female athlete triad: Causes, diagnosis, and treatment. *Physician and Sportsmedicine* 24, (July), 67–76, 86.

Spirduso, W. W. (1995). *Physical Dimensions of Aging*. Champaign: Human Kinetics.

Stager, J. M. & Hatler, L. K. (1988). Menarche in athletes: the influence of genetics and prepubertal training. *Medicine and Science in Sports and Exercise* 20, 369–73.

Stephens, T., Jacobs, D. R., & White, C. C. (1985). A descriptive epidemiology of leisure-time physical activity. *Public Health Reports* 100, 147–58.

Sundgot-Borgen, J. & Larsen, S. (1993). Preoccupation with weight and menstrual function in female elite athletes. *Scandinavian Journal of Medicine and Science in Sports* 3, 156–63.

Surbey, M. K. (1990). Family composition, stress, and the timing of human menarche. In *Socioendocrinology of Primate Reproduction*, ed. T. E. Ziegler & F. B. Bercovitch, pp. 11–32. New York: Wiley Liss.

Tanner, J. M. (1962). *Growth at Adolescence*, 2nd edn. Oxford: Blackwell Scientific Publications.

United Nations Development Programme (1993). *Human Development Report 1991*. New York: Oxford University Press.

US Bureau of the Census (1992). *Poverty in the United States: 1991*. Washington, DC: US Government Printing Office.

US Department of Health and Human Services (1996). *Physical Activity and Health: A Report of the Surgeon General*. Washington, DC: US Department of Health and Human Services.

Van Camp, S. P., Bloor, C. M., Mueller, F. O., Cantu, R. C. & Olson, H. G. (1995). Nontraumatic sports deaths in high school and college athletes. *Medicine and Science in Sports and Exercise* 27, 641–7.

Veldhuis, J. D., Evans, W. S., Demers, L. M., Thorner, M. O., Wakat, D. & Rogol, A. D. (1985). Altered neuroendocrine regulation of gonadotropin secretion in women distance runners. *Journal of Clinical Endocrinology and Metabolism* 61, 65–80.

Warren, M. P. (1983). Effects of undernutrition on reproductive function in the human. *Endocrine Reviews* 4, 363–77.

Warren, M. P. (1995). Amenorrhea in ballet dancers. *Eleventh International Jerusalem Symposium on Sports Injuries*, abstracts, p. 18. Tel Aviv: Israel Society of Sports Medicine.

Wolf, A. M., Gortmaker, S. L., Cheung, L., Gray, H. M., Herzog, D. B. & Colditz, G. A. (1993). Activity, inactivity, and obesity: Racial, ethnic, and age differences among schoolgirls. *American Journal of Public Health* 83, 1625–7.

Women's Sports Foundation (1993). *Sport and Fitness in the Lives of Working Women*. New York: Women's Sports Foundation.

11 Biological correlates of social and geographical mobility in humans: an overview

C. G. NICHOLAS MASCIE-TAYLOR

Introduction and historical perspective

The study of the relationship between biological characters and migration has a murky history in human biology and can be traced back to the era when every characteristic of the skeleton was considered to be an inherent racial hallmark such that their geographical distribution was assumed to result solely from the movement of peoples of different 'races'. However the pioneering work of Franz Boas (1910) on migration to the USA of Old World Jews and Sicilians demonstrated that stature and other anthropometric measurements are modified in the offspring of migrants and instead of 'diluting the American stock' they became more like the Americans, that is taller.

Basler (1927) showed that it made a difference to the shape of the head (as measured by the cephalic index) if an infant had been swaddled and placed on its back or on its side. Ewing (1950) studied the skull shape of a religious sect, the Lebanese Maronites. He found those living in Lebanon were characterized by hyperbrachycephaly (wide, broad skulls) while those who had migrated to the USA no longer showed this characteristic. Like stature, early anthropologists had thought hyperbrachycephaly to be an entirely racial/inherited character. Ewing also studied the cradling practices of the sect in Lebanon and the USA. In Lebanon a baby would be placed in a wooden cradle for 18 hours or more each day. The skull being soft and pliable at that age came into contact with the hard wooden cradle for long periods and the back of the skull became flattened leading to hyperbrachycephaly. The Lebanese migrants to the USA ceased to use this cradling practice and so their children had normal skull shapes. Thus the abnormal skull shape was not due to genes at all but was simply the result of an environmental practice.

As these examples show, the early work on migration was concerned with a study of the impact of geographic migration. However, the genetic structure of populations can be modified by both geographic and social migration and the first part of this review is devoted to an examination of the role of social migration or social mobility as it is more often called.

The concept of social mobility

Social mobility is a major area of sociological inquiry with both theoretical and practical applications. Ideas about movement between occupational and social groups can be traced back to the theory of social selection and recruitment first put forward by Plato in the *Republic* over 2000 years ago. Plato, was a strong hereditarean, and proposed that in the ideal republic there are three classes of citizen – the guardians or rulers, the soldiers (auxiliaries) and the remainder comprising farmers, builders, weavers, etc. He believed that children will usually have the same nature as their parents and hence mobility between the classes will only occur rarely (see Chapter 1).

In reality, in modern populations there is a reasonable amount of mobility as Glass (1954) and his colleagues have demonstrated in the classic work on the extent of mobility in Britain. They conducted a study on a random sample of 10,000 adult civilians (18 years of age or older) from Britain in 1949 and collected data on age, sex and marital status, schools attended, qualifications obtained and their own and their fathers' occupations. From this they were able to construct the extent of inter-generational mobility. The main conclusions from the 1949 study were: (1) there was a considerable amount of short-range mobility (i.e. small movements between adjacent occupational groups) while long-range mobility, the so called 'rags to riches' or 'riches to rags' mobility, was very rare; (2) there was a barrier to movement across the non-manual/manual divide; and (3) there was a high degree of self-recruitment at the top of the social scale (the closure thesis or elite self-recruitment thesis).

Shortly after the British mobility data were published, information on the extent of inter-generational mobility were collected from a number of European countries as well as the USA. Lipset and Bendix (1959) re-analysed data from nine industrialized countries (France, Germany, Sweden, Switzerland, USA, Japan, Denmark, Italy and UK) and reclassi-fied the occupations into non-manual, manual and farming categories in order to make comparisons between countries. They then determined the amount of upward and downward mobility across the non-manual/man-ual divide and summed it to give figures for the total mobility, which they termed the 'total vertical mobility'. Lipset and Bendix found that the nine countries showed very similar high rates of total vertical mobility, ranging

Table 11.1. *Goldthorpe and Hope's classification of occupations*

Class	Occupations
I	Higher-grade professionals, administrators, managers and large proprietors
II	Lower-grade professionals, administrators and managers, high-grade technicians, and supervisors of non-manual employees
III	Routine clerical workers, sales personnel and other non-manual workers
IV	Farmers, small proprietors and self-employed workers
V	Supervisors of manual workers and lower grade technicians
VI	Skilled manual workers
VII	Semi- and unskilled manual workers

between 27% (France and Japan) and 31% (Germany), but contrary to their expectation, the USA did not show any higher rates than the supposedly more 'traditional' European countries.

Classification of occupations

Unfortunately there is no agreed or 'correct' classification of occupations and those used reflect a mixture of political and theoretical preferences. Even so, there is some agreement and most classifications are broadly similar at the top (professional and high administrative occupations) and the bottom (unskilled manual workers). Most problems arise in the middle at the manual/non-manual divide. The classification used by the Oxford group the seven categories set out in Goldthorpe and Hope's (1974) classification (Table 11.1).

This classification places clerical workers, foremen, shopkeepers and skilled manual workers in four different social classes (classes III to VI). However, Heath (1981) argues that these four social classes are on the same level and therefore inappropriate for differentiating between upward and downward mobility (see next section). Goldthorpe and Hope's scheme differs from the Registrar-General's classification which is based on a five point scale with professional (I), managerial (II), non-manual and skilled manual (IIINM and IIIM, semi-skilled manual (IV) and unskilled manual (V) groups. Detailed discussions on the types of classifications can be found elsewhere (Hope, 1972).

The extent of social mobility

More recent work by the Oxford group of sociologists has provided further insight into the type and extent of mobility and in particular whether the three main types of movement reported by Glass were still occurring. They used a database of some 10,000 adult males who con-

Table 11.2a. *Intergenerational mobility (outflow)*

Father's class	Respondent's class							Total %	Total N
	I (%)	II (%)	III (%)	IV (%)	V (%)	VI (%)	VII (%)		
I	48.4	18.9	9.3	8.2	4.5	4.5	6.2	100.0	582
II	31.9	22.6	10.7	8.0	9.2	9.6	8.0	100.0	477
III	19.2	15.7	10.8	8.6	13.0	15.0	17.8	100.1	594
IV	12.8	11.1	7.8	24.9	8.7	14.7	19.9	99.9	1223
V	15.4	13.2	9.4	8.0	16.6	20.1	17.2	99.9	939
VI	8.4	8.9	8.4	7.1	12.2	29.6	25.4	100.0	2312
VII	6.9	7.8	7.9	6.8	12.5	23.5	34.8	100.2	2216
Total %	14.3	11.4	8.6	9.9	11.6	20.8	23.3	99.9	8343

See Table 11.1 for class classification.

Table 11.2b. *Intergenerational mobility (inflow)*

Father's class	Respondent's class							Total (%)
	I %	II (%)	III (%)	IV (%)	V (%)	VI (%)	VII (%)	
I	23.6	11.6	7.5	5.8	2.7	1.5	1.9	7.0
II	12.7	11.4	7.1	4.6	4.5	2.7	2.0	5.7
III	9.5	9.8	8.9	6.1	7.9	5.1	5.5	7.1
IV	13.1	14.3	13.3	36.7	10.9	10.4	12.5	14.7
V	12.1	13.1	12.2	9.0	16.1	10.9	8.3	11.3
VI	16.3	21.6	26.9	19.6	29.2	39.4	30.2	27.7
VII	12.7	18.1	24.1	18.1	28.6	30.0	39.7	26.6
Total %	100.0	99.9	100.0	99.9	99.9	100.0	100.1	100.1
Total N	1197	948	721	830	969	1734	1944	8343

See Table 11.1 for class classification.

stituted a representative sample of those aged between 20 and 64 years resident in England and Wales. Table 11.2a shows the extent of inter-generational mobility (i.e. an outflow table of the destinations of men from different social classes) using the Goldthorpe scheme. It can be seen that just under half of the sons (48.4%) who were in social class I had fathers in the same social class (the father's social class was defined as the occupation of the father when the respondent was aged 14 years). Even so there has been a considerable amount of inter-generational mobility and the diag-onal line provides information on non-migrants (28% overall), while 72% were migrants. If one assumes that classes III to VI are on the same level, then 31% were upwardly mobile and 18% downwardly mobile. The excess of upward over downward mobility is due to the expansion of the professional and managerial occupations.

What do these data tell us about short-range and long-range movement? There is clearly more short-range than long-range movement. Only about 6% of men of class I origin had dropped into the semi-/unskilled manual category (class VII), whereas over 25% had dropped the shorter distance into classes III to VI. A similar picture is seen starting from class VII – about 15% were upwardly mobile to classes I and II, whereas over 50% were mobile into classes III to VI. There is evidence therefore that the 'rags to riches' and 'riches to rags' scenarios are rarer than 'rags to moderate affluence' or vice-versa (Heath, 1981). These data also provide information on the extent of movement between manual and non-manual groups (the manual/non-manual barrier). Assuming that classes I–III are definitely in the non-manual group then 25% of men originating in classes VI and VII families were upwardly mobile into a non-manual class. However, although rather fewer men with class I fathers were likely to finish up in the manual group the data do suggest that the idea of a 'barrier' is incorrect.

Finally we turn to the notion of the elite self-recruitment. Table 11.2a shows that the highest levels of occupational stability are found in the extremes of classes I and VII, with class I showing the higher level – 48% compared with 35% in class VII. Westergaard and Resler (1975) using the 1949 data, calculated that if parental origins played no part, only 3% of sons would have remained in social class I compared with the observed 40%. The index of observed to expected (Index of Association or mobility index) was 40:3 or about 13:1. Applying the same calculation to the more recent data yields an index of association of 48:14 or less than 4:1. However the index is sensitive to size and the more recent data show that class I is much bigger (14%) compared with the 1949 (3%) data. The dramatic increase in class I also illustrates an important paradox; although nearly half of the men whose fathers were in class I followed in their fathers' footsteps, 75% of the men in class I had come from a lower class. This can be seen from Table 11.2b which shows inflow (the difference between Tables 11.2a and 11.2b is that an outflow table calculates the percentages across rows whereas an inflow table calculates percentages down the columns). Thus the majority of class I are newcomers and the upwardly mobility is a consequence of the expansion of the professional and managerial groups mentioned earlier.

Relationship between social mobility and psychometric characters

Most of the literature focuses on the relationship between IQ and social class, between social class and education, and the interactions between IQ, class and education. However, much less attention has been paid to the relationship between social mobility and IQ and there are virtually no data dealing with the relationship between mobility and other personality traits.

In the late 1950s and early 1960s there was considerable debate on the causes of social class differences in intelligence. In essence there were two schools of thought – one group held the view that a large part of the heterogeneity in intelligence between social classes was due to innate or inherited differences, the other that the class differences were the result of differences in environmental conditions. Burt was the main protagonist of the hereditarean view and Floud and Halsey (see Halsey, 1959) for the environmental standpoint. In 1961 the British Journal of Statistical Psychology published an article by Burt (Burt, 1961) on intelligence and social class. In it Burt reported on two quite separate sets of analyses. In the first set of analyses he examined the relationship between social mobility and the intelligence of father–child pairs. He reported on the distribution of intelligence of adults in six occupational classes. There was a gradation in mean IQs from class I (higher professional) to class VI (unskilled) with an overall difference of nearly 55 IQ points between the means of classes I and VI. He then reported on the distribution of children based on the occupational classes of their fathers. Although there was still a gradation in means, the range for children was much smaller and the difference was only just over 28 IQ points between the means of classes I and VI. Furthermore, the distribution of scores within each occupational class was wider than that of their fathers.

Burt drew particular attention to this increased variability as evidence in support of the polygenic hypothesis. He wrote, 'Consider, for example, the lowest occupational class of all. Among the adults only 20 persons out of 261 have an intelligence above the general average: among the children as many as 76, nearly four times as many – a discrepancy of 56. Dr Floud and others, who hold as she does that differences in intelligence are due wholly to environmental advantages or disadvantages, can hardly maintain that the high level reached by these boys – all children of unskilled workers – results from the superior advantages which their home environment confer... On the Mendelian hypothesis, however, such apparent anomalies are exactly what we should anticipate, if the child's intelligence is determined mainly, or at any rate largely, by his genetic constitution, and if that in turn is the result of chance recombination of parental genes.'

Burt went on to calculate, for both adults and children, the maximum amount of social mobility which could occur if occupational class depended solely on intelligence. He calculated that if the distribution of intelligence of children was to replicate that of the adults, steady state mobility must take place and Burt calculated this basic mobility as 22%. In the second set of analyses Burt set out to see how intelligence, motivation, home background and educational achievement influenced inter-generational occupational mobility. He concluded that intelligence and motivation were the most important causal factors in defining individual social mobility, although education and home background also made some contribution.

Burt's paper led to increased research interest both in Britain and in the USA on the relationship between IQ scores of fathers and their adult sons and occupational mobility. However, by the mid-1970s many aspects of Burt's work, particularly that on separated identical twins, were coming under close scrutiny. Hearnshaw (1979) who wrote the official biography on Burt, was critical of the 1961 paper, criticizing the quality of the data and their use for elaborate statistical analysis. He wrote 'it was a dubious exercise, though perhaps "fraud" is too strong a word to use'. Dorfman (1978) was not so reticent and in the lead article in *Science* (September, 1978) he accused Burt of fabrication. Dorfman's paper was entitled 'The Cyril Burt question: new findings' with a subtitle 'The eminent Briton is shown beyond reasonable doubt, to have fabricated data on IQ and social class'. The summary to the article said, 'A detailed analysis of these data reveals, beyond reasonable doubt, that they were fabricated from a theoretical normal curve, from a genetic regression equation, and from figures published more than 30 years ago before Burt completed his surveys.'

Dorfman's allegations did not go unchallenged and there was correspondence in Science refuting Dorfman's claim of fabrication as well as Dorfman's replies (Dorfman 1979a,b,c; Rubin, 1979; Rubin & Stigler, 1979; Stigler, 1979). Two recent books on Burt (Joynson, 1989; Fletcher, 1991) both suggest that Dorfman's evidence was defective, a conclusion recently accepted by Jensen (1995). For a full discussion of the issues see Mascie-Taylor (1995). Here I will address only two issues. The first is that Dorfman argued that the intelligence scores of fathers and children were remarkably close to a theoretical normal distribution. He showed that the fit based on a mean of 100 and standard deviation of 15 was 'extraordinarily good'. Then he calculated the goodness of fit for a further 105 published distributions of which 33 were for IQ, 42 for height and 30 for weight and found that Burt's father and child data were significantly closer to a theoretical normal curve than the other 105 distributions. In order to calculate the goodness of fit we need to know the sample size. Nowhere in the 1961 paper does Burt reveal how many father–son pairs were studied, the only information given refers to the fathers' data 'the actual number in class I was nearer to one hundred and twenty than to three'. Taking the upper figure of 120 gives a multiplier of 40 (120/3). Hence $40 \times 1000 = 40,000$ father–son pairs. It is this number that Dorfman used in his calculations. But of course, the number could be anywhere between 62 and 120 and there is no reason to believe that Burt used the same weighting for all cells! If the total sample size was only 1000, then Burt's data are not so exceptional.

Dorfman is on much firmer ground when he predicted the mean IQs of sons in a class from the mean IQ of the fathers in the same class. He showed that the regression coefficient, after rounding to two decimal places, was 0.50 for all six occupational groups and furthermore the product-moment correlation between the six means was 0.999. These results are very unlikely to occur if the data are from what Burt defined as a *'pilot study'* where the data are 'crude' (M-T's emphasis). Mascie-Taylor (1995) concluded that even if Burt did not fabricate the data he was deliberately deceptive.

Since Burt's data cannot be trusted what have other surveys found? Young and Gibson (1963) published the results of a pilot study of 47 father–son pairs living in Cambridge, UK. They found a significant relationship between social mobility and IQ with the upwardly mobile sons having higher IQs than their fathers and downwardly mobile sons having lower IQs than their fathers. Later Gibson (1970) examined the IQ scores of Cambridge University scientists (aged between 25 and 34 years), their fathers and brothers. He found that the differences in IQ between the scientists and their fathers in each social class were related to the distance the scientists had moved up the social scale (Table 11.3). In the 22 families in which the IQs of the father and two male siblings were known, the upwardly mobile siblings tended to have higher IQs than the non-mobile or downwardly mobile siblings. In a subsequent paper, Gibson and Mascie-Taylor (1973) examined the relationship between IQ components and mobility and showed that both verbal and visuo-spatial IQ components were important in social mobility although verbal differences were greater.

Waller (1971) examined the IQ scores of 173 sons and their 131 fathers who were representative of the non-farming white population of Minnesota, USA . He was able to show a clear relationship between father–son IQ differences and social mobility and the correlation between father–son difference in IQ score and father–son difference in social position was positive ($r = +0.368$). Mascie-Taylor and Gibson (1978) examined the IQ scores of 193 male householders living in a Cambridge suburb. They found that the correlation coefficients between IQ scores (verbal, visuo-spatial and total) and present occupational status were significantly higher than the correlations between IQ scores and social class of origin, suggesting that intra-generational social mobility is positively related to IQ (Table 11.4). Parent–offspring data were available for 85 father–son pairs. Analysis of the IQ differences between fathers and sons in relation to their social class differences provided evidence for selective migration related to both

Table 11.3. *IQs of scientists and their fathers in different social classes*

Fathers' social class (Scientist's initial class)	N	Scientists (class I)		Fathers	
		Mean	SD	Mean	SD
I	18	129.7	4.24	130.2	8.00
II	36	128.6	6.25	122.6	10.90
IIINM	12	125.7	5.97	121.6	9.08
IIIM	10	123.4	5.93	113.1	10.72

See Table 11.1 for class classification; IIINM: non-manual; IIIM: manual.

Table 11.4. *Correlation coefficients between IQ components and social class in the Cambridge (UK) study*

	Verbal IQ	Visuo-spatial IQ	Total IQ
Based on father's class	−0.41	−0.29	−0.41
Based on present class	−0.58	−0.52	−0.64
Difference between correlation coefficients	$p < 0.05$	$p < 0.01$	$p < 0.01$

IQ components and total IQ scores. They also found a simple linear relationship between the extent of social mobility and the degree of difference between father's and son's IQ scores (Table 11.5) and this held for both upward and downward mobility. These calculations assumed that the distance between adjacent social classes was equal.

Not all studies have shown a relationship between IQ and social mobility. In their study of the Otmoor villages in Oxfordshire, UK, Gibson *et al.* (1983) also used the 'step' model described above. They divided the sample into locally-born and non-locally born groups and found little or no relationship between IQ scores and the extent of mobility for the locally-born. However, the non-locally born males demonstrated a similar pattern to that found in Cambridge with the upwardly mobile having, on average, higher mean IQs than the non-mobile or downwardly mobile groups. In addition, IQ scores were available on 85 father–son pairs and the relationship between father–son IQ difference and social mobility showed that the number of steps moved up or down the social scale by the sons is positively related to the magnitude of the difference between their IQ and that of their fathers.

Anderson *et al.* (1952) estimated the proportion of social mobility which was associated with differences in IQ score. Their estimates of about 22% were based upon an ideal situation in which the sons' social mobility restored the IQ/social class relationship to the paternal level.

Table 11.5. *Social mobility of sons and mean differences in IQ between them and their fathers in the Cambridge (UK) study*

Number of steps	N	Verbal IQ		Visuo-spatial IQ		Total IQ	
		Mean	SD	Mean	SD	Mean	SD
Upwardly mobile							
+1	15	+4.27	6.78	+4.47	14.49	+3.87	7.82
+2	16	+7.14	7.27	+10.00	14.33	+8.71	7.02
+3	6	+11.67	10.42	+14.17	13.66	+13.67	12.11
Downwardly mobile							
−1	10	+0.40	6.67	−9.80	13.47	−4.80	7.57
−2	6	−8.67	7.17	−15.83	10.83	−13.33	5.75
−3	1	−10.00	—	−22.00	—	−15.00	—

Although these studies suggest a strong relationship between IQ and mobility, other factors are involved. American sociologists, in particular, have been concerned with determining to what degree 'circumstances of birth condition subsequent status' (Duncan, 1966) or as Heath (1981) puts it 'who gets ahead and who stays behind?' Duncan used the technique of path analysis to examine the son's attainment. The path takes into account attainment at different stages of the life-cycle. It assumes that social origins and family circumstances – i.e. the father's educational and occupational status – will influence the son's educational level and that together they will impact on the kind of occupation the son will achieve when he first enters the labour market; his first occupation, educational level and social origins may all be expected to influence his subsequent career. Figure 11.1 presents some data from the Oxford University mobility study (Heath, 1981). The numbers (path coefficients) represent the magnitude of the direct effect of one variable on another after controlling for other variables already in the model. Figure 11.1 shows that the son's education has a major impact on his first occupation (0.50) and the first job influences his current job. In addition, education has a direct influence on the son's present job. Education therefore has a double impact, one acting directly, the other indirectly. Caution should be exercised in interpreting this model. Although there are arrows from the father's education and the father's occupation leading to the son's education there is also a third arrow which comes from 'out of the blue'. This arrow represents the residuals, that is all the unknown factors which affect a man's education but which have not been included in the model. These unknown factors include genetic and environmental factors (e.g. how well a person gets on with the teachers). Unfortunately the unknown factors are more important determinants of educational and

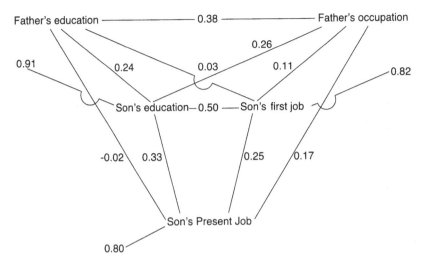

Figure 11.1. Path analysis of determinants of son's present job (Oxford University Mobility Study). Co-efficients represent the magnitude of direct effects of each variable on another after controlling for other variables in the model. (*Source:* Heath, 1981.)

occupational achievements than the known factors. Given these results it is hardly surprising that authors come to different conclusions as to the role of intelligence in shaping class structure.

Relationship between social mobility and anthropometric characters

Social class differences in height and weight of children and adults have been reported in many studies from a wide variety of populations (Goldstein, 1971; Bogin and MacVean, 1978; Bielicki *et al.*, 1981; Mascie-Taylor, 1984; Johnston, 1986; Cernerud, 1995). For example, Schreider (1964) showed that French peasants differed from other workers in many characters and Cliquet (1968) found that certain physical varaibles were associated with upward mobility. Social class differences were observed in all children regardless of whether they were from developed, socialist or capitalist countries, but in all studies, the higher the social class, the taller and heavier, on average the children.

Although most of the work on social mobility has involved males, a study conducted in Aberdeen in the 1950s suggested that taller women raised their social class on marriage (Illsley, 1955). In this study, classification of womens' social class before marriage was on the basis of their

Table 11.6a. *Mean stature and weight by social class for British boys at ages 7, 11 and 16 from the National Child Development Study*

Variable	Social class				
	I	II	III	IV	V
Stature (cm)					
7 years	124.4	124.2	122.6	122.1	120.6
11 years	146.1	145.2	143.8	143.0	141.7
16 years	172.5	171.7	170.0	169.4	167.8
Weight (kg)					
7 years	24.5	24.8	24.0	23.9	23.2
11 years	36.7	36.7	36.0	35.4	34.7
16 years	59.4	59.9	58.9	58.4	57.0

See Table 11.1 for class classification.

father's occupation, and their class at the time of the interview, on their husband's occupation. Data from a 1984 study ($N = 10,000$) in Britain, (Knight and Eldridge, 1984) of adult heights and weights lend some support to the earlier finding with the shortest women (those less than 155 cm tall) least likely to raise their social class through marriage, while, to a lesser extent, the tallest were more likely to raise their class in this way.

Despite the wealth of information, little is known of the influence of social class from longitudinal data or of the effect of social mobility on growth. In order to redress this balance, Lasker and Mascie-Taylor (1989) used data from the National Child Development Study (NCDS; Butler and Bonham, 1963; Davie *et al.* 1972; see Chapter 3), a longitudinal study of all children born in England, Wales and Scotland during March 1958 who were followed up when they were aged 7, 11 and 16 years, to attempt to answer these questions.

Table 11.6a presents the mean stature and weight for boys at ages 7, 11 and 16 by social class (Registrar-General's classification). There is a clear downward trend in means from class I (professional) to class V (unskilled) at all three ages for stature but not quite so obviously for weight. The impact of social mobility was quantified by the number of social class changes between each interval, i.e. between birth and 7 years, 7 to 11 years and 11 to 16 years, using the occupation of the male head of household. Analyses of variance for height and weight were run which took into account the social class of origin, the sex of the child and the extent of social mobility (Table 11.6b). About two-thirds of the families remained in the same social class from each study to the next. Those children who rose in social class, for example from V to IV or III to II, have, on average, lower stature and weight for their new social class but are taller and heavier

Table 11.6b. *Three way ANOVA (analysis of variance) of social class, change in social class, and sex on stature and weight of British children at 7, 11 and 16 years of age from the National Child Development Study*

	17 years		11 years		16 years	
	F	p	F	p	F	p
Stature						
Social class	60.5	< 0.001	57.3	< 0.001	39.3	< 0.001
Social mobility	11.7	< 0.001	9.3	< 0.001	8.3	< 0.001
Sex	55.8	< 0.001	28.0	< 0.001	3311.3	< 0.001
Weight						
Social class	24.7	< 0.001	13.8	< 0.001	4.2	< 0.002
Social mobility	5.3	< 0.005	0.6	NS	4.0	NS
Sex	37.3	< 0.001	69.9	< 0.001	443.0	< 0.001

NS = not significant

Table 11.6c. *Three way ANOVA (analysis of variance) of effects of social class, changes in social class and sex on growth in stature and weight of British children between 7 and 16 years of age from the National Child Development Study*

	Growth in stature				Growth in weight			
	7–11		11–16		7–11		11–16	
	F	p	F	p	F	p	F	p
Class	2.6	< 0.03	0.9	NS	4.1	< 0.002	2.2	< 0.07
Mobility	1.2	NS	0.5	NS	0.6	NS	1.3	NS
Sex	320	< 0.001	6666	< 0.001	267.3	< 0.001	1560	< 0.001

NS: not significant

than children in their old social class. (Cernerud, 1995, reported similar findings based on a representative sample of children in Stockholm, Sweden).

However, a large fraction of the differences in size according to social class seen at later ages (11 and 16 years) is already achieved by age 7. Analyses of variance showing the effects of social class, changes in social class and sex (Table 11.6c) on growth in stature and weight demonstrate this point (changes in height and weight for the two periods, 7–11 years and 11–16 years were computed as measures of growth in height and weight respectively). Growth in stature and weight is only significant in the period 7–11 years and not thereafter, and social mobility is not important in either time period.

Bielicki and Waliszko (1992) recently examined the relationship between stature and upward mobility in a large study of Polish military conscripts (aged 19 years). They created 10 different, non-overlapping social groups of subjects. Each group was homogeneous with respect to four criteria – parental education, parental occupation, urban–rural residence and number of siblings. The 10 groups are shown in Table 11.7. Within each social group conscripts were selected who differed in educational level. Level A conscripts were those who, at the time of examination were secondary school students or graduates. Level B conscripts comprised those who had terminated their education at the level of primary school or had gone to basic trade schools, i.e. had not continued into secondary education. Finally they compared the statures of conscripts from the two educational levels within each social group. The results showed that within all 10 social groups the 19-year-olds who were in level A, were, on average, taller than those who were in level B. Indeed in 8 of the 10 groups the differences between levels A and B conscripts were significant.

Social mobility and genetic variation

Bielicki and Waliszko (1992) put forward three hypotheses to account for their results. The first, called the 'genetic linkage' hypothesis, supposes that tallness is linked to some intellectual attribute which enhances educational attainment or that there is some pleiotropic effect of polygenes affecting stature and IQ. They accept that these explanations are theoretically possible but do not know of any evidence to support such a conclusion. The second hypothesis they call the 'trump card' hypothesis. Tallness is a desirable attribute in many social situations and tallness may 'aid in climbing the social scale'. Although plausible they discount this hypothesis as explaining their results. Instead they propose the 'third factor' hypothesis, whereby the observed differences are brought about by a 'third factor' of a purely environmental nature.

These three hypotheses, as Bielicki and Waliszko point out, have implications for the social class differences in stature. If the first hypothesis, and to a lesser extent, the second hypothesis were correct then there would be genetic implications since the observed phenotypic differences between classes would also reflect genotypic differences too. However, the 'third factor' hypothesis would have no genetic consequences.

What is the evidence for genetic differences between social classes? Most genetic studies in the past have failed to record social class data but there are a few exceptions. Dawson (1964) studied the occupational classes and ABO and Rhesus blood groups in Ireland and found little evidence for

Table 11.7. *Stature (means and standard deviations) of 19-year-old Polish conscripts from 10 socially homogeneous groups, by education of the conscript*

| Social Group | Education of conscript | | | | | | Difference |
| | Level A | | | Level B | | | A–B |
	Mean	SD	N	Mean	SD	N	
1. Urban, parents secondary school, two children	177.31	6.28	456	176.05	6.15	202	+ 1.26*
2. Urban, parents secondary school, three children	177.17	6.04	125	175.06	5.66	97	+ 2.14*
3. Urban, parents primary or trade school, father skilled worker, grandfather worker, two children	176.61	5.81	311	175.30	6.37	658	+ 1.31**
4. Urban, parents primary or trade school, father skilled worker, grandfather worker, three children	176.24	6.23	136	173.84	6.24	395	+ 2.40**
5. Urban, parents primary or trade school, father skilled worker, grandfather peasant, two children	176.58	6.62	242	175.30	6.13	446	+ 1.28*
6. Urban, parents primary or trade school, father skilled worker, grandfather peasant, three children	176.21	6.06	146	175.41	6.36	309	+ 0.80
7. Rural, parents primary or trade school, father skilled worker, grandfather peasant, two children	176.50	6.14	98	174.94	6.02	331	+ 1.56*
8. Rural, parents primary or trade school, father skilled worker, grandfather peasant, three children	175.49	6.55	114	173.87	5.86	469	+ 1.62**
9. Rural, father primary school, father and grandfather peasant, two children	174.53	5.37	102	174.20	5.90	450	+ 0.33
10. Rural, father primary school, father and grandfather peasant, three children	175.23	5.92	138	173.67	5.96	822	+ 1.56**

Urban residence: cities, population 100,000–500,000; rural residence: villages.
* $p < 0.05$; ** $p < 0.01$; N: number of individuals.
Level A: conscripts at time of examination who were secondary school students or graduates.
Level B: conscripts at time of examination who had terminated education at primary school or who had attended basic trade school, but not secondary school.

differences between classes. Wheatcroft (1973) did, however, find some class differences for the phenyl-thio-carbamide (PTC) tasting status although it is difficult to assess the statistical significance of her results because numerical data were not provided.

The most compelling evidence for genetic differences between social classes comes from the study of Cartwright and his colleagues (1978). They examined five polymorphic systems (serum haptoglobin (Hp), serum transferrrin (Tf), red blood cell isoenzymes of acid phosphatase (AP), adenylate kinase (AK) and esterase D (EsD)) in a sample of 999 individuals from Nottinghamshire, UK and found evidence of social class differences between polymorphic markers. They also examined the effects of social mobility and found upward social movement was associated with genotypic groupings (Table 11.8a). Further evidence of a relationship between polymorphic marker frequencies and social classes was provided by Mascie-Taylor and colleagues (1985) who examined the associations between eight polymorphic markers and mobility in the Otmoor villagers and found a significant relationship with haptoglobin. There were higher frequencies of Hp1-1 homozygotes and lower frequencies of Hp2-2 homozygotes in the downwardly mobile group (Table 11.8b) but no differences were observed in the upwardly mobile group. In addition Mascie-Taylor and others (1985) went on to examine the relationship between IQ and the polymorphic markers. They found an association between haptoglobin and visuo-spatial IQ which became stronger after allowing for the possible effects on IQ of sex, geographic location, social class and family structure. This procedure removed the evidence for all the other associations except for one, between Kell blood group and IQ.

In 1983, Beardmore and Karimi-Booshehri reported an association between the ABO blood group phenotype and social class in Britain. They found that in both sexes the A phenotype was significantly more frequent than the O phenotype in social classes I and II, the converse being true for classes III, IV and V. However, attempts to replicate their findings using other British datasets failed (Mascie-Taylor and McManus, 1984; Golding *et al*, 1984). Mascie-Taylor and McManus went on to examine the relationship between social mobility of women and ABO phenotype using the NCDS dataset. They classified the women (mothers of children in the follow-up) on the basis of her own and her father's social class, as either upwardly mobile, downwardly mobile or non-mobile. They found no evidence that inter-generational social mobility was related to ABO phenotype. Given these apparent contradictions between studies, the journal, *Nature* concluded 'the general conclusion seems to be that there can be no general conclusion'.

Table 11.8a. *Gene frequencies and social class movement*

Polymorphism	Up 1 + classes	No change	Down 1 + classes
Hp 1	0.4300 ± 0.036	0.3915 ± 0.043	0.4100 ± 0.052
2	0.5700	0.6085	0.5900
Tf C	0.9925 ± 0.006	0.9910 ± 0.007	0.9790 ± 0.015
B/D	0.0075	0.0090	0.0210
AP Pa	0.3590 ± 0.027	0.3215 ± 0.028	0.3320 ± 0.038
Pb	0.5970	0.6340	0.6035
Pc	0.0440	0.0445	0.0645
AK 1	0.9645 ± 0.0101	0.9825 ± 0.008	0.9560 ± 0.016
2	0.0355	0.0175	0.0440
EsD 1	0.8740 ± 0.018	0.8555 ± 0.021	0.9030 ± 0.023
2	0.1260	0.1445	0.0970
Maximum numbers	338	287	159

Hp: serum haptoglobin; Tf: serum transferrin; AP: red blood cell isoenzymes of acid phosphatase; AK: adenylate kinase; EsD: esterase D.

E^2 *statistic for social movement data*

	1	2
1 Up 1 + classses		
2 No change	0.04013	
3 Down 1+ classes	0.04001	0.01682

Table 11.8b. *Social mobility and haptoglobin genotypes among males*

Social mobility	Haptoglobin			χ^2
	2–2	2–1	1–1	
Up	33	49	14	Up/stable = 0.3 ns
Stable	40	51	17	
Down	12	31	18	Stable/down = 7.8 < 0.05

Relationship between social class, social mobility and other characters

Jenner *et al.* (1980) found differences in catecholamine excretion rates of men living in the Otmoor region according to day of the week and occupation. Urinary catecholamines have been used as indicators of job stress. In their populational study Jenner and his colleagues found that adrenaline excretion rates were higher on work days than rest days. In addition, rates for non-manual occupations were higher than for manual occupations and the rates for professional and managerial occupations were, in general, the highest of all. Brown (1981, 1982) compared the 24-hour urine catecholamine/creatinine ratios among Filipino immigrants in Hawaii in relation to their degree of acculturation and found significantly higher noradrenaline (norepinephrine) and adrenaline (epinephrine) ratios among individuals classified as intermediately acculturated.

Geographic migration

As migration is a means of transferring genes from one population to another, it has been a much studied area in genetics. Such a transfer (gene flow) can lead to changes in gene frequencies and thus it is one of the major mechanisms of evolution (Roberts, 1988). The effects of immigration and emigration can be demonstrated.

The Black Caribs of Central America provide a good example of the genetic effects of immigration (Crawford, 1984). Crawford has used genetic survey data, census and historical records to show how the present population evolved. The earliest records go back to the start of the 1500s and the population at that time comprised Carib and Arawak peoples located on St Vincent Island of the Lesser Antilles. An African component was added to the gene pool between 1517 and 1646 (through, for example slaves). Some of the population, now known as the Black Caribs, settled in British Honduras, Guatemala and Nicaragua. Thereafter they dispersed along the coast of Central America and further admixture took place with input of European genes through the Creoles. Consequently the current Black Carib population contains evidence of an African (through haemoglobin S allele, the Fy^4 gene and $Gm^{z,a;b}$ haplotype), European (Gm haplotype $Gm^{f;b}$) and American Indian (Diego blood group and albumin Mexico) contributions. Crawford has estimated that overall the Africans have contributed between 60–80% of the gene pool, Europeans, 16% and American Indian, 17–40%. Thus the gene pool appears to be predominately from immigrant rather than original sources (this conclusion ignores the fact that some polymorphisms are adaptive, and they cannot provide a good measure of origins, since gene frequencies would have been changing in the last 400 years).

The population shows a very high level of polymorphic variability resulting from the varied ancestry of the migrants. In addition, the migrants have brought with them genes which have enhanced the genetic adaptability of the population. The area where the Black Caribs live was, until recently, a heavily malarious area and the migrants brought with them genes that confer some advantage against malaria (genotypes AS and AC for haemoglobin, FyFy homozygote of the Duffy blood group) which were not present in the American Indian population.

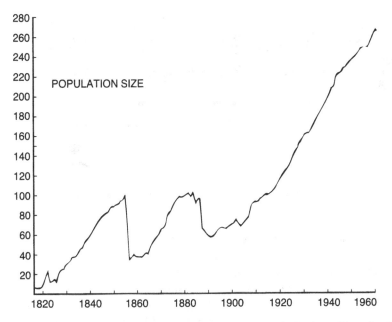

Figure 11.2. The size of the population of Tristan da Cunha on December 31 of each year from 1816 to 1960. (*Source:* Roberts, 1968.)

The classic work on the islanders of Tristan da Cunha (Roberts, 1968) can be used to show the genetic effects of emigration. The population on Tristan da Cunha was founded in 1812 and as Figure 11.2 shows the population increased steadily until it reached 103 in 1855. Then, with the death of William Glass one of the founder members and with the persuasion of the pastor, 70 inhabitants left in the space of 15 months and the population was reduced to 33. The population increased again until an accident at sea killed 15 adult males on November 28, 1885. Many of the widows and their families subsequently left the island and by 1891 the population fell to 59. The impact of these two bottlenecks on the genetic constitution of the population can be seen from Figure 11.3 when the contributions to the gene pool change substantially before and after the bottlenecks.

The impact of immigration and emigration in changing gene frequencies can also be seen in Britain where there is marked geographical heterogeneity for the ABO blood groups (Kopec, 1970). There are north–south and east–west clines with the frequency of blood group O tending to decline from Scotland to southern England and from Wales to East Anglia. Blood group A shows the opposite geographical trend.

Britain experiences a large amount of geographical migration. If migra-

Contribution to Gene Pool

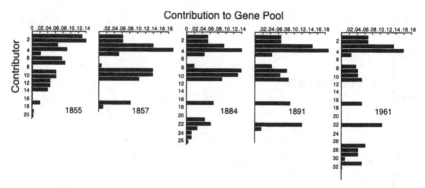

Figure 11.3. Contributions of particular ancestors to the gene pool in Tristan da Cunha before and after 1855 and 1885. (*Source*: Roberts, 1968.)

tion is at random with respect to blood groups then over time the present clinal pattern will disappear and genetic homogeneity will result. The NCDS provides information on the ABO and Rhesus (positive and negative only) blood groups of mothers since all of them were typed in 1958 when their children were born. The children and their families were restudied periodically and in 1974 the geographical location was known for 8850 mothers. The data were grouped by county and the centre of population ascribed to each. The grid coordinates in kilometres, easting and northing, for each centre were determined and the frequency of each blood phenotype A, O, B + AB (they were merged since there were so few ABs), Rh + ve and Rh-ve regressed by a routine which weighted for county sample size against longitude (easting) and latitude (northing). The results tend to show a flattening of the curves from 1958 to 1974 (e.g. A phenotype on easting) thus showing the homogenizing effect of migration (Mascie-Taylor and Lasker, 1987).

Assuming this pattern of migration continues, how long will it take to reach homogeneity for the ABO blood group phenotypes? In order to answer the question the same dataset was used but the geographical location of the mother was reassigned to one of the eleven standard regions as defined by the Registrar-General. Knowing the mother's region of birth and her residence in 1958 a migration matrix as shown in Table 11.9 was drawn up. Using a mathematical model, the time in generations (using 25 years to represent a generation) required to reduce the existing genetic differences between any two regions to homogeneity was calculated (based on 99% similarity between regions). A cluster analysis of times to homogeneity based on the inter-regional migration matrix is presented in the form of a dendrogram (Figure 11.4). The analyses of movement patterns thus demonstrate that migration is gradually eroding the observed clinal variation for ABO blood groups and that provided current movement

Table 11.9. *Inter-regional maternal migration matrix*

		NW	N	EWR	NM	E	SE	S	SW	MID	WA	SC
							Region of residence in 1958					
	NW	566	11	17	4	5	12	4	3	9	9	9
	N	17	571	41	3	2	11	2	2	1	4	17
	EWR	26	27	542	15	6	11	5	6	4	7	18
	NM	15	11	28	407	17	33	7	7	24	6	15
Region	E	13	12	22	14	303	159	16	6	8	8	9
of	SE	36	31	23	19	58	1002	40	30	25	19	19
birth	S	17	4	10	7	17	98	240	22	14	14	3
	SW	5	4	13	4	9	48	15	330	18	14	14
	MID	24	11	9	15	9	27	6	11	550	18	11
	WA	17	4	3	6	2	13	4	8	7	359	2
	SC	5	7	6	7	3	12	3	3	4	7	766

Regions: 11 standard regions as defined by the Registrar-General.

patterns continue, within eight generations, or approximately 200 years, virtually all genetic heterogeneity for ABO blood groups within England and Wales will have disappeared. However, another 12 generations or 300 years will elapse before Britain as a whole reaches genetic homogeneity (these conclusions do not take into account possible new immigration during this period).

There is also evidence of differences in average IQ between regions of Britain (Lynn, 1979). The 1946 cohort (children born in the first week of March 1946) of children tested at 8, 11 and 15 years of age showed highly significant differences between regions with higher means in London and South-East England (101.9) and the lowest mean in Scotland (97.3). The 1958 cohort confirmed these findings. Lynn (1977) has also suggested selective emigration as the principal cause of the 3–4 IQ points difference between Scotland and England. Mascie-Taylor (1985) analysed the movement patterns of children in the 1958 cohort between birth and 11 years of age in relation to the 11 standard regions described earlier. He compared the IQ scores of out-migrants (emigrants) from a region with the non-

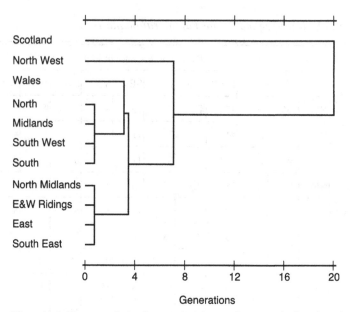

Figure 11.4. Cluster analysis of generation times to homogeneity based on the ABO blood group system in the UK.

migrants (sedentes) and found that in all regions the out-migrants had, on average, a higher IQ than the sedentes. Next he compared the IQ of in-migrants (immigrants) to their new region with the IQ of that region and showed that in all regions immigrants had a higher IQ. Finally the net effect of migration (out–in) was computed. For Scotland there was a difference in IQ between out and in of − 3.9 IQ points which supports Lynn's conclusion.

The impact of migration on anthropometric variables has been studied. Shapiro (1939) showed that the Japanese who migrated to Hawaii were taller and heavier than sedentes living in the same village from which the migrants originated. Furthermore the Hawaiian born offspring of the migrants (most of whom had married Japanese women) grew up to be taller and more linear in build than their migrant parents or the sedentes. This study is an example of rural–urban migration and Bogin (1988) has shown that such migrations result in biological changes in the growth and development, fertility, morbidity and mortality of migrants and their descendants. In general, children from developing countries living in cities are taller and mature earlier than do children from rural areas. However, migration to a slum or squatter area is not beneficial and studies have shown that the growth of children living in slums is not significantly different from children living in impoverished rural areas.

Finally, there has been considerable discussion about the impact of migration on health and disease but since other chapters in this volume are devoted to health, only brief mention will be made here. Studies include the impact of migration as a source of disease transmission and the relationship between migration, industrial diseases and mental health. The extensive anthropological studies on the Tokelauan population have shown that migrants to New Zealand tend to have an excess prevalence and/or incidence of many chronic diseases including diabetes, gout, bronchitis, varicose veins and asthma compared to their island dwelling compatriots (Ostbye *et al.*, 1989). Migrants were also at higher risk for coronary heart disease than non-migrants.

In Britain an unusual set of circumstances led to a dangerous and potentially debilitating deficiency disease in migrants from South Asia to Britain. In the early 1970s reports began accumulating of cases of rickets and osteomalacia among immigrants from India and Pakistan. Serum assays of 25-hydroxycholecalciferol were used to estimate that 21% of South Asian children were affected and that vitamin D deficiency was the major factor leading to the bone disorders. What were the reasons for the vitamin D deficiency? Inadequate dietary vitamin D intake and failure to synthesize cholecalciferol in the skin due to the inadequate sunlight and the heavily pigmented skin were suggested. Although these may be contributory factors, it is now recognized that removing chapattis from the diet produced a dramatic improvement. The high phytate content of unleavened chapattis inhibited calcium mobilization and thus increased the likelihood of developing rickets (Figure 11.5).

Interaction between social and geographic mobility

There have been a few instances where the joint effects of geographical and social migration have been studied. Mascie-Taylor (1984) used the British NCDS to examine migrational changes in relation to height and weight. The results show that there was a tendency for simple additivity of the effects of social mobility and geographic migration on height and weight and no indication of any significant interactive effects. For instance the geographically mobile are, on average, 11.2 mm taller and the upward socially mobile + 3.3 mm above the grand mean. Therefore the mean of those who are both geographically and upwardly mobile should be $11.2 \pm 3.3 = 14.5$ mm whereas the observed result was 13.2 mm. In a subsequent paper, Mascie-Taylor (1985) related IQ differences to the dual effects of social and geographical migration. He also showed that simple additivity occurred.

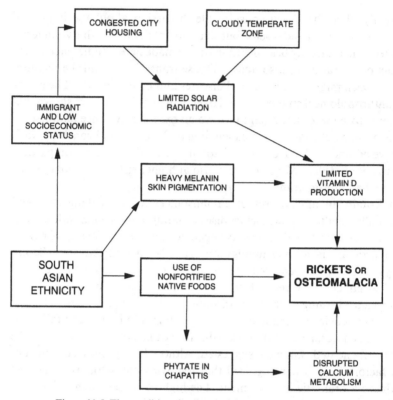

Figure 11.5. The conditions leading to rickets and osteomalacia in South Asian migrants to the UK.

Conclusion

This review has attempted to bring together various studies which have looked at the relationships between biological characteristics and migration. The impact of migration varies – on the one hand, it can promote and maintain phenotypic differences between groups, as the examples of height and IQ with social class have demonstrated. Such migration is selective. On the other hand, if migration is random, then homogeneity can occur, as the eroding of the clinal variation in ABO phenotypes in Britain demonstrates. Understanding of the role of migration is therefore an important element in adaptation, genetic and evolutionary studies.

References

Anderson, C. A., Brown, J. C. & Bowman, M. J. (1952). Intelligence and occupational mobility. *Journal of Political Economy* **40**, 218–39.

Basler, A. (1927). Ueber die Einflusse der Lagerung von Sauglingen auf die Bleibende Schadelform. *Zeitschrift für Morphologie und Anthropologie* **26**, 247–55.

Beardmore, J. A. & Karimi-Booshehri, F. (1983). ABO genes are differentially distributed in socio-economic groups in England. *Nature* **303**, 522–4.

Bielicki, T. & Waliszko, H. (1992). Stature, upward social mobility and the nature of statural differences between social classes. *Annals of Human Biology* **19**, 589–93.

Bielicki, T., Szczotka, H. & Charzewski, J. (1981). The influence of three socio-economic factors on body weight in Polish military conscripts. *Human Biology* **53**, 543–55.

Boas, F. (1910). Changes in bodily form of descendants of immigrants. *Senate Document 208, 61st Congress, 2nd Session, Washington, DC.*

Bogin, B. (1988). Rural-to-urban migration. In *Biological Aspects of Human Migration*, ed. C. G. N. Mascie-Taylor & G. W. Lasker, pp. 90–129. Cambridge: Cambridge University Press.

Bogin, B. & MacVean, R. B. (1978). Growth in height and weight of urban Guatemalan primary school children of low and high socioeconomic class. *Human Biology* **50**, 477–87.

Brown, D. E. (1981). General stress in anthropological fieldwork. *American Anthropologist* **83**, 74–92.

Brown, D. E. (1982). Physiological stress and culture change in a group of Filipino-Americans, a preliminary investigation. *Annals of Human Biology* **9**, 553–63.

Butler, N. R. & Bonham, D. G. (1963). *Perinatal Mortality*. Edinburgh: Livingstone.

Burt, C. (1961). Intelligence and social mobility. *British Journal of Statistical Psychology* **14**, 3–24.

Cartwright, R. A., Hargreaves, H. J. & Sunderland, E. (1978). Social identity and genetic variability. *Journal of Biosocial Science* **10**, 23–33.

Cernerud, L. (1995). Height and social mobility. A study of the height of 10-year-olds in relation to socio-economic background and type of formal schooling. *Scandinavian Journal of Social Medicine* **23**, 28–31.

Cliquet, R. L. (1968). Social Mobility and the anthropometric structure of populations. *Human Biology* **40**, 17–24.

Crawford, M. H. (1984). *Black Caribs: Case study in Biocultural Adaptation*. New York: Plenum.

Davie, R., Butler, N. R. & Goldstein, H. (1972). *From Birth to Seven*. London: Longman.

Dawson, G. W. P. (1964). The blood group frequencies in some occupational groups in County Dublin. *Annals of Human Genetics* **22**, 315–20.

Dorfman, D. D. (1978). The Cyril Burt Question: new findings. *Science* **201**, 1177–86.

Dorfman, D. D. (1979a). Correspondence, *New Statesman* (Feb.), 150.

Dorfman, D. D. (1979b). Reply to Stigler and Rubin. *Science* **204**, 246–54.

Dorfman, D. D. (1979c). Burt's data: Dorfman's Analysis. *Science* **206**, 142–4.

Duncan, O. D. (1966). Path analysis: sociological examples. *American Journal of Sociology* **72**, 1–16.

Ewing, J. F. (1950). Hyperbrachycephaly as influenced by cultural conditioning.

218   *C. G. N. Mascie-Taylor*

Papers of the Peabody Museum of American Archaeology and Ethnology, Harvard University **23** (2).

Fletcher, R. (1991). *Science, Ideology and the Media: the Cyril Burt Scandal.* New Brunswick: Transacation Publishers.

Gibson, J. B. (1970) Biological aspects of a high socio-economic group. I. IQ, education and social mobility. *Journal of Biosocial Science* **2**, 1–16.

Gibson, J. B. & Mascie-Taylor, C. G. N. (1973). Biological aspects of a high socio-economic group. II. IQ components and social mobility. *Journal of Biosocial Science* **5**, 17–30.

Gibson, J. B., Harrison, G. A., Hiorns, R. W. & Macbeth, H. M. (1983). Social mobility and psychometric variation in a group of Oxfordshire Villages. *Journal of Biosocial Science* **15**, 193–205.

Glass, D. V. (1954). (Ed). *Social Mobility in Britain.* London: Routledge and Kegan Paul.

Golding, J., Hicks, P. & Butler, N. R. (1984). Blood groups and socio-economic class. *Nature* **309**, 396–7.

Goldstein, H. (1971). Factors influencing the height of seven year old children – results from the National Child Development Study. *Human Biology* **43**, 92–111.

Goldthorpe, J. H. & Hope, K. (1974). *The Social Grading of Occupations: A New Approach and Scale.* Oxford: Oxford University Press.

Halsey, A. H. (1959). Class differences in general intelligence. *British Journal of Statistical Psychology* **12**, 1–4.

Hearnshaw, L. S. (1979). *Cyril Burt Psychologist.* London: Hodder and Stoughton.

Herrnstein, R. J. & Murray, C. (1994). *The Bell Curve.* New York: Free Press.

Heath, A. (1981). *Social Mobility.* Glasgow: William Collins.

Hope, K. (1972). *The Analysis of Social Mobility.* Clarendon Press: Oxford.

Illsley, R. (1955). Social class selection and class differences in relation to stillbirths and infant deaths. *British Medical Journal* **24**, 1520–4.

Jenner, D. A., Reynolds, V. & Harrison, G. A. (1980). Catecholamine excretion rates and occupation. *Ergonomics* **23**, 237–46.

Jensen, A. R. (1995). IQ and science: the mysterious Burt affair. In *Cyril Burt, Fraud or Framed*, ed. N. J. Mackintosh, pp. 1–12. Oxford: Oxford University Press.

Johnston, F. E. (1986). Somatic growth of the infant and preschool child. In *Human Growth*, ed. F. Falkner & J. M. Tanner, vol. 2, pp. 3–24. New York: Plenum Press.

Joynson, R. B. (1989). *The Burt Affair.* London: Routledge.

Knight, I. & Eldridge, J. (1984). *The Heights and Weights of Adults in Great Britain.* HMSO, London.

Kopec, A. C. (1970). *The Distribution of the Blood Groups in the United Kingdom.* Oxford: Oxford University Press.

Lasker, G. W. & Mascie-Taylor, C. G. N. (1989). Effects of social class differences and social mobility on growth in height, weight and body mass index in a British cohort. *Annals of Human Biology* **16**, 1–8.

Lipset, S. M. & Bendix, R. (1959). *Social Mobility in Industrial Society.* Berkeley: University of California Press.

Lynn, R. (1977). Selective emigration and the decline of intelligence in Scotland. *Social Biology* **24**, 173–82.

Lynn, R. (1979). The social ecology of intelligence in the British Isles. *British Journal of Social and Clinical Psychology* **18**, 1–12.

Mascie-Taylor, C. G. N. (1984). The interaction between geographical and social mobility. In *Migration and Mobility*, ed. A. J. Boyce, pp. 161–78. London: Taylor and Francis.

Mascie-Taylor, C. G. N. (1985). Biosocial correlates of IQ. In *The Biology of Intelligence*, ed. C. Turner, pp. 99–127. Nafferton Books.

Mascie-Taylor, C. G. N. (1995). Intelligence and social mobility. In *Cyril Burt, Fraud or Framed*, ed. N. J. Mackintosh, pp. 70–94. Oxford: Oxford University Press.

Mascie-Taylor, C. G. N. & Gibson, J. B. (1978). Social Mobility and IQ components. *Journal of Biosocial Science* **10**, 263–76.

Mascie-Taylor, C. G. N., Gibson, J. B., Hiorns, R. W. & Harrison, G. A. (1985). Associations between some polymorphic markers and variation in IQ and its components in Otmoor Villagers. *Behavior Genetics* **15**, 371–83.

Mascie-Taylor, C. G. N. & Lasker, G. W. (1987). Migration and changes in ABO and Rh blood group clines in Britain. *Human Biology* **59**, 337–44.

Mascie-Taylor, C. G. N. & McManus, I. C. (1984). Blood groups and socio-economic class. *Nature* **309**, 395–6.

Ostbye, T., Welby, T. J., Prior, I. A. M., Salmond, C. E. & Stokes, Y. M. (1989). Type-2 (non-insulin-dependent) diabetes mellitus, migration and westernization: the Tokelaua Island migration study. *Diabetologia* **32**(8), 585–90.

Roberts, D. F. (1968). Genetic effects of population size reduction. *Nature* **220**, 1084–8.

Roberts, D. F. (1988). Migration in the recent past: societies with records. In *Biological Aspects of Human Migration*, ed. C. G. N. Mascie-Taylor & G. W. Lasker, pp. 41–69. Cambridge: Cambridge University Press.

Rubin, D. B. (1979). Reply to Dorfman. *Science* **204**, 245–6.

Rubin, D. B. & Stigler, S. J. (1979). Dorfman's Data Analysis. *Science* **205**, 1204–6.

Schreider, E. (1964). Investigations into social stratification of biological characters. *Yearbook of Physical Anthropology* **12**, 184–201.

Shapiro, H. L. (1939). *Migration and Environment*. Oxford: Oxford University Press.

Stigler, S. J. (1979). Reply to Dorfman. *Science* **204**, 242–5.

Waller, J. H. (1971). Achievement and Social Mobility: relationships among IQ score, education, and occupation in two generations. *Social Biology* **18**, 252–9.

Westergaard, J. & Resler, H. (1975). *Class in a Capitalist Society: A Study of Contemporary Britain.* London: Heinemann.

Wheatcroft, P. (1973). Biological variables and social class in Birmingham. In *Genetic Variation in Britain*, ed. D. F. Roberts & E. Sunderland, pp. 277–85. London: Taylor and Francis.

Young, M. & Gibson, J. B. (1963). In search of an explanation of social mobility. *British Journal of Statistical Psychology* **16**, 27–36.

12 How female reproductive decisions cause social inequality in male reproductive fitness: evidence from eighteenth- and nineteenth-century Germany

ECKART VOLAND AND ATHANASIOS CHASIOTIS

Introduction

Social inequality between men leads to their inequality in mating success in traditional (Borgerhoff Mulder, 1987), historical (Voland and Engel, 1990) and modern (Pérusse, 1993) societies. Socially successful men are married to more women in polygamous societies (Borgerhoff Mulder, 1987) and have more extramarital affairs (Flinn, 1986). They have higher chances of remarriage after becoming widowers (Boone, 1986) or following divorce (Haskey, 1987) and marry the youngest and most fertile brides (Røskaft et al., 1992). Clearly, differential male mating success must necessarily be reflected by women's mating preferences in favour of socially successful men which in fact is well confirmed by psychological research (Buss, 1994; Waynforth and Dunbar, 1995; Bereczkei and Csanaky, 1996). These female preferences cannot be completely without any influence on male reproductive success, which is why the question arises of how and to what extent female mating and reproductive decisions cause social inequality in male reproductive fitness.

This question is not easy to answer, because reproductive fitness is hard to measure, for both theoretical and practical reasons. Human conduct can lead to extremely long lasting fitness effects. If, for example, fitness enhancing resources are passed from generation to generation, this can easily exert an impact on the reproductive performance of future descendants yet to come – long after one's own death. Clearly, it is highly questionable when and how to measure reproductive fitness. The second problem that biologists face when trying to deal with fitness differentials is the indirect fitness that is allocated by kin selection and that is even more difficult to handle.

220

Social group differences in reproductive fitness and the reasons for their origin are, however, not merely of empirical interest since they also play a predominant role in the socio-biological and behavioural ecology theory. William Irons (1979) proposed a hypothesis according to which there should regularly be a correlation between cultural and reproductive success in human populations as a consequence of which a demographically more or less constant and pressing over-reproduction of the socially elite classes occurs. Ever since Irons (1979) published this hypothesis, which he was able to substantiate by data on the positive correlation of personal wealth with individual reproductive success among Yomut Turkmen, major research efforts have been undertaken to verify or dismiss the widely disputed assumption of the general applicability of Irons' main findings to other societies. Recent research now suggests that in many historical and contemporary non-industrial populations, the cultural success/reproductive success link actually has a functional background (for reviews, see Betzig, 1986; Cronk, 1991; Irons, 1993; Low, 1993; Pérusse, 1993). The doubts raised are however justified as to their applicability to industrialized societies (Kaplan et al., 1995; Pérusse, 1993; Vining, 1986).

Even in pre-modern Germany there are indications for the validity of Irons' hypothesis from the Ostfriesland region known as the Krummhörn (Voland, 1990; Klindworth and Voland, 1995). However, it is unclear as to what extent this individual finding can be generalized with respect to other regions and epochs. In this chapter, therefore, we have collected the available empirical evidence on social group differences in reproductive fitness for eighteenth- and nineteenth-century Germany. Due to the difficulties mentioned, we have attempted this by means of a disaggregation of fitness into its individual measurable components – fertility, infant survivorship and social placement of one's adult offspring. Social group differences with regard to these three components should deliver reliable indications for differential reproductive fitness. Data sets usable for this purpose are provided by Historical Demography with its family reconstitution studies. On the basis of church records and other historical sources, numerous vital and social statistical data can be linked to individual lives and to the histories of whole families and lineages, as well as to village genealogies and even the genealogical networks of entire regions.

In pre-modern rural Germany, the social position of a family within the social system was completely characterized by its property or its property rights to estates. Accordingly, in the following presentation we are going to differentiate between relatively prosperous full-time farmers, i.e. owners or leaseholders of 'full-time position', as a rule, and landless workers. The contrast of a two-class society suggested by this method can be more or less mitigated though by agrarian or non-agrarian middle classes (farmers with

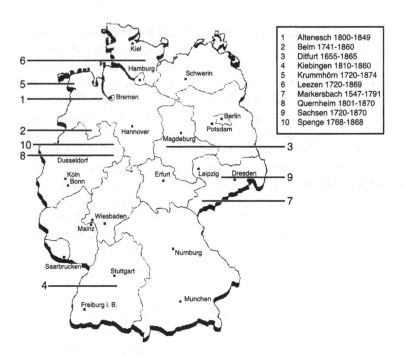

1	Altenesch 1800-1849
2	Belm 1741-1860
3	Ditfurt 1655-1865
4	Kiebingen 1810-1860
5	Krummhörn 1720-1874
6	Leezen 1720-1869
7	Markersbach 1547-1791
8	Quernheim 1801-1870
9	Sachsen 1720-1870
10	Spenge 1768-1868

Figure 12.1. Localities of the populations referred to in the text.

medium-sized holdings and tradesmen) depending on the region. This applies in particular to those regions in which a division of property had led to a progressive fragmentation of land ownership. The reproductive behaviour of these middle classes will not be taken into consideration here for reasons of clarity. The rule of thumb seems to be that their data tend to fall in the middle of the range, with the extremes being provided by the farmers with large holdings on the one hand and the landless workers on the other. More precise descriptions of the populations under study with regard to their historical, cultural, economic and ecological profiles are to be found in the literature cited, together with the precise criteria for their allocation to the respective social groups. The localities from where the samples have been drawn are shown in Figure 12.1.

Social inequality in reproductive fitness

Fertility

Figure 12.2 shows the average number of live-born children in farming families with land (upper class) and in landless worker families (lower

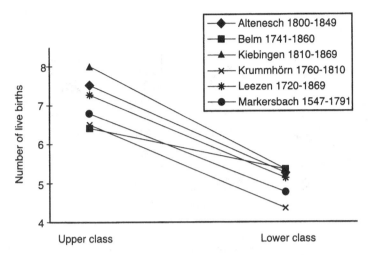

Figure 12.2. Number of live births per family according to social class for six
rural populations.
Sources: Altenesch: Hinrichs *et al.* (1981: table 4, p. 57); Belm: Schlumbohm
(1994: table 3.15b, p. 150); Kiebingen: Kaschuba and Lipp (1982: table II.79,
p. 524); Krummhörn: Klindworth and Voland (1995: table 4, p. 232); Leezen:
Gehrmann (1984: p. 253); Markersbach: Weiss (1981: p. 28).

class) for six early modern, rural German populations. Although the
values are only comparable with each other to a limited degree because the
methods of computation may have varied somewhat, the results are unam-
biguous. In all six populations, farmers' wives had more children than the
wives of landless workers. The difference is so pronounced that the two
data sets do not even overlap. Similar positive correlations between the
number of births and property were also found in other pre-modern
European populations: in England (Hughes, 1986), in Norway (Røskaft *et
al.*, 1992) and in Sweden (Low and Clarke, 1991).

There were two proximate reasons for these differences in the numbers
of children. In the first place, upper-class women began reproducing
earlier, because their age at marriage was generally lower than that of
lower-class women. So on average, they gained two to four years of fertile
marriage years depending on the population and their cohort, independent
of the overall mean age at marriage. This can be shown, for example, with
regard to the Krummhörn population (Figure 12.3). The average age
among Krummhörn men (1720–1874) at their first marriage was about 30
years and there were virtually no social differences. However, in the case of
women, the mean age at marriage of the brides of farmers with large
holdings was 24.9 years and thus 2.3 years lower than those of the brides of
landless workers.

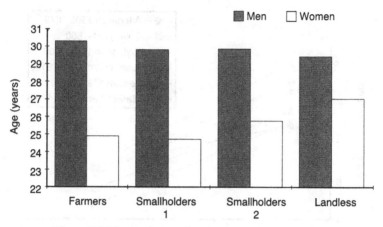

Figure 12.3. Mean age at marriage for men and women according to the bridegroom's social status (Krummhörn, 1720–1874). (*Source:* Voland and Engel, 1990, Fig. 1, p. 147.)

These widespread social status differences in the marriage age of women could possibly be the result of a conditional female partner choice strategy. The maxim of mate selection would then read, 'If you are young, you can be demanding and only marry a man who promises to provide you with above-average social standing. The older you become, the more you should give up your choosiness!'. And in fact the Krummhörn data suggest that women of different ages did show striking differences in their choice of husbands – the younger the women were when they married, the more likely it was to a well-off man. Almost one-third of the women under 20 years of age (compared to fewer than 10 per cent of the women over 30 years of age) married a farmer; conversely, one in five of farmers, but only one in every 25 of the landless men, married a woman under 20 years of age (Voland and Engel, 1990). If one assumes that young women were equally attractive as marriage partners for men from all social groups, then the variation in the marriage age of women can only be seen as the outcome of systematic mate choice on the part of the women with preference for socially successful men. Incidentally, the social status of a bride's natal family had no significant influence on her marriage age, despite pronounced social endogamy (Voland and Dunbar, 1997).

This interpretation leads to a more general hypothesis – hypergamous women, i.e. those who successfully pursued a higher standard when selecting a mate generally married at a younger age as compared to those women who married within their own social class or even experienced a social downward mobility with marriage. And data do support this general hypothesis. Among the daughters of the landless, social climbers were 2.3

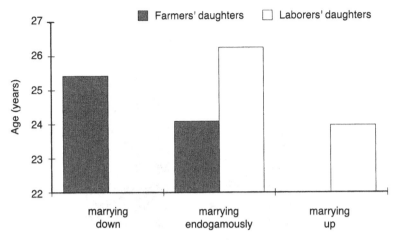

Figure 12.4. Age at marriage according to social mobility (Krummhörn 1720–1874). (*Source:* Voland and Engel, 1990: Fig. 2, p. 149.)

years younger on average than socially endogamous brides at the time of their marriage. The converse was true for upper-class daughters – if they married beneath their social status, they were 1.3 years older on average, than their sisters, who had succeeded in preserving their social standing by marriage (Figure 12.4).

The social-group difference in female marriage age appears to have been unrestrictedly characteristic for pre-modern peasant Germany. Schlumbohm (1992) examined the demographic data on the mean marriage age for a total of 17 populations from various regions in Germany and found not one single case in which the average marriage age of farmers' wives was higher than that of the wives of landless workers. Weiss (1993) also obtained comparable results. The conclusion is that in German farming societies of the eighteenth and nineteenth centuries, women preferred men with higher social standing as spouses, whatever the proximate motives for these preferences may have been – emotions, material expectations or normative rules.

With regard to the age of the women at the last birth, on the other hand, there do not seem to be any systematic differences between the social groups. In some populations, upper-class women had a longer period of child-bearing (as in Kiebingen), in others, the lower-class women did (as in the Krummhörn), and again in others, there was no notable difference between the two groups (as in Belm oder Leezen) (see also Knodel, 1988).

The second reason for the increased number of children among farmers with large holdings lies in the shorter mean inter-birth intervals (Hinrichs *et al.*, 1981; Kaschuba and Lipp, 1982; Gehrmann, 1984; Ebeling and

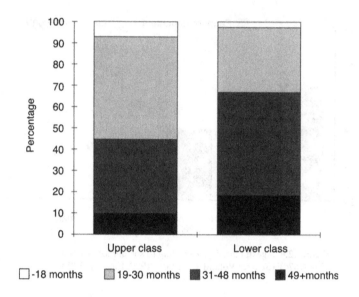

Figure 12.5. Distribution of mean birth intervals (Dupâquier and Lachiver categories) within elite and non-elite 'biologically completed' first marriages (Krummhörn 1760–1810).
(*Source:* Klindworth & Voland, 1995: Fig. 3, p. 228.)

Klein, 1988; Klindworth and Voland, 1995). This has automatically led to an increase in age-specific marital fertility rates. Figure 12.5 shows the distribution of the birth-interval categories, using the Krummhörn case (1760–1810) as an example.

Infant and child survivorship

Although land ownership in pre-modern Germany may have determined the increased fertility, this did not automatically lead to increased fitness, because it was possible for the reproductive advantage held by the social upper classes to have been exhausted by an increase in infant and child mortality. In fact, the data do show that the chances that children from prosperous families would become adults were by no means above average (Figure 12.6). On the contrary, in Altenesch and in the Krummhörn, the survival chances of both sexes combined were much better for children of workers. This result is a bit counter-intuitive, because one should be able to expect that materially better living conditions would be more of an advantage than a disadvantage for the survival of infants and children. The fact that this is not the case, however, lies in the obvious manipulations with which the survival chances of the newborn were influenced in

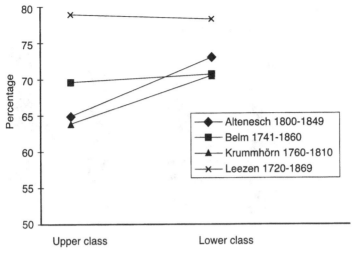

Figure 12.6. Survivorship to age 15 according to social group.
Sources: Altenesch: Hinrichs *et al.* (1981: table 5, p. 60); Belm: Schlumbohm (1994: table 3.16c, p. 154); Krummhörn: Klindworth and Voland (1995: Fig. 5, p. 230); Leezen: Gehrmann (1984: table 11, p. 96).

farmer families. In the families of farmers with large holdings in the Krummhörn, boys suffered a rate of mortality that was unexpectedly high (Voland, *et al.*, 1991; Voland and Dunbar, 1995). Their mortality rose especially in relation to the number of living brothers (Figure 12.7). Manipulative 'family planning', the goal of which was to limit the number of male heirs in farmer families, was obviously responsible for this observation.

For geographic and ecological reasons, the Krummhörn region did not provide any opportunities to expand the local resource base, thereby preventing any significant population growth during early modern times. This meant that reproductive displacement competition favoured tactics that primarily relied on wealth preservation as the safest protection against one's lineage being socially and genetically driven out. Conventionally, in this region, the youngest son inherited the undivided farm and its inventory from his father (this is called 'ultimogeniture'). His siblings had to be compensated by cash and other forms of moveable wealth with daughters only receiving half of what each son could expect. Depending on the number of offspring, a more or less substantial amount of wealth was thus drawn out of the business with every transfer of inheritance, sometimes burdening the heir with great economic hardship that often took a lifetime to recoup. Under these conditions rational economic behaviour would aim

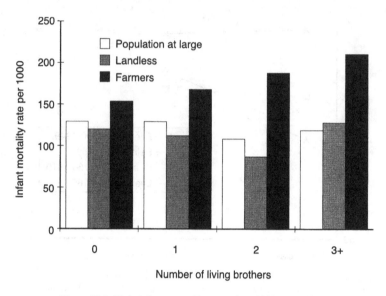

Figure 12.7. Male infant mortality rates (per 1000) according to number of living brothers (Krummhörn 1720–1874). (*Source:* Voland and Dunbar, 1995: Fig. 1, p. 38.)

at limiting the number of potential heirs as well as reducing the rate of inheritance transfers. Indeed, the tendency for ultimogeniture to lengthen the generation time can be interpreted as an attempt to resolve the latter problem.

Female infant mortality shows some social group pecularities as well. The landless daughters' probability of dying shows little variation with respect to the number of living sisters. In contrast, the farmers' daughters' fate is highly dependent on the number of living sisters. Although there is some suggestion that having one or two living sisters increases female infant survivorship in comparison to having no sister, it became clear that farmer parents drastically reduced their investment in their newborn daughters if they already had three or more girls. Thus, these later-born girls faced a risk of mortality that was approximately threefold of what could be expected (Voland and Dunbar, 1995).

Although in the Krummhörn this kind of parental discrimination against later born daughters did not lead to an overall female infant over mortality in farmer families, in Leezen and in Ditfurt, however, the daughters of farmers generally suffered from an infant mortality that was greater than that of the sons (Gehrmann, 1984; Stephan, 1993). Again, this pattern of female-bias infant mortality can reasonably be understood as an outcome of differential parental investment behaviour and may be interpreted as the adaptive answer on the part of farmer families to the prevailing local ecological and economic circumstances (Voland, 1984, 1995; Voland *et al.*, 1997). Interestingly enough, the regional comparisons show that in particular, upper-class families were very flexible in adapting their 'family planning' to the respective local conditions, whereas no similar manipulations were observed in the family formation processes of lower-class families. In an exaggerated and pointed comparison one could, therefore, characterize the reproductive strategies of the farmers as 'manipulative', and those of the landless workers as 'opportunistic' (Voland, 1995; Voland *et al.*, 1997). In summary, it is worth recording the fact that infant mortality in upper-class families was pushed up in part by 'avoidable deaths'. This explains why the survival chances of the children in farmer families were not better than those of others.

Social placement

After fertility and survivorship, there is a third component of reproductive fitness – that of the probability of adult children becoming reproductive and possibly able to assume socially privileged positions, which in turn contributed to increased reproductive fitness. For this one would have to know how the occupational and marital prospects of young adults differed depending on their social origin. Historical demographic research has previously been unable to answer this question of differential marriage probabilities satisfactorily, because some of the adult children left the parishes of their birth as single persons and the church records naturally do not contain any information about the fate of these emigrants in their new homes. Thus, the data provided in Figure 12.8 on social group differences refer only to the marriage probabilities in a parish (Leezen) or in a more closely defined regional environment (Krummhörn) and therefore have to be interpreted with more care. Accordingly in the two single populations for which – to our knowledge – there are reliable data, surviving upper-class children had by far better marriage prospects (Figure 12.8).

There are of course interesting differentiations which are concealed – at least in the Krummhörn – behind the overall increased probability of marriage. The daughters in particular, rather than the sons of farmers were more likely to marry. Furthermore, children with a large number of surviving same-sex siblings were at a disadvantage in comparison with children of smaller families (Figure 12.9). Increased fertility in farmers with large holdings fanned the flames of intra-familial competition for marriage and reproductive opportunities with the consequence that the reproductive advantage in families of farmers with large holdings was again neutralized by increased celibacy and emigration rates (Voland and

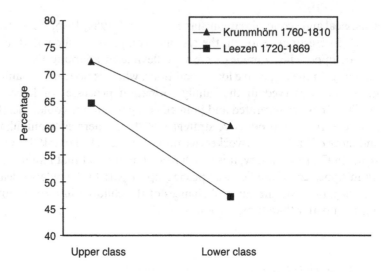

Figure 12.8. Local marriage probabilities of 15 year olds (both sexes combined) according to their fathers' social class. *Sources:* Krummhörn: Klindworth and Voland (1995: table 2, p. 230); Leezen: Gehrmann (1984: table 11, p. 96); Voland (1984: Tab. 3, p. 104).

Dunbar, 1995). In lower-class families this effect was barely noticeable (Figure 12.9).

The advantages of upper-class children with respect to their marriage probabilities were paralleled by considerably improved social status prospects (Figure 12.10). In early modern rural Germany it was practically impossible for children from the agrarian lower classes to rise to the leading class of farmers. Although a certain degree of social mobility upwards is observed in all populations, for members of the landless class this rise is successful, as a rule, only into the milieu of the smallholders (Figure 12.11). If there was any social rise into the group of the farmers with large holdings at all, it took several generations to accomplish, with each generation only able to take one step on the way while rising from being workers to becoming farmers.

However, farmers' children – not unexpectedly – had a much better chance of themselves being able to live as farmers (Figures 12.10 and 12.11). However, the statistics show that there was no guarantee of being able to preserve one's status – only one-half to about two-thirds of the married farmers' children were able to preserve their natal status. This meant that more than half (as in the Krummhörn) or about one-third (as in Quernheim) of the farmers' children lost social status with their marriages (Figure 12.10). This social decline led them more frequently to join the group of smallholders than to the group of workers (Figure 12.11). In

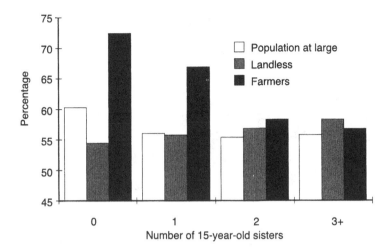

Figure 12.9. Probability of 15-year-old daughters marrying according to the number of 15-year-old sisters (Krummhörn 1720–1874). (*Source:* Voland and Dunbar 1995: Fig. 4, p. 41.)

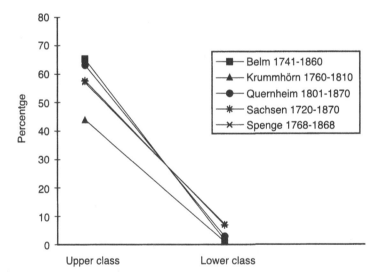

Figure 12.10. The probability of becoming a member of the upper class for married children (both sexes combined) whose fathers are either upper class members or lower class members. *Sources:* Belm: Schlumbohm (1994: table 6.03.d, p. 378), Krummhörn: Voland *et al.* (1991: table 4, p. 112); Quernheim: Mooser (1981, table 1, p. 185, sons only); Sachsen: Weiss (1993: table 14, p. 127; table 20, p. 137); Spenge: Ebeling and Klein (1988: table 2, p. 33).

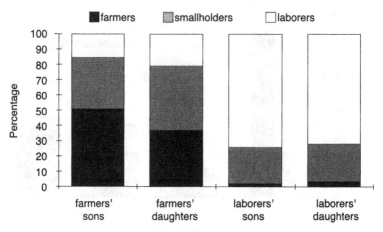

Figure 12.11. Intergenerational social mobility: percentages of married farmers' and labourers' sons and daughters, who lived as farmers, smallholders or labourers (Krummhörn 1720–1874). (*Source:* Voland *et al.*, 1991: table 4, p. 112.)

summary it remains to be stated that the German farming village communities of the eighteenth and nineteenth centuries were societies characterized by an overall social downward mobility. To be a member of one of the privileged social classes, one had to be born within that class. However, a high-ranking birth did not necessarily guarantee permanent high-ranking social status as adults.

Social differences in men's average reproductive careers

In summary it can be said that for the two components of reproductive fitness, namely fertility and social placement of one's grown-up offspring, there are significant social group differences, which resulted in an over reproduction of the local elite as a rule. With respect to the rearing success of the children born to these groups, however, no basic advantage was seen among the upper classes – on the contrary, the upper class farmers' children tend to die more frequently due to parental manipulation.

The reasons for and the degree of over reproduction among the local wealthy elites can be examined by using the situation in Krummhörn as an example. Table 12.1 summarizes the reproductive performance of the 60 wealthiest men in 1812, and compares it with the contemporary non-elite population mean (Klindworth and Voland, 1995), keeping in mind that the reproductive data of married men were compared. If the lifelong celibates had been taken into consideration, this contrast could possibly be even more pronounced. On average, elite men had roughly two children more than ordinary men. Even if an increased infant and child mortality

Table 12.1. *Average reproductive careers for always-married elite and non-elite men (Krummhörn, Germany, 1760–1810)*

	Elite ($n = 60$)	Non-elite ($N = 954$)	$p <$
Children	6.58	4.59	0.001
stillbirth rate	0.015	0.037	0.05
Livebirths	6.48	4.42	0.001
infant mortality rate	0.195	0.134	0.002
1-year olds	5.22	3.83	0.001
child mortality rate	0.201	0.185	ns
15-year olds	4.17	3.12	0.001
probability of celibacy	0.112	0.120	ns
probability of 'emigration'	0.156	0.272	0.001
Married children	3.05	1.90	0.001

ns: not significant
(From Klindworth and Voland, 1995.)

(including striking sex differentials) tended to reduce the rich group's initial reproductive 'advantage' they were able to raise one child more to adulthood than the average man. Moreover, children from the elite group had markedly better local marriage and thus reproductive prospects than the offspring of ordinary families. Whereas an average man could expect to see two of his children married and thus only just replace himself in genetic terms, a rich farmer, who ended up with three married children on average, succeeded in contributing 50% more to the local gene pool.

Differential reproductive performance accumulated to yield pronounced social status differences in Darwinian fitness. This can be shown, for example, by comparing the long-term reproductive success of an upper-class farmer family with the population mean by adding together all age- and sex-specific 'reproductive values' (Fisher, 1930) of all the living descendents at defined time intervals (Figure 12.12). In short, a prosperous farming couple of the eighteenth century had almost twice as many gene replicates in the local population 100 years after marrying than an average family. Hence, wealth was correlated with reproductive fitness within the Krummhörn population.

Conclusion

Two things now become clear. Firstly, ownership of land was a component of natural selection. This in turn was associated with an important consequence – striving to maintain or even increase the size of the property and striving for social upward mobility, i.e. striving for cultural success, was

Figure 12.12. The ratio of farmers' fitness vs. population mean (Krummhörn 1720–1750). (*Source:* Voland, 1990: Fig. 3, p. 69.)

tantamount to striving for reproductive fitness, regardless of whether this correlation was actually recognized or not. The surmise first voiced by Irons (1979) that there might be a positive correlation between social and reproductive success applies to the Krummhörn case, and with regard to the demographic data given above, it almost certainly applies to other pre-modern farming populations as well. In this sense, social competition for privileged positions in the social hierarchy is to be understood as a cultural reflex of a basic biological process, namely that of reproductive competition.

Secondly, the psychological traits underlying male socio-economic competition appear to have been selected by sexual selection, since the over reproduction of social elites turns out to be largely a female driven phenomenon. Ultimately it is the women who solely or decisively control some of the components responsible for the above-average reproduction of the upper class (Table 12.2). The contribution of upper-class men to their over reproduction lies primarily in their efforts to achieve economic and social success. Ultimately, a family's economic background, on the one hand, decides the level of inheritance and dowry payments and in this way the marriage and status prospects of the children, thus deciding the long-term existence of a lineage. On the other hand, women control the direct generative components of reproductive fitness.

Upper-class women were willing to marry at an earlier age and thus allocated more fecund years to their husbands by this means. In the

Table 12.2. *Components of social inequality in reproductive fitness and their control by males and females*

Component	Control by males	by females	Through
Fertility			
Fecund years spent in marriage	+	+ +	Female age at marriage
Rate of reproduction	+	+ +	Breast-feeding patterns
Infant and child survivorship	+	+ +	e.g. breast-feeding
Social placement of offspring			
Allocation of marriage chances	+ +	+	Dowry and inheritance payments
Status preservation	+ +	+	Economic power

+ + decisive control; + some control.

Krummhörn, for example, a married upper-class man spent an average of 21.4 years in a fecund marriage, a worker spent much less time, with only 17.3 years (Klindworth and Voland, 1995). Furthermore, upper-class women had shorter inter-birth intervals and therefore higher fertility rates. In times without any technical means of birth control, inter-birth intervals were decisively controlled by breast-feeding patterns (Knodel and Van de Walle, 1967; Margulis *et al.*, 1993) – this naturally being the sole domain of the women.

Finally, upper-class women had presumably the main responsibility for differences in the allocation of living and survival chances to their children. It does not take much imagination to see how different assessments of a child's desirability can result in differences in maternal responsiveness and care, in the provision of medical and caloric resources, in work loads, in the strictness of physical and mental punishment and other intra-familial transactions. The most significant transaction by the help of which parents were able to consciously or unconsciously allocate their investment is undoubtedly breast-feeding – again, exclusively the domain of the mothers.

Thus the question raised at the beginning about the correlation between female reproductive preferences and social inequality in reproductive fitness during eighteenth-century and nineteenth century Germany can clearly be answered. Reproductive fitness differentials are mediated by female mating and reproductive decisions, and these are contingent on male social status. Hence female mating and reproductive decisions are a decisive hinge in the correlation between male social and reproductive inequality.

References

Bereczkei, T. & Csanaky, A. (1996). Mate choice, marital success, and reproduction in a modern society. *Ethology and Sociobiology* **17**, 17–35.

Betzig, L. (1986). *Despotism and Differential Reproduction – A Darwinian View of History*. Hawthorne: Aldine de Gruyter.

Boone III, J. L. (1986). Parental investment and elite family structure in preindustrial states: a case study of late medieval-early modern Portuguese genealogies. *American Anthropologist* **88**, 859–78.

Borgerhoff Mulder, M. (1987). On cultural and reproductive success: Kipsigis evidence. *American Anthropologist* **89**, 617–34.

Buss, D. M. (1994). *The Evolution of Desire – Strategies of Human Mating*. New York: Basic Books.

Cronk, L. (1991). Human behavioral ecology. *Annual Review of Anthropology* **20**, 25–53.

Ebeling, D. & Klein, P. (1988). Das soziale und demographische System der Ravensberger Protoindustrialisierung. In *Bev'lkerungsgeschichte im Vergleich: Studien zu den Niederlanden und Nordwestdeutschland*, ed. E. Hinrichs & H. van Zon, pp. 27–48. Aurich: Ostfriesische Landschaft.

Fisher, R. A. (1930). *The Genetical Theory of Natural Selection*. Oxford: Clarendon.

Flinn, M. V. (1986). Correlates of reproductive success in a Caribbean village. *Human Ecology* **14**, 225–43.

Gehrmann, R. (1984). *1720–1870 – Ein historisch-demographischer Beitrag zur Sozialgeschichte des ländlichen Schleswig-Holstein*. Neumhnster: Wachholtz.

Haskey, J. (1987). Social class differentials in remarriage after divorce: Results from a forward linkage study. *Population Trends* **47**, 34–42.

Hinrichs, E., Liffers, R. & Ziegler, J. (1981). Sozialspezifische Unterschiede im generativen Verhalten eines Wesermarsch- Kirchspiels (1800–1850) – Ergebnisse der Auswertung des Familienregisters von Altenesch. In *Sozialer und politischer Wandel in Oldenburg – Sudien zur Regionalgeschichte vom 17. bis 20. Jahrhundert*, ed. W. Gunther, pp. 49–73. Oldenburg: Holzberg.

Hughes, A. L. (1986). Reproductive success and occupational class in eighteenth-century Lancashire, England. *Social Biology* **33**, 109–15.

Irons, W. (1979). Cultural and biological success. In *Evolutionary Biology and Human Social Behavior – An Anthropological Perspective*, ed. N. A. Chagnon & W. Irons, pp. 257–272. North Scituate: Duxbury.

Irons, W. (1993). Monogamy, contraception and the cultural and reproductive success hypothesis. *Behavioral and Brain Sciences* **16**, 295–6.

Kaplan, H. S., Lancaster, J. B., Bock, J. A. & Johnson, S. E. (1995). Fertility and fitness among Albuquerque men: A competitive labour market theory. In *Human Reproductive Decisions – Biological and Social Perspectives*, ed. R. I. M. Dunbar, pp. 96–136. Houndsmills: MacMillan. St. Martin's.

Kaschuba, W. & Lipp, C. (1982): Dörfliches Überleben – Zur Geschichte materieller und sozialer Reproduktion ländlicher Gesellschaft im 19. und frühen 20. Jahrhundert. Tübingen: Tübinger Vereinigung für Volkskunde.

Klindworth, H. & Voland, E. (1995). How did the Krummhörn elite males achieve above-average reproductive success? *Human Nature* **6**, 221–40.

Knodel, J. E. (1988). *Demographic Behavior in the Past – A Study of Fourteen German Village Populations in the Eighteenth and Nineteenth Centuries*.

Cambridge: Cambridge University Press.

Knodel, J. & Van de Walle, E. (1967). Breast feeding, fertility and infant mortality: an analysis of some early German data. *Population Studies* 21, 109–31.

Low, B. S. (1993). Ecological demography: a synthetic focus in evolutionary anthropology. *Evolutionary Anthropology* 1, 177–87.

Low, B. S. & Clarke, A. (1991). Family patterns in nineteenth-century Sweden: impact of occupational status and landownership. *Journal of Family History* 16, 117–38.

Margulis, S. W., Altmann, J. & Ober, C. (1993). Sex-biased lactational duration in a human population and its reproductive costs. *Behavioral Ecology and Sociobiology* 32, 41–5.

Mooser, J. (1981). Soziale Mobilität und familiale Plazierung bei Bauern und Unterschichten – Aspekte der Sozialstruktur der ländlichen Gesellschaft im 19. Jahrhundert am Beispiel des Kirchspiels Quernheim im östlichen Westfalen. In *Familie zwischen Tradition und Moderne – Studien zur Geschichte der Familie in Deutschland und Frankreich vom 16. bis zum 20. Jahrhundert*, ed. N. Bulst, J. Goy & J. Hoock, pp. 182–201. Göttingen: Vandenhoeck & Ruprecht.

Pérusse, D. (1993). Cultural and reproductive success in industrial societies: Testing the relationship at the proximate and ultimate levels. *Behavioral and Brain Sciences* 16, 267–322.

Røskaft, E., Wara, A. & Viken, C. (1992). Reproductive success in relation to resource-access and parental age in a small Norwegian farming parish during the period 1700–1900. *Ethology and Sociobiology* 13, 443–61.

Schlumbohm, J. (1992). Sozialstruktur und Fortpflanzung bei der ländlichen Bevölkerung Deutschlands im 18. und 19. Jahrhundert – Befunde und Erklärungsansätze zu schichtspezifischen Verhaltensweisen. In *Fortpflanzung: Natur und Kultur im Wechselspiel*, ed. E. Voland, pp. 322–46. Frankfurt/M.: Suhrkamp.

Schlumbohm, J. (1994). *Lebensläufe, Familien, Höfe – Die Bauern und Heuerleute des Osnabrückischen Kirchspiels Belm in proto-industrieller Zeit, 1650–1860.* Göttingen: Vandenhoeck & Ruprecht.

Stephan, P. (1993). Sterben in früheren Jahrhunderten – Versuch einer Begründung von Unterschieden in der Sterblichkeit von 1655 bis in das 20. Jahrhundert in einer d'rflichen Population. *Biologisches Zentralblatt* 112, 28–81.

Vining Jr., D. R. (1986). Social versus reproductive success: The central theoretical problem of human sociobiology. *Behavioral and Brain Sciences* 9, 167–216.

Voland, E. (1984). Human sex-ratio manipulation: historical data from a German parish. *Journal of Human Evolution* 13, 99–107.

Voland, E. (1990). Differential reproductive success within the Krummhörn population (Germany, 18th and 19th centuries). *Behavioral Ecology and Sociobiology* 26, 65–72.

Voland, E. (1995). Reproductive decisions viewed from an evolutionarily informed historical demography. In *Human Reproductive Decisions – Biological and Social Perspectives*, ed. R. I. M. Dunbar, pp. 137–59. Houndsmills: MacMillan.

Voland, E. & Dunbar, R. I. M. (1995). Resource competition and reproduction – the relationship between economic and parental strategies in the Krummhörn population (1720–1874). *Human Nature* **6**, 33–49.

Voland, E. & Dunbar, R. I. M. (1997). The impact of social status and migration on female age at marriage in an historical population in north-west Germany. *Journal of Biosocial Science* **29**(3), 355–60.

Voland, E., Dunbar, R. I. M., Engel, C. & Stephan, P. (1997). Population increase and sex-biased parental investment in humans: Evidence from 18th- and 19th-century Germany. *Current Anthropology* **38**(4), 129–35.

Voland, E. & Engel, C. (1990). Female choice in humans: a conditional mate selection strategy of the Krummhörn women (Germany, 1720–1874). *Ethology* **84**, 144–54.

Voland, E., Siegelkow, E. & Engel, C. (1991). Cost/benefit oriented parental investment by high status families – The Krummhörn case. *Ethology and Sociobiology* **12**, 105–18.

Waynforth, D. & Dunbar, R. I. M. (1995). Conditional mate choice strategies in humans: evidence from 'Lonely Hearts' advertisements. *Behaviour* **132**, 755–79.

Weiss, V. (1981). Zur Bevölkerungsgeschichte des Erzgebirges unter frühkapitalistischen Bedingungen vom 16. bis 18. Jahrhundert. *Sächsische Heimatblätter* **27**, 28–30.

Weiss, V. (1993). *Bevölkerung und soziale Mobilität – Sachsen 1550–1880*. Berlin: Akademie-Verlag.

13 Environmental constraints, social inequality and reproductive success. A case-study in Morocco

EMILE CROGNIER

Introduction

The impact of social inequalities on people's health has been thoroughly observed for more than one century in industrialized countries and is still being documented, both in developed or in developing nations (see several contributions in this volume). The available data consistently suggest that social inequalities are a major cause of differences in many aspects of human biology and health, from the primary stages of foetal life to adulthood and senescence, including differences in life expectancy and health status at each stage in life.

What is less obvious, however, is whether this relationship still holds as far as poorly stratified societies are concerned. For example, to earn one's living means, in many peasant groups living in traditional societies, an equal effort to extract resources from an adverse environment whatever one's social position. Is it not possible, in such conditions, that environmental variations prevail over social inequalities as differentiating factors in biology and health?

The economic conditions in rural Morocco offer good examples of this situation. Although undergoing industrial growth, the major part of the population of this country live in rural areas and are principally occupied in agriculture. In the southern province of Marrakesh whose territory is partitioned between dry plains and a steep mountain ridge, the high Atlas mountains, cultivation either faces dryness in the lowlands or small and difficult land areas in the highlands. Excluding the city of Marrakesh, by far the greatest number of inhabitants in the province are peasants. Until recently, many were still following the traditions of a millenary society, organized in *quasi* autarkic economy. This traditional society, which has been the backbone of Moroccan identity, was associated with the high fertility and mortality rates common to many developing countries, but

239

still more typical of Muslim nations (Kirk, 1968; United Nations, 1971; Nagi, 1983; Gadalla *et al.*, 1984).

In 1984, a survey designed by the Public Health Agency of the Province of Marrakesh to evaluate the impact of a family planning campaign implemented several years before, investigated 6000 nuclear Berber families chosen at random in the rural province of Marrakesh (Crognier, 1989). Information on reproductive events and socio-economic characteristics was systematically collected; the data considered in this study represent those of agriculturist women 50 to 70 years of age at the time of the survey. The oldest had entered the reproductive process in 1930–35, whereas the youngest did so in 1950–55. These two periods are prior to changes in reproductive behaviour, which from the late 1960s began to alter the previous equilibrium in cities and, later on, in rural areas (Ministère du Plan, 1983, 1984; CERED, 1986; Enquête Démographique Nationale, 1988).

Though Berbers do not aim to build an egalitarian society, social differences and their specific corollaries are not evident in the daily life of the majority of peasants. Everybody works hard; all houses whatever their size share more or less the same Spartan comfort; meals are either rich or poor, according to the season of the year rather than a reflection of incomes. In the traditional context of peasant life, and considered from a general standpoint, appreciable variations do not seem significantly to affect the behaviours associated with education, family life or health care.

In contrast to this apparent evenness of living conditions, there is an evident environmental heterogeneity, assignable to five main subdivisions:

(1) The irrigated plain in the eastern part of the province, close to the city of Marrakesh.
(2) The dry plain in the northern and the western parts.
(3) The low and middle mountain, from 600 to 1000 m altitude.
(4) The high moutain, above 1000 m.
(5) The urban or suburban environments of small towns several shops, local administrative services and a medical centre gathered around the market place (the 'souk').

A previous study comparing highlanders to lowlanders and to town inhabitants of a particular Berber group, the Guedmiua (Crognier *et al.*, 1993; Crognier, 1994, 1996), showed that environmental conditions powerfully influenced fertility, infant/child mortality and subsequently the behavioural components of reproductive success, resulting in contrasted patterns among groups. Hence the present hypothesis is that environmental strain could surpass social inequalities in modulating the expression of reproductive success.

Table 13.1. *Chronology of reproductive life (years), total fertility rate and number of offspring surviving to 15 years of age, among samples of Berber women living in contrasted environments*

Variable	ir. pl. ($N = 60$)	dry pl. ($N = 189$)	low mt. ($N = 136$)	high mt. ($N = 82$)	town ($N = 49$)	F	P
Mean age at:							
menarche	14.6	14.5	14.7	15.1	15.6	6.55	***
marriage	17.6	18.0	18.6	18.6	16.0	8.17	***
1st birth	19.1	19.7	20.5	20.6	18.3	5.16	***
2nd birth	22.3	22.7	23.1	22.9	22.8	0.49	ns
3rd birth	24.7	25.4	25.5	25.8	25.1	0.45	ns
4th birth	27.3	27.8	27.8	28.6	26.5	1.24	ns
last birth	36.3	35.7	35.1	37.0	35.2	0.48	ns
Mean:							
int. mar.-last b.	22.4	21.4	19.9	21.3	23.5	1.97	ns
age at menopause	49.2	45.7	45.8	48.7	49.8	10.6	***
int. last b.-men.	13.6	10.3	11.3	12.2	14.6	2.51	*
Mean number of:							
live births	8.3	7.7	7.4	7.6	8.7	1.32	ns
offspring > 15 yrs	5.1	4.6	4.3	4.9	5.1	1.80	ns

ir. pl.: irrigated plain; dry pl.: non-irrigated plain; low mt.: low mountain; high mt.: high mountain; int mar.-last b.: mean interval between marriage and last birth; int. last b.-men.: mean interval between last birth and menopause; ns: not significant.
* $p < 0.05$; ** $p < 0.01$; *** $p < 0.001$.

Patterns of fertility and environmental variation

Table 13.1 expresses the main chronological features of female reproductive life in the various ecosystems. The general pattern is in agreement with earlier studies (Crognier et al., 1992, 1993; Varea, 1993). The mean differences in ages at menarche and at marriage do not result in differences in the chronology of births or the age at which women's reproductive period ends, whatever the environment. Finally, their reproductive scores do not differ significantly, either in the number of live births or in that of offspring surviving until sexual maturity. This second characteristic gives a crude evaluation of reproductive success which is close to the initial Darwinian concept of 'success in leaving progeny', i.e. the number of adult offspring that a sexually mature individual is able to produce.

In humans however, the number of progeny is no longer the mere outcome of the balance between carrying capacity and predation. Reproductive strategies occur (Daly and Wilson, 1978; Irons, 1979; Betzig *et al.*, 1988), in which cultural and social projections shape the reproductive pattern. In particular, with developing living standards, the flows of en-

Table 13.2. *Offspring survival in contrasted environments, estimated by the Kaplan-Meier method on right censored data.*

Variable	ir. pl. (N = 497)	dry pl. (N = 1296)	low mt. (N = 633)	high mt. (N = 616)	town (N = 334)
Number surviving to 15 yrs	339	925	437	455	234
Percentage surviving	68.2	71.4	69.0	78.9	70.0

Log-rank test $\chi^2 = 7.46$ DF = 4 $P \chi^2 < 0.1136$
Wilcoxon test $\chi^2 = 9.48$ DF = 4 $P \chi^2 < 0.0500$
$- 2\log(\text{LR})$ test $\chi^2 = 5.58$ DF = 4 $P \chi^2 < 0.2325$
ir. pl.: irrigated plain; dry pl.: non-irrigated plain; low mt.: low mountain; high mt.: high mountain.

ergy, nutrients and care demanded by parenthood often conflict with other necessities or choices (Handwerker, 1983; Hill and Randall, 1984; Lancaster *et al.*, 1987; Betzig, 1988). A more informative variable is therefore relating the number of liveborn offspring to the number of those who survive until sexual maturity. This 'relative reproductive success' is more a measure of efficiency in the process of reproduction than an appreciation of the 'reproductive value' (Fisher, 1930), *sensu stricto*.

In these Berber samples, the comparison of survival function estimates versus time computed for each environment, does not show evidence of environmental influence upon survival until the age of 15 years, in so far as the log-rank test probability of distinct survival in the highest ages fits the null hypothesis (Table 13.2). The Wilcoxon test, which places more weight on early survival times, however, indicates a probable environmental influence during the early stages of life. The computation of survival function estimates up to one year of age (Table 13.3), shows effectively differences, with the urban and irrigated plain environments showing a lower level of survival than the others.

The causes of death recorded may help the interpretation of this result. Gastro-intestinal diseases and tetanus which, both occur soon after birth, predominate in both places and are the cause of 68 and 70% of deaths respectively, whereas they are less frequent in other environments. Conversely, the other main causes of death from communicable diseases, as for example some respiratory infections, occur later (Baudot and Bley, 1992).

Patterns of fertility and economic strata

The appraisal of economic conditions is not easy in rural Morocco as there is no standard evaluation of income. Indirect factors were therefore used in this study:

Table 13.3. *Offspring's causes of death and differential survival to the age of 1 year, as a function of the diverse environments*

Variable	ir. pl. ($N = 558$)	dry pl. ($N = 1470$)	low mt. ($N = 671$)	high mt. ($N = 686$)	town ($N = 359$)
Cause of death:					
gastro-intest. disease	44 (35%)	70 (28%)	21 (24%)	22 (22%)	23 (31%)
other infectious disease	25 (20%)	130 (52%)	53 (60%)	52 (52%)	13 (18%)
tetanus	42 (33%)	14 (6%)	5 (6%)	10 (10%)	28 (39%)
congenital and obstetric	15 (12%)	37 (15%)	10 (11%)	16 (16%)	10 (13%)

$\chi^2 = 118.1$ DF12 $P < 0.0001$
χ^2Likelihood ratio $\chi^2 = 117.8$ DF12 $P < 0.0001$
χ^2Mantel-Haenszel $\chi^2 = 1.131$ DF12 $P < 0.288$

Number surviving to 1 year	479	1342	584	625	309
Percentage surviving	85.8	91.3	87.0	91.1	86.1

Log-rank test $\chi^2 = 23.48$ DF $= 4$ P $\chi^2 < 0.0001$
Wilcoxon test $\chi^2 = 24.10$ DF $= 4$ P $\chi^2 < 0.0001$
$-2\log(LR)$ test $\chi^2 = 19.65$ DF $= 4$ P $\chi^2 < 0.0006$

ir. pl.: irrigated plain; dry pl.: non-irrigated plain; low mt.: low mountain; high mt.: high mountain

- The number of rooms in the house.
- The presence/absence of latrines.
- The possession of a radio or a TV set.
- The type of domestic light.
- The origin of domestic water.
- The means of transportation.
- The number of adults in the household.
- The number of wage earners.
- The agricultural area owned.

Iterated factor analyses reduced these to six factors without a substantial loss of information:

- The number of rooms in the house.
- The possession of a radio or a TV set.
- The type of domestic light.
- The means of transportation.
- The number of wage earners.
- The agricultural area owned.

Then, a correspondence analysis was performed on this set of estimators and the individual parameters of each woman interviewed was overlaid on the frame made by the first two axes, thus allowing classification with respect to economic characteristics. The bounds of the contiguous areas were delineated by hand screening of initial questionnaires.

Four economic strata were finally distinguished:

(1) The lowest typically includes families living in small houses without latrines, lit by candles or oil lamps, possessing no radio, no bicycle and no mule, living from a single wage earner and possessing no land.

(2) The second group includes small farmers, still with very poor household equipment but generally owning a mule and about 1 ha of land.

(3) This category consists of farmers who, though bigger, were still observing a traditional way of life. Houses are large, most include latrines, there is a radio, but no mechanical means of transport, instead there are several mules, and the fields owned range from 4–5 ha in irrigated places to 20–30 ha in dry zones.

(4) This group is similar to the third as far as economical level is concerned. However, it is more modern. Electricity and tap water, TV sets and motor-bikes are frequent, all features which generally indicate a peri-urban or urban location.

The Table 13.4 shows the chronology of women's reproductive life according to these economic strata. Compared to other groups of women, the poorest group are younger when their reproductive life ends as well as having a shorter reproductive life span as estimated by the interval between marriage and last birth. An expected consequence is the smaller number of live births and the subsequent smaller number of offspring surviving until reproductive maturity. In a crude Darwinian interpretation, the reproductive success of the poor stratum is clearly inferior to that of other strata, the highest scores being those of the two wealthiest groups, whatever their traditional or modern life style.

To which causes is the discrepancy between the poorest strata and the others assignable? Table 13.5 details more parameters which are disposed to influence the reproductive pattern – those which refer to the tempo of births do not appear to differ (waiting time to first birth and intervals between subsequent offspring; duration of breast-feeding), whereas discrepancies are found in the frequencies of childless women, that of the poorest stratum being 10 times higher than that of the urban sample. Two characteristics might explain this result:

(1) The wealthiest couples could consult a doctor if pregnancy does not occur soon after the wedding (Gharbi *et al.*, 1997).

Table 13.4. *Chronology of reproductive life (years), total fertility rate and number of offspring surviving to 15 years old, among samples of Berber women from different economic strata*

Variable	ECO1 (N = 112)	ECO2 (N = 145)	ECO3 (N = 137)	ECO4 (N = 120)	F	P
Mean age at:						
menarche	14.8	14.8	14.7	14.8	0.42	ns
marriage	17.9	18.3	18.6	17.3	4.76	***
1st birth	20.2	20.1	20.1	18.8	5.00	***
2nd birth	23.1	22.6	22.5	21.7	1.82	ns
3rd birth	26.2	25.2	25.0	24.2	2.12	ns
4th birth	28.4	27.9	27.3	26.3	2.92	*
lst birth	33.1	36.8	36.8	36.3	3.48	**
Mean:						
int. mar.-last b.	17.3	21.2	19.9	21.3	4.44	***
age at menopause	47.3	47.3	47.5	47.9	1.86	ns
int. last b.-men.	13.7	10.5	11.0	12.1	2.30	ns
Mean number of:						
live births	6.1	7.9	8.5	8.4	12.7	***
offspring > 15 yrs	3.7	4.6	5.0	5.4	10.5	***

ECO1 – 4: poverty – wealth; int. mar.-last b.: mean interval between marriage and last birth; int. last b.-men.: mean interval between last birth and menopause; ns: not significant.
$*p < 0.05$; $**p < 0.01$; $***p < 0.001$.

Table 13.5. *Possible factors of the differences observed between reproductive scores according to economic strata*

Variable	ECO1 (N = 112)	ECO2 (N = 145)	ECO3 (N = 137)	ECO4 (N = 120)	statistical significance
Mean (months):					
wait t. 1st birth	26.5	23.5	19.8	22.8	$F = 2.27$ ns
interval between births	38.1	37.5	33.7	38.1	$F = 1.76$ ns
lactation extent	17.5	18.4	17.8	16.7	$F = 0.86$ ns
Number and percentage:					
with infant death	53 (50)	56 (39)	44 (34)	40 (31)	$\chi^2 = 5.18$ ns
childless	11 (9.8)	9 (6.1)	4 (2.9)	1 (0.8)	$\chi^2 = 31.2$ ***
widows	48 (43)	20 (14)	19 (14)	21 (18)	$\chi^2 = 41.3$ ***
Mean effective conjugal life	29.2	30.2	30.8	31.3	(Lsd 1–4 *)

ECO1 – 4: poverty – wealth; ns: not significant.
$*p < 0.05$; $**p < 0.01$; $***p < 0.001$.

Table 13.6. *Offspring survival in contrasted socio-economic strata computed by Kaplan-Meier method on right censored data*

Variable	ECO1 (N = 684)	ECO2 (N = 1092)	ECO (N = 1128)	ECO4 (N = 840)
Number surviving to 15 yrs	447	669	699	575
Percentage surviving	65.3	61.3	62.0	68.40

Log-rank test $\chi^2 = 13.6$ DF = 3 $P\chi^2 < 0.0035$
Wilcoxon test $\chi^2 = 13.6$ DF = 3 $P\chi^2 < 0.0035$
$-2\log(LR)$ test, $\chi^2 = 19.6$ DF = 3 $P\chi^2 < 0.0002$
ECO1 – 4: poverty – wealth.

(2) The wealthiest men would not hesitate to divorce if their wife remained childless, thus introducing a bias in the distribution of fertility among economic classes.

Widowhood also shows significant differences. It is more than twice as frequent in the poorest group than in the others and it consequently affects the span of effective conjugal life by allowing for a shorter exposure to the risk of pregnancy.

If child survival from birth to maturity is considered in association with economic levels (Table 13.6), significant differences also appear as indicated by homogeneity tests, though the order of relative efficiency is not maintained. The poorest group appear to provide better care for the offspring than the two next and wealthier strata. The highest percentage of survivors is, however, found in the fourth group, i.e. those who are more integrated in modern life.

These survival differentials suggest the need for further investigation – why and how do poor families better protect their progeny than those enjoying better life conditions? Is the proximity of medical assistance provided by town dwelling the main cause of improved offspring survival in the fourth group, or is this attributable to other factors (education, sanitary level, etc.)?

Discussion and conclusion

The part played by economic and social factors in the determination of health and general well-being in human societies was not indeed questioned in this study. Our inquiry was directed to testing the model at its boundaries, i.e. in a society expressing weak social differences.

The results obtained bring out a clear answer to our inquiry – in spite of contrasted ecosystems, both crude reproductive success, and its assessment in terms of the relative survival of offspring up to sexual maturity, do not seem to be affected by environmental causes among the Berber agriculturists of Southern Morocco. On the contrary, economic status clearly influences the two aspects of reproductive success.

Is this surprising? Though exhibiting attenuated wealth contrasts, this peasant society does not by any means aim to be egalitarian. Furthermore, as is observed in many traditional societies (e.g. Caldwell and Caldwell, 1987; Draper, 1989), the true wealth is that of progeny, whose extension is an important measure of social success. Hence the expression, through reproductive success, of socio-economic disparities that are otherwise concealed.

The effect of environmental diversity is not altogether absent, as seen in the analysis, it would at least entail differential survival in infancy. Its absence at the age of evaluation of reproductive success, however, would argue for the intervention of compensatory mechanisms thus far unravelled.

Humans, it would appear, are more able to cope with environmental difficulties, than with the forces stemming from their own social culture.

References

Baudot, P. & Bley, D. (1992). Climat et structure par âge de la mortalité infantile. Le cas du Maroc. In *Risques pathologiques, rythmes et paroxysmes climatiques*, ed. J. P. Besancenot, pp. 39–48. Paris: John Libbey Eurotext.

Betzig, L. (1988). Mating and parenting in Darwinian perspective. In *Human reproductive behaviour*, ed. L. Betzig, M. Borgerhoff Mulder & P. Turke, pp. 3–20. Cambridge: Cambridge University Press.

Betzig, L., Borgerhoff Mulder, M. & Turke, P. (1988). *Human Reproductive Behaviour*. Cambridge: Cambridge University Press.

Caldwell, J. & Caldwell, P. (1987). The cultural context of high fertility in sub-Saharan Africa. *Population and Development Review* 13(3), 409–37.

CERED (1986). *Analyses et tendances démographiques au Maroc*. Rabat: Centre d'Etudes et de Recherches Démographiques.

Crognier, E. (1989). La fécondité dans la province de Marrakech (Maroc): enquête anthropologique. *Anthropologie et Préhistoire (B. Soc. Roy. Belge Anth. Prehist.)*, **100**, 113–23.

Crognier, E. (1994). Darwinian fitness and reproductive strategies in human populations. The case of Berber groups of Morocco. *Journal of Human Ecology* 4(1), 77–88.

Crognier, E. (1996). Behavioral and environmental determinants of reproductive success in traditional Moroccan Berber groups. *American Journal of Physical Anthropology* **100**, 181–90.

Crognier, E., Bernis, C., Elizondo, E. & Varea, C., (1992). Reproductive patterns as environmental markers in rural Morocco. *Collegium Anthropologicum*

16(1), 89–97.

Crognier, E., Bernis, C., Elizondo, E. & Varea, C., (1993). The patterns of fertility in a Berber population from Morocco. *Social Biology* **3–4**, 192–9.

Daly, M. & Wilson, M. (1978). *Sex, Evolution and Behavior*. North Scituate, Mass: Duxbury Press.

Draper, P. (1989). African marriage systems: perspectives from evolutionary ecology. *Ethology and Sociobiology* **10**, 145–69.

Enquête Démographique Nationale (ENDPR, 1986–88), *Rapport préliminaire*. Rabat: Direction de la Statistique.

Fisher, R. A. (1930). *The Genetical Theory of Natural Selection*. New York: Dover.

Gadalla, S., McCarthy, J. & Kak, N. (1987). The determinants of fertility in rural Egypt: a study of Menoufia and Beni-Suef Governorates. *Journal of Biosocial Science* **19**, 195–207.

Gharbi, M., Bachtarzi, T., Nacer, M. T., Bouzidi, Z. & Soukhal, A. (1997). Le recours au système de soins dans les premiers mois de la vie conjugale. In *Conception, naissance et petite enfance au Maghreb*, pp. 155–62. Cahiers de l'IREMAM no. 9/10. Aix en Provence.

Handwerker, W. P. (1983). The first demographic transition: an analysis of subsistence choice and reproductive consequences. *American Anthropologist* **85**, 4–27.

Hill, A. G. & Randall, S. (1984). Différences géographiques et sociales dans la mortalité infantile et juvénile au Mali. *Population* **40**, 921–46.

Irons, W. (1979). Human female reproductive strategies. In *Evolutionary Biology and Human Social Behavior*, ed. N. A. Chagnon & W. Irons, pp. 169–213. North Scituate: Duxbury Press.

Kirk, D. (1968). Factors affecting Moslem fertility. In *Population and Society*, ed. C. B. Nam, pp. 235–43. Boston: Houghton Mifflin Co.

Lancaster, J. B., Altmann, J., Rossi, A. S. & Sherrod, L. (1987). *Parenting Across a Life Span*. New York: Aldine de Gruyter.

Ministère du Plan, Royaume du Maroc (1983). *Population légale du Maroc, d'après le recensement général de la population et de l'habitat de 1982*. Rabat: Direction de la Statistique.

Ministère du Plan, Royaume du Maroc (1984). *Caractéristiques socio-économiques de la population d'après le recensement général de la population et de l'habitat de 1982*. Rabat: Direction de la Statistique.

Nagi, M. H. (1983). Trends in Moslem fertility and the application of the demographic transition model. *Social Biology* **30**(3), 245–62.

United Nations (1971). *Human Fertility and National Development: A Challenge to Science and Technology*. New York: United Nations Department of Economic and Social Affairs.

Varea, C. (1993). Marriage, age at last birth and fertility in a traditional Moroccan population. *Journal of Biosocial Science* **25**, 1–15.

14 The emergence of health and social inequalities in the archaeological record

MARK NATHAN COHEN

Historians and philosophers have long debated the meaning of civilized progress for human health and the quality of human life. However, the popular presumption which dominates most Western thinking and teaching about our history is that history is synonymous with progress and progress means improvement in health. A corollary of this presumption is that inequality or class stratification, which is a relatively recent phenomenon intimately associated with the rise of civilization, actually benefits even the poor because the rich lead the poor 'upwards' so that, despite disparities in wealth, the poor actually do better than they would if left uncivilized. However, we now know a good deal about the modern significance of 'civilized' behavioural patterns for health and nutrition, and a great deal of data are now available about the comparative health, nutrition and demography of the 'primitive' *versus* the 'civilized' poor. Moreover, patterns of health through prehistory can be traced in skeletons from prehistoric cemeteries. All three sets of data tell a story very different from the image of civilized 'progress'.

Anthropologists commonly recognize three broad categories of human society with special reference to inequality – those described as *egalitarian*, those which display *ranking* and those which are socially *stratified*. We have found that an enormous variety of societies, across both time and space, fits reasonably well in these broad categories. The three categories have two broad, major correlates. Firstly, they are clearly and conspicuously related to the size and density of the populations involved. The smallest societies are generally the most egalitarian; the largest are the most sharply stratified.

Secondly, larger and denser societies have, as a rule, progressively replaced smaller societies in human prehistory (although only in a statistical sense, not in the sense of a universal, unilinear sequence or law) so human societies have tended to move from egalitarianism to stratification.

The three types of society fall crudely, statistically, along a time line. We think that most of the human societies of early prehistory were egalitarian although there are some notable exceptions. Societies which display ranking become more common beginning 10,000 to 15,000 years ago. Stratification is more or less synonymous with the origins of civilization and 'the state' beginning about 5000 years ago, although a dwindling number of egalitarian and ranked societies persisted well into the twentieth century.

Stratified societies, which can wield a great deal of political power whether or not they are otherwise superior in any adaptive or moral sense, have replaced more egalitarian societies through growth, incorporation or competitive elimination, or have driven them to extreme environments such as deserts, the Arctic or central rainforests in which the more powerful groups have (until recently) had little interest. Human 'progress' is synonymous with population growth and the displacement of smaller societies by larger ones; and it is therefore also synonymous with increasing stratification. For anthropologists, 'civilization' is not merely coincident with the emergence of stratification, it is defined by the emergence of stratification and coercive power as much as, or more than, by any other quality.

The density of human population and the size of human groups are in turn closely correlated with food-getting technology. It has been customary to assume that the correlation is Malthusian in nature: that population hovers below a limit set by technology until a new invention raises the limits of the food supply. In contrast, I (Cohen, 1977, 1989) among others, following Ester Boserup (1965), have argued that it is the growth of population itself that stimulates both technological and social reorganization. That point is moot; but the correlation itself is not.

The smallest and most egalitarian groups typically live by foraging or hunting and gathering wild resources. Under reasonable circumstances, hunting and gathering provide a rich and varied diet with a moderate workload. But under all but the richest circumstances it only does so when human population density is low and groups are small. Such groups typically average no more than 30–50 people and exploit territories averaging at least one square mile per person. Such egalitarian hunting and gathering communities also tend to be relatively open with individuals coming and going at will and this freedom to move is one of the major underpinnings of their egalitarian structure. At the same time, low population densities and the absence of any form of transport mean that the number of people who are in direct or even indirect contact in any defined time period is relatively small.

The adoption and subsequent intensification of farming increases human efficiency in the use of land and permits more people to feed themselves in a given territory (although farming probably does not improve the efficiency of labour). Agriculture began, I believe, only as some combination of the growth of the human population, and the decline of preferred wild resources (particularly large wild game) made hunting and gathering increasingly labour intensive and unattractive. Farming techniques, in turn, also demand sedentism and the storage of foods since harvests tend to be seasonal and crops need to be tended and then stored.

Sedentism and farming also appear to increase human reproductive rates for several reasons further stimulating population growth and density. Natural human fecundity may increase with sedentism possibly because of reduced strain on women carrying children, changes in diet and food processing, and the ready availability of weaning foods, although the reasons are hotly debated (see Cohen, 1989). Farming economies also utilize the labour of children more readily than foraging economies, affecting birth control decisions in the direction of larger families (Hassan, 1981).

But farming, storage and sedentism not only *permit* people to aggregate in larger groups at higher densities, they also *encourage* or even *require* people to choose larger population aggregates for two reasons. Firstly, sedentary farming populations, stored foods and towns or cities are vulnerable to attack, conquest, expropriation of stored resources, and even enslavement and there is safety in numbers. Hunter–gatherers are notoriously difficult to conquer or enslave although they can be driven out or killed. One can take their territory but not their stored resources. Secondly, economic or craft specializations, which may begin to emerge, encourage people to congregate because specialization reduces the need to spread out over the land and because proximity improves the efficiency of specialized tasks.

Agriculture (and certain kinds of space-intensive foraging with storage) and/or the larger denser populations which cause or accompany them, also change social rules slightly. They broaden the definition of ownership of resources so that at least farmed and stored foods are private property and planted trees or improved fields may also be. Sedentism and prior investment in resources also mean that the mobility that foragers use to resolve conflicts and resist social pressures are no longer available. Groups are less likely to split and they are less likely to welcome new members.

The combination of sedentism and vulnerability in turn makes stratification and central government control possible. Earlier in human history people simply moved away from anyone 'who would be king' or who simply wanted a greater share of the food, just as members of modern foraging societies did in the twentieth century. Stratification and civilization did not emerge because someone invented them, they emerged because the conditions enabling people to resist being stratified were gradually undermined.

Civilizations in turn are defined by several principles:

(1) They are often heterogeneous assemblages of specialists who are functionally interdependent – but often with conflicting interests – rather than just neighbours, friends and relatives farming in parallel to one another.

(2) Stratification involves marked class-based inequalities in which one class permanently owns *all* the natural resources or means of production and members of the other class(es) own nothing but their labour.

(3) Civilizations employ real physical coercive power (execution, forced exile, maiming, police, jails, armies) to reinforce the system of heterogeneous occupations and classes with their conflicting needs since the correct participation of the lower class is no longer likely to be maintained by individual self interest.

(4) Civilization involves urbanization to concentrate specialized functions for efficiency and defence.

(5) Civilizations are often defined at least in part by fixed spatial boundaries (lines on a map) and not by the (often temporary) location of groups of people.

Civilizations appear to have emerged independently in several areas of the world. (The Middle East and Egypt, India, China, sub-Saharan Africa, Peru and Mexico are often mentioned but the exact number of truly independent centres is in dispute.) From these centres, civilization then spread outward. The presence of a civilized state appears to act as a catalyst for the development of others.

While increases in group size and the incorporation of smaller groups have resulted in *internal* stratification of individual societies, competition between such groups has also resulted, often through conquest, in a second kind of inequality, inequality of power and exchange *between* societies. Differentiation and class stratification refer not only to people within a community but to differences among communities themselves, since civilization implies a loss of community self-sufficiency and participation in networks of trade (and therefore of politics) which are inevitably hierarchical (see Strickland and Shetty, Chapter 1). In recent years it has become common to refer to the unequal world distribution of wealth and power as the modern 'world system' but similar power networks are quite literally as old as civilization itself, i.e. as much as 5000 years old in some parts of the world. For example, regional satellite communities of ancient Egypt can be identified throughout most or all of the history of that civilization.

Civilizations are typically built in part by outright conquest, which, as Robert Carneiro pointed out (1970), involves both power and *circumscription* or limits on people's ability to expand their resource base or move away resulting from physical barriers to movement, sharply declining resource potential in possible areas of expansion, or the proximity of other people. Warfare did not begin with civilization, but warfare for conquest of territory and/or enslavement of people is a civilized phenomenon.

Archaeologists identify these trends in the archaeological record in a number of ways. The increasing number and size of habitation sites testify to increasing density of population and the increasing size of individual communities. The increasing depth of habitation refuse (thickness of cultural strata), the presence or absence of permanent construction of houses and storage facilities and preservation and recovery of resources representing all of the seasons of the year indicate permanence or sedentism. The increasing local differentiation in art and artefact styles reflects the closing of social boundaries. The elaboration of non-economic or symbolic goods indicates social differentiation and ranking. The presence or absence of particularly demanding crafts and their distribution in archaeological sites indicate the presence of specialists. The scale of corporate construction implying massive amounts of labour, evidence of large scale inter-regional trade, and the elaboration of special tombs implying massive differences in wealth, are characteristics that signify civilized society and stratification. The most visible trappings of civilization in the archaeological record are monumental constructions requiring co-ordinated labour well beyond the command of any tribal 'chief'. I often ask my students what it takes to build a pyramid like those of ancient Egypt, one of the first civilizations, and they always talk about architectural design, stone cutting tools, water power, pulleys, ramps and perhaps a stable food supply or even engineering. What they fail to realize is that pyramids are, more than anything else, a sign of a major social and political transformation. What it really takes to build one is the ability to put thousands of people to work on a basically questionable task and keep them there for periods of decades or more. That requires coercive power. And it is that demonstrated ability that is the keynote of civilization in the archaeological record.

The satellite communities of such civilizations can be identified by the presence of shared artefact styles and the movements of raw materials which can often be traced to their source. Archaeologists are also aided in their reconstructions by the predictability of certain aspects of ethnographic and historical data in relationship to group size, house size, labour costs, etc., which permits us cautiously to fill in or flesh out the archaeological record. And of course, once civilization emerges written records may describe the structure of the society, although they are often as incomplete or biased in their own way as the archaeology itself.

The combination of these trends towards stratification occurring repeatedly in various regions of the world (the regions at least partly independent of one another) suggests that they represent independent parallel evolution of common solutions to common social and economic needs, and not simply invention, historical chance or cultural diffusion. It has been suggested (Cohen, 1977) that, like agriculture, they primarily represent common solutions to the problems of increasing population density. That interpretation remains controversial although the correlation with population density is unmistakable.

What does 'egalitarian' mean?

Individual members of 'egalitarian' societies are not necessarily equal in their wealth or power – and ethnographic descriptions make clear that the individual inequalities can have dramatic effects on the health, nutrition and survival of individuals, primarily those who cannot fend for themselves. However, in such societies, all able bodied persons (including children from a relatively young age and the surviving, capable elderly) are free to obtain food and have the same access to resources as others. Individual ownership is recognized only after labour has been invested in specific resources in the act of gathering or processing – or, more rarely, ownership may be invested in tending selected long-lived resources such as trees. But the rights of such 'ownership' are typically far more limited than they are in our own system. No one (other than some foetuses, some infants, and, sometimes the crippled or incapacitated elderly) is denied access. Moreover, such groups lack storage of food or accumulated possession of other types of symbolic wealth (because they move frequently and so minimize the accumulation of 'things') so that inequalities resulting from superior performance tend to be temporary. Prevailing ethics tend to make hoarding or private accumulation difficult, and there is no mechanism for storing wealth for the long-term. Those ethics are not really very different from ours – or at least not very different from the rules we obey when we are dealing with family, friends and kin. The difference is that in egalitarian societies everyone falls into the categories of family, friends and kin. So sharing is common. Your family, kin and friends have a 'right' to a share of what you have, in much the way that we cannot really refuse to let a friend or kinsman come to Christmas dinner and we cannot refuse to serve as much food as people want to eat. Moreover they know they have the right so they take what they want. A hunter who gets a large animal will expect to be met by kin, friends and neighbours wielding knives to take a share. And the prevailing ethic, which is the real 'power' in the group, supports the knife-wielders rather than defending the 'property interests' of the hunter.

It is important to note that, in such groups, aggressive sharing refers not only to essential resources such as food, it also extends commonly to symbolic wealth items such as jewellery. This is one major reason why storing wealth is difficult which in turn makes it difficult to accumulate or demand disproportionate shares of real resources. An anthropologist friend of mine had a prized necklace 'stolen' in the field by a member of such a group who simply made admiring noises and took it off her neck like an aggressive sister might in our own culture. The 'thief' wore it quite openly for a brief period until it was taken by the next person (following fairly strict rules of etiquette!) and so on until everybody had had a chance to take/share/'steal' it. After this, having understood the pattern, my friend 'stole' it back again mouthing the appropriate admiring noises. It is clear that status differences based even on symbolic wealth would be hard to maintain in such a society.

In such a system differential success implies not differential wealth but differential obligation to share and to be rewarded with friendship, respect, reciprocity or sexual favours. Respect is a wonderful commodity because it is diffuse, poses no economic threat, and is potentially infinite in its expansion so that unlike exchanges for other commodities, exchanges for respect never become too one-sided to function. Moreover, respect dissipates and requires constant renewal, motivating the successful to continue their participation. In short, strong pressures reinforce equal access to economic goods.

Anthropologists describe such economic exchange as 'reciprocity' or even as 'tolerated theft'. Rather than being exchanged with strangers in a market for profit, goods are exchanged through pre-existing social ties, thus reinforcing those ties. Much like our exchange of presents within a family, there is an unspoken ethic of 'from each according to ability, to each according to need', and repayment may be in the form of small tokens, respect and prestige or awareness of future obligations.

Such societies are also 'egalitarian' in their leadership. Differences in ability and success certainly exist and leaders emerge from among the successful. But five qualities distinguish such leadership from that of more complex societies and limit the potential for inequality. Firstly, such leadership tends to be achieved by success and is not ascribed. Secondly, it tends to be ephemeral, lasting only as long as one is successful. Thirdly, it tends to be specialized, applying only in the context of success and not extending to other spheres (so a good hunter will be asked to lead the hunt but not necessarily to settle disputes, etc.). Fourthly, leadership implies authority (judgment which is respected) but not power (i.e. the ability to coerce through physical force.) A good hunter can direct others because his judgment is respected, but they are free to stop obeying whenever they wish. Fifthly, leadership conveys no economic privilege. Such leaders, in short, are very like the leaders we choose or elect within our own small groups.

If an individual tries to be too 'big' he is likely to face the ridicule of the group, as Richard Lee demonstrated graphically in an oft-cited paper entitled 'Eating Christmas in the Kalahari' (1969b); and if he tries to enforce his leadership or privilege he will find that others in his group exercise the ultimate freedom of hunter–gatherers. They leave, taking their individual self-sufficiency with them to forage somewhere else. Groups commonly split after disputes which get more common as the groups get larger and the splinter groups may coalesce in new patterns through a pattern known as group 'flux', a fluidity reflected in the relative absence of stylistic boundaries in the archaeological record

Ranking

At this second level of size and complexity (i.e. in ranked societies), it is common for people to tolerate private ownership and unequal ownership of forms of symbolic wealth (e.g. jewellery) and of economic goods including produced food and improvements to resources. But the inequalities tend to be on a graded scale of individuals, small and temporary, rather than sharply bounded categories; and they do not imply ownership of (or even privileged access to) basic economic resources such as food, land, water, etc., or to other 'means of production'. One's crop can be owned, and storage makes possible the perpetuation of unequal success – but there is nothing but prestige and reproduction to reward unequal success beyond a certain point. Moreover, everyone has the right to farm some land and to own what is produced and there are no striking contrasts in the equipment with which people farm.

The elaboration of symbolic wealth and the tolerance of unequal symbolic wealth results at least in part from problems of information processing as the number of 'other' people gets larger (Cohen, 1985). Since people no longer recognize all of their fellows as individual friends and kin, they begin to classify and stereotype others into groups, based on large scale kin affiliations (clans or lineages) or on craft specialities, and they begin to use various types of symbolic wealth as visual signs and symbols which allow discrimination and stereotyping. I refer to discrimination and stereotyping here only in their most benign sense, sorting people into categories for the sake of organization but without judgments about inequality or rank. A genuine need for classification and stereotyping to simplify information processing underlies the secondary development of discrimination in its more malignant sense since classifications, initially involving relatively benign, parallel or segmentary divisions, lead easily to invidious comparisons.

As the number of people becomes too large to co-ordinate informally through *ad hoc* leaders, ascribed leaders marked by symbolic wealth also emerge because of the need for predictable, permanent foci of communication and decision making. And, since the group is too large to accomplish all necessary sharing on a face to face basis, centralized sharing or redistribution emerges in which the central person ('big man', 'chief', etc.) acts as a central repository for produce and redistributes it to members of the community. There is the obvious possibility here for the differential control of wealth and in fact such a 'chief' may use such wealth for political favours or for small corporate projects (e.g. by assigning earth moving and building tasks and feeding the workers). But often chieftainship may be more of an economic drain than a privilege to an individual. Just as the treasurer of one of our own informal associations or projects who is acting as a redistributor, often ends up making up a shortfall from his own pocket, the redistributor chief may end up poor. Such chiefs still generally lead from authority, not power, although their authority is now more permanent and more reinforced by symbolic reminders of office – and people are not so free to move away as they once were.

Stratification

The real watershed in social organization and inequality occurs with the appearance of civilizations. Unlike what came before it, class-based stratification is permanent, ascribed, inflexible and largely unrelated to individual capabilities. It no longer refers to inequality in personal ability or individual investment but to prior ownership of essential resources and to capital goods produced by prior investment. Moreover, we are no longer talking about slight inequality on a graded scale in which a successful individual or one with a better piece of land, has 110%, 120% or even 200% of what a less successful individual has. Class stratification is a system in which members of a privileged class may own even up to or over 10,000% of what a poor person owns. Members of the lower class have no rights to those resources – unless they trade their labour or their daughters, or are fed by benign masters.

Civilization is related to potential inequality in another way. Whereas smaller societies are based on the fact (or sometimes the fiction) of kinship relationships and presume an inherently positive relationship towards family, friends and kin who are essentially like oneself, civilizations are commonly built of heterogeneity – different ethnic groups and different craft specialists whose needs may be (or may be perceived to be) in conflict, but who are actually more dependent on one another than are farmers because specialization reduces self-sufficiency. The combination of inter-dependence and perceived difference and conflict of interest makes all individuals more vulnerable, but particularly those of the lower classes. The modern economist Amartya Sen (1981; see also Pelto and Pelto, 1985) has described how social and political 'entitlements' (the obligations of others) increasingly replace individual self-sufficiency in the food quest as civilization becomes more complicated. For present purposes, Sen's key point is that individuals become increasingly dependent on economic, social and political relationships, socially defined rights to food (entitle-ments) and the obligations of others to provide food which once people obtained for themselves. Overall efficiency may increase through specializ-ation of labour, but each individual becomes more vulnerable because he/she risks not only the failure of food supplies *per se* but also the failure of the entitlement rights. The problem is that the most reliable entitlements – those involving the obligations of family, friends and kin – gradually lose their power, leaving people to the mercies of fickle governments or equally fickle commerce. Famine can result, even when food is plentiful, from, among other things, the fact that trade routes are blocked, that govern-ments cut off supplies or stop providing food stamps, or that one's own economic speciality, required to enable one to earn the money needed to buy food is no longer viable.

The benefits and costs of civilization and stratification for health

Civilized organization can make major contributions to human well-being by allowing professional managerial skills and other specialized skills to emerge, by permitting stored foods to be used for emergencies, by permit-ting organized labour to be directed at essential large tasks like building irrigation ditches, and by permitting corporate investment of centralized wealth, potentially for benign ends like medical research, weather forecast-ing (or forecasting the Nile's annual flood). Civilized size and organization also enhance the aggressive and defensive power of groups, which clearly and significantly contributes to the biological welfare of their citizens although of course it commonly acts to the detriment of others. Civiliza-tion, of course, also makes it possible to withhold rights and privileges from one's own citizens to an unprecedented degree (Wittfogel, 1957). Anthropologists and archaeologists agree that civilization is both exploitative and managerial. We disagree as to which function came first and which has prevailed or in what balance they occur.

It is a common part of our historical mythology to assume that these managerial contributions have offset the disadvantages of class stratification and enhanced the well-being of all the people, reducing the burden of disease and malnutrition and increasing life expectancies. But in fact the advantages must be balanced against the disadvantages inherent in stratification, in the spatial and economic organization of civilizations, and in governmental ability to deprive people of the means of survival. It becomes an empirical question whether, and for whom, life has been enhanced (see chapters by Macintyre, Wilkinson, Stephens, and Ben-Shlomo and Marmot). This is a question that can now be addressed using three types of data:

(1) 'Uniformitarian (natural science or "natural law") predictions' about the probable behaviour of parasites, nutrients and human physiology under different conditions in prehistory based on contemporary observations and epidemiological knowledge, models, computer simulations, etc.

(2) Study of the health of ethnographic (i.e. living) members of small societies that, despite their inevitable participation in the modern world system, retain some features of behaviour and health associated with smaller, more egalitarian groups.

(3) The skeletons of prehistoric populations unearthed through archaeology. Human skeletons respond to the presence or absence of various kinds of insult – wounds, workload, infection and malnutrition and episodes of unspecified stress leaving non-specific and specific scars on the skeleton. They also display variations in growth and mortality.

The empirical analysis of data on health in small 'uncivilized' contemporary groups and in recorded history as well as in the skeletons of prehistoric populations suggests that, in fact, throughout most of history and prehistory, the rise of class stratification, with all its enhanced power for managing investments and solving human problems has not offset problems inherent in larger populations and class stratification itself. Rather, civilization has exacerbated health and nutritional problems for most people throughout most of history. There is little if any evidence of 'permanent' improvement in health, nutrition or longevity anywhere in the world until sometime in the nineteenth century, and in fact there is a great deal of evidence to the contrary. Even in the nineteenth century, improvements were largely confined to politically privileged upper classes or rich countries and may have resulted more from world dominion than from scientific progress. The benefits did not reach most contemporary populations until well into the twentieth century. Moreover, the clearly progressive history of the last 50 years may well be partly a fluke. It certainly is not the culmination of a steady trend. It would appear to be related more than anything else to the fact that we are enjoying a relative hiatus in the onslaught of new epidemic diseases that has little relationship to anything people have done. And in fact the hiatus has been so brief, in historical terms, that it may not be a hiatus at all, rather just a minor blip on a curve. We have done very little that could discourage the spread of the next world pandemic by an unknown disease and we have done a great deal to encourage one (Fenner, 1970; Cohen, 1989; Garrett, 1994; Mims, 1995; Fenner, pers. comm.; see also chapter by Porter and Ogden).

There is a great deal of evidence to suggest that lower class status and participation of satellite communities in the networks of civilizations have had negative effects beginning with Ancient Egypt and its Nubian satellites (Martin *et al.*, 1984), and extending well into the twentieth century.

The pattern of data from skeletal pathology suggests a widespread, parallel, increase in many common indicators of pathology as hunter–gatherers become farmers and as they participated in larger scale civilizations, although the latter data are less complete (Cohen and Armelagos, 1984; Cohen, 1989; Larsen, 1995). Caution needs to be observed in inferring community health from skeletons because a cemetery may not be a representative sample of the living population and because several factors, other than disease itself, affect whether or not a given skeleton records an insult. For example, an individual who dies from an infection too quickly will not display the insult in the skeleton because skeletal lesions develop slowly. An individual who lives longer with the insult – presumably because the individual is actually healthier in other ways – is more likely to display a skeletal scar. So skeletal pathology and health may actually be *positively* correlated. An increase in the frequency of visible skeletal lesions such as that which commonly occurs as groups become more sedentary could theoretically reflect an improvement in community health in which more people lived long enough to develop skeletal lesions. (For a full discussion see Wood *et al.*, 1992; Harpending, 1990; Ortner, 1992.) However, it is possible to show that these other factors, though unquestionably real, cannot account for the pattern of increasing pathology we observe (Cohen, 1989, 1997; Cohen *et al.* 1994). Briefly, this paradoxical interpretation can be ruled out in the data discussed here for several reasons. Firstly of course, the other lines of evidence – ethnographic descriptions of health and uniformitarian 'predictions' from the known behaviour of parasites and nutrients – suggest that infection and malnutrition should and do increase when human groups become denser and more sedentary so concurring patterns in the skeletal pathology should hardly surprise us. But, secondly, and more importantly the interpretation of pathology as reflecting improved survivorship can simply be ruled out on quantitative grounds. The trends occur widely, in parallel, ruling out purely local or random problems in cemetery sampling (who gets buried and who does not). Archaeological preservation can often be ruled out since (with the obvious problem of poor preservation of infants and the elderly) most of the data being compared come from the relatively well preserved core of the population and since differences in preservation do not clearly parallel the trends seen. And the arguments based on improved survivorship in later populations can also *as a rule* be ruled out because we can show quite easily that survivorship cannot have improved to any significant degree across the board after the adoption of farming or through the period of the early civilizations. Simply summarized, the known pattern of population growth rates over the past 100,000 years of prehistory makes it impossible to contend that there was a general improvement in human survival which could account for the widespread prehistoric increase in visible pathology. Population growth did not accelerate (and could not have accelerated using any reasonable population estimates) to any significant degree during the period of prehistory under discussion; and what did occur probably reflects increased fertility, for which there is good evidence, not increased survival, for which there is little if any such evidence.

The data are clearest with regard to the quality and quantity of human nutrition. Our own experience suggests that an array of animal and vegetable foods eaten fresh with minimal processing is the best way to achieve a balanced diet, so we ought to expect that hunter–gatherers who eat that way would be well nourished. And, the new field of evolutionary medicine can be, and has been, used that way to suggest that our bodies are probably designed for the kind of diet hunter–gatherers enjoy (Eaton *et al.*, 1988). A comprehensive review of the world's few remaining foraging communities (Cohen, 1989) suggests that their nutrition is, in fact, clearly superior to that of modern satellite communities of the 'Third World', and of the modern poor of any country, despite the fact that those modern foragers now live in some of the world's poorest environments. Reports of vitamin deficiencies among hunter–gatherers are rare, except under extreme physical conditions where individuals are starving – either in extreme, central deserts or the Arctic where no one else even tries to live. Secondary deficiencies due to parasites are reported in some tropical rainforest groups; but most hunter–gatherers have only light parasite loads. Even in the tropics nutrition is generally good; and the problems increase rather than decrease as people grow in numbers or become sedentary, and as parasite loads increase. Protein deficiency is almost never reported at all among hunter–gatherers and we now wonder whether protein deficiency was ever a problem in the hunter–gatherer world (Speth, 1988.) Mineral deficiencies (iron, calcium, magnesium) are almost unknown outside the arctic or in rainforest populations with heavy parasite loads. Caloric intakes in all but the most extreme cases average at or above modern Third World averages (Cohen, 1989). People like the Hadza in East Africa, in environments which are still relatively rich in game, have caloric intakes similar to those of affluent Westerners. But even the !Kung, whom we now think of as a relatively impoverished foraging group, who live in a hot dry desert where heat and the lack of water restrict their activities, and who have few resources, and are legally forbidden to hunt some large animals, have caloric intakes that match Third World averages, for example equalling the *average* caloric intakes estimated for India and China. And, they do far better than the Third World poor (see Cohen, 1989 for an overview and other citations; for specific studies Jelliffe *et al.*, 1962 (San); Lee, 1969a (San); Truswell and Hansen, 1976 (San); Jones, 1980 (Anbarra); Sen Gupta, 1980 (Onge); Ichikawa, 1981 (Pygmy); Hill *et al.*, 1984 (Ache)).

Moreover, optimal foraging studies or studies of foraging efficiency in various activities (e.g. Winterhalder and Smith 1981; others summarized in Simms, 1987; Russell, 1988; Cohen, 1989) provide ample evidence that prehistoric hunter–gatherers in a world rich in large game, which is the most efficiently exploited of wild resources when plentiful, would have fared far better than later groups. In any case, it is important to note that the natural ratio of calories to other nutrients is lower in hunter–gatherer diets than in modern diets (and 'standard' dietary studies) in which calories are more common. For hunter–gatherers, calories and fat are the elements most likely to be in short supply – that is not true of the modern world. Hunter–gatherers in fact may discard hunted game that has no fat presumably because fat is scarce but protein is not. So a hunter–gatherer's diet with a caloric intake matching that of the modern Third World is actually a far superior diet.

Farming, particularly as it intensifies gradually eliminates dietary variety. It results in loss of freshness, and therefore of water soluble vitamins, as reliance is placed on stored foods. It often reduces the nutritional quality of individual foods such as wheat, and it requires heavy dependence on foods chosen for the density of calories per unit area and for the ability to be stored, not for nutritional value. Large scale transport of food, without expensive modern refrigeration, further reduces the variety and quality of foods consumed by putting a premium on 'shelf life' and packaging.

In addition, and again with the exception of extreme environments in which many hunting and gathering groups now live and where reported starvation biases our picture of hunter–gatherer life, there is little evidence that extreme hunger or starvation is a more significant problem for these populations than for more 'modern groups' (sources have been reviewed in Cohen, 1989). Although farmers can store food, their domesticated and concentrated crops, which have often been removed from their natural environment, are more vulnerable to crop failure than wild ones. Storage has always been imperfect; and it has always increased the danger of human predation.

Trade networks are also a double-edged sword. Modern satellite populations may benefit from world trade networks when those systems work. But people's inability to generate economic demand for food, political boundaries, politically motivated or required specialization in failed crops or disfavoured industries, political with-holding of food, the ability of the rich to exert disproportionate demand on resources even from far away, failure of transport and economic institutions, fluctuations in world markets, cheating and market manipulation by managers, and the movement of resources away from populations in need, are all factors which apparently make modern Third World groups more, not less, vulnerable to malnutrition and starvation. The USA, for example, imports beef from areas of Central America where peasants have no land on which to feed themselves, including from the Maya (with whom I have worked). In short, where natural disasters may cause hunger or starvation in primitive groups, the very political mechanisms that supposedly protect civilized people from starvation can actually harm them. The rich clearly benefit, the poor do not. It seems very clear to me that the problem in the modern world is not a lack of supply of food, it is lack of economically viable *demand* for food on the part of the poor, which no new food technology will correct.

Skeletons of early, Palaeolithic hunter–gatherers indicate that they were commonly rather large people. The overall trend in human stature until fairly recently, if mixed, has been downward although it has been reversed for affluent people and nations in the last 100 years (Cohen and Armelagos, 1984; Fogel, 1984; Fogel *et al.*, 1985). The seventeenth- and eighteenth-century Europeans against whom we proudly measure ourselves to demonstrate our progress, were actually some of the smallest people who ever lived. The skeletons of hunter–gatherers also commonly show fewer stigmata of malnutrition than later populations. For example, rates of porotic hyperostosis or iron deficiency anaemia are generally lower for hunter–gatherers than in later populations. However, it is not clear whether the widespread temporal increase in anaemia reflects reduced dietary iron intake or secondary loss to parasites, or even the body's own tendency to with-hold iron from circulation to fight parasites and infections (Stuart-Macadam and Kent, 1992). Where comparisons are possible, childhood growth patterns and signs of prenatal stress, assessed through growth disruption markers of teeth and the length of long bone diaphyses compared to dental ages, suggest that childhood was healthier among hunter–gatherers than among later populations. Rickets is found almost exclusively in the skeletons of relatively recent populations.

Infectious disease also seems to have increased through human history not only with population growth, sedentism and urbanization, but also with such class-based factors as trade patterns serving primarily the wealthy, conquest, troop movements, the imposing of 'civilized' lifestyles and the resettlement of conquered populations. Tuberculosis is very clearly a disease of crowded urban poor and of ghettoes and reservations. Bubonic plague spread over large scale trade routes (Biraben, 1968), and international military operations are clearly implicated in the spread of influenza and cholera.

This increase lasted at least until the end of the nineteenth century, and, after a brief respite, may well be increasing again. This involves not only AIDS and tuberculosis (see Porter and Ogden), but also a myriad of regional infections, as for example Ebola, which may now spread through modern trade networks. This raises the spectre that our 'progress' is not the historical norm but a brief deviation from the norm fuelled by privilege. Contemporary hunter–gatherers, for example, display low rates of infantile and other diarrhoea, and of anaemia related to parasitic worms, as a function of their small group size, low population density and mobility. It is thought that the range of infections which could have plagued such populations before civilization would have been comparatively limited. We usually assume that only two classes of disease could have afflicted such populations: (1) zoonotic and soil borne diseases which do not depend on people for survival and transmission; and (2) chronic diseases able to survive in small human populations because they remain in each individual for long periods.

In contrast, the major epidemic diseases are commonly thought to require urban civilizations for their survival and to require civilized trade and military manoeuvres for their dissemination. One model of the history of such epidemic diseases suggests that, because they kill or immunize each victim, depriving themselves of fuel, any local epidemic in a small and isolated group of people (or groups connected only by speed of foot) would burn out before it could spread very far. Therefore, although mini-epidemics might have occurred in prehistory as the result of local mutations in the structure of common or zoonotic viruses, the modern epidemic or pandemic diseases in widespread, ongoing form must have originated from mutations which occurred after human beings were living in large dense populations with trade networks so that new victims could be recruited as fast as old victims were eliminated. According to this scenario, these are entirely diseases of civilization (Black *et al.*, 1974).

An alternative model suggests that the epidemic diseases could have existed and snaked their way slowly around populations in the prehistoric world with devastating consequences in each location (CIBA, 1977). But even by that model they would have moved faster and wreaked havoc more often, presumably contributing more to overall mortality in later, denser populations, until they reached such high frequency in civilized populations that they returned while survivors of previous epidemics were still numerous and only young children were vulnerable, making them diseases of children who could survive because they were cared for by healthy parents as occurred in recent history.

In prehistoric skeletons, epidemic diseases (except possibly smallpox) are rarely visible by present techniques, although new techniques of DNA analysis are promising (Spigelman and Lemma, 1993; Rafi *et al.*, 1994). But periostitis and osteomyelitis – non-specific signs of infection – almost invariably increase in frequency as group sizes increase, as would be expected from basic epidemiological principles. Even yaws, a chronic disease handed directly from person to person, and therefore considered a disease likely to have plagued human populations throughout our evolutionary history, clearly increases in frequency through time in most regions as a function of group size and aggregation (Cohen and Armelagos, 1984; Cohen, 1989; Larsen, 1995). New skeletal evidence suggests that yaws was present in both the New World and the Old long before Columbus (DuTour *et al.*, 1995). But congenital syphilis, which is identifiable through congenital malformations of teeth, and therefore also presumably venereal syphilis resulting from the same (?) organism as yaws, appears only after Columbus on both sides of the Atlantic; and this disease thus appears to be associated either with his voyage, or with newly emerging civilized social patterns of the sixteenth century (Baker and Armelagos, 1988; DuTour *et al.*, 1995). Similarly, tuberculosis appears in the skeletal record with a clear concentration in late, dense, civilized settlements. Leprosy was once widely considered the 'older' form of mycobacterial infection, predating tuberculosis. But it, too appears in skeletons only in fairly recent (pre-) history.

Historically, cholera, which is not visible in the skeleton and therefore in archaeological evidence, but which does not commonly attack small isolated groups unless they are connected to an urban centre, appears to have been spread around the world by British occupation troops. Bubonic plague (also not visible in the skeleton but known only as a rare zoonosis in primitive groups) may originally have spread by caravan routes and later certainly spread by shipping. As late as the eighteenth century in France, it was still primarily a plague of major cities and seaports, not of rural communities (Biraben, 1968). As is well known, many diseases spread to the New World by European and African colonization and conquest (Crosby, 1972; McNeill, 1976).

William McNeill (1976) has pointed out that disease has even acted as an agent of inequality. Epidemic diseases which circulated through European civilizations eventually attacking only those children born since the last epidemic, killed all or almost all individuals in 'virgin soil' populations without previous exposure. The physical decimation of those populations accompanied by the apparent immunity of Europeans (who had been infected as children) was a powerful demonstration of the superiority of Europeans and their gods and contributed to the collapse of American Indian resistance as well as population. The archaeological record of this episode is now being pursued actively (Verano and Ubelaker, 1992; Cohen *et al.*, 1994; Larsen and Milner, 1994).

The teeth of human skeletons display scars of growth disruption which act as markers of biological stress on the individual during childhood, i.e. when teeth are forming. These indicators are 'non-specific' in the sense that they are known to be caused by an array of insults ranging from starvation to severe disease. It is not clear whether or not they refer specifically to famine or epidemics, or even that they have the same cause in different cases (even on the same tooth) or different populations. But their record is one of increase through time in most regions of the world. Neither these nor any other indicator provide any suggestion or indication that individuals came to be in a general way better 'buffered' against episodes of stress as they became more civilized (Cohen and Armelagos, 1984; Cohen, 1989; Larsen 1995).

There is some direct evidence suggesting the declining health of satellite communities in prehistory. Teams working with skeletons of ancient Nubians now suggest that nutrition and health fluctuated in response to the changing political climate, and that the quality of life tended to decline when political connections and trade with Egypt were at their most active. Ancient Nubia was apparently one of those satellite regions where trade moved resources *from* rather than *to* people in need (Martin *et al.*, 1984). A sequence of declining health associated with agricultural intensification, at the American site of Dickson Mounds in Illinois, was originally interpreted as a function of agricultural intensification but may (also) reflect a sequence of progressive economic exploitation by a nearby major population centre (Goodman *et al.*, 1984).

The most pathology-ridden skeletal population that has been described is not any prehistoric population at all but a free black American population from the state of Arkansas early in the twentieth century (see Rose, 1985). This population may be skewed by the effects of the documented northward migration of young, healthy adults, although there is not much reason historically to think that 'real' health and nutrition in the group were particularly good, but it nonetheless calls into question the sense that pathology is most common among 'primitives'.

Although reconstruction of life expectancy from (undocumented) cemeteries is very difficult (Sattenspiel and Harpending, 1983), the best estimates of prehistoric life expectancies based on such cemeteries, in combination with calculated rates of population growth and the observed fertility and mortality of living hunter–gatherers, suggest that hunter–gatherer groups probably had average life expectancies of about 25–30 years. This is a poor figure by affluent twentieth century standards (see Wilkinson, Chapter 4), but one which compares quite reasonably with figures from Europe as late as the eighteenth and early nineteenth centuries. The figure is better than those from many European urban centres from earlier periods and better, particularly, than recorded figures for lower classes of those cities. It is also conspicuously better than the figures for India, one of the few satellite countries for which good data are available, prior to about 1920.

There is no evidence from either ethnography or from archaeology that primitive populations lost a higher proportion of their infants or their children than populations as late as the nineteenth or even early twentieth centuries. For example, infant mortality rates in many major cities in the USA at the beginning of the twentieth century were worse than the average of the smallest groups. My own collection of available data on living hunter–gatherers and skeletal populations (Cohen 1989: 195–201, 215–22) suggests that infant mortality rates of about 200 per 1000, and child mortality rates (death prior to age 15 including infant mortality) of 400–500 per 1000, are typical of hunter–gatherers. Most of Europe did not improve 'permanently' on that until sometime in the nineteenth century, and most of the world did not improve on it until well into the twentieth century.

But what of the record of population growth? Why did the growth of world population accelerate at all after the Neolithic adoption of farming and the rise of civilization, as we normally assume (see Hassan, 1981, for a fairly standard reconstruction)? One interesting new possibility is that it did not. Work by geneticists (Harpending *et al.*, 1993) now suggests that the major increase in prehistoric population occurred much earlier, and that we are underestimating the size of the human population which existed at the dawn of farming. In this case, in keeping with the data presented here, the real acceleration in growth may have occurred at the time of efficient and relatively well nourished hunters. But that remains speculative.

Returning to more traditional perceptions, it is assumed that, after the Neolithic and through the period of early civilizations, population growth (the average for the species, worldwide) accelerated from near zero to about 0.1 or 0.15% per year. But, as mentioned above, there is a good deal of evidence, both from ethnography and from archaeological samples, that human *fertility* increased when people became sedentary. Fertility probably increased again as agriculture intensified and women's roles changed (Ember, 1983), and it probably increased again under European colonization, as several ethnographers have suggested (e.g. O'Brien, 1994). As such, it is possible and, given the evidence, probable that population growth accelerated despite the fact that life expectancy was actually reduced.

In summary, the available evidence from prehistory and from the comparison of 'primitive' and 'civilized' contemporary societies does not support the image that civilization has brought progress in the form of improved health. The structural changes inherent in civilization – large cities, dense populations, complex patterns of economic interdependence and large scale trade, actually threaten health by introducing new threats of infectious disease and reducing the probabilities that individuals will get varied diets of fresh food. The patterns of economic interdependence may also *increase*, rather than reduce, the possibility that some people will get no food at all. Class stratification, which is also an inherent, defining part of what we call civilization, exacerbates these problems by further reducing the economic demand which the poor can exert while simultaneously allowing the rich to exert disproportionate demand for luxury goods. Stratification therefore reduces the resources available to the poor far below the level that world crowding would otherwise create. It also increases population density artificially in some portions of the civilized society ('inner cities') resulting in crowding and poverty which may reach such heights that the areas become breeding grounds for infectious disease which in turn can affect the whole society. The only advantage which civilization confers for health is its capacity to invest accumulated wealth to help alleviate some of these problems. We certainly need to invest to maintain our vigilance against new epidemic diseases. We also have a moral obligation to invest in the health and nutrition of those whom class stratification and civilization otherwise harm.

References

Baker, B. & Armelagos, G. J. (1988). The origin and antiquity of syphilis. *Current Anthropology* **29**, 703–22.

Biraben, J-N. (1968). Certain demographic characteristics of the plague epidemic in France, 1720–22. *Daedalus* **97**, 536–45.

Black, F., Hierholzer, W., Pinheiro, F., Evans, A., Woodall J., Opton E., Emmons, J. West, B. Edsall, G. Downs, W. & Wallace G. (1974). Evidence for persistence of infectious agents in isolated human communities. *American Journal of Epidemiology,* **100**, 230–50.

Boserup, E. (1965). *The Conditions of Agricultural Growth.* Chicago: Aldine.

Carneiro, R. (1970). A theory of the origin of the state. *Science* **169**, 733–8.

CIBA (1977). *Health and Disease in Tribal Societies.* New York: Elsevier.

Cohen, M. (1977). *The Food Crisis in Prehistory.* New Haven: Yale University Press.

Cohen, M. (1985). Prehistoric hunter gatherers: the meaning of social complexity. In: *Prehistoric Hunter Gatherers: the Emergence of Cultural Complexity,* ed. T. Price & J. Brown, pp. 99–122. New York: Academic Press.

Cohen, M. (1989). *Health and the Rise of Civilization.* New Haven: Yale University Press.

Cohen, M., O'Connor, K., Danforth, M., Jacobi, K. & Armstrong, C. (1994). Health and death at Tipu. In *In the Wake of Contact* ed. C. Larsen & G. Milner, pp. 122–34. New York: Wiley-Liss.

Cohen, M. (1997). Does paleopathology measure community health: a rebuttal of the 'osteological paradox' and its significance for the interpretation of human history. In *Integrating Archaeological Demography: Multidisciplinary Approaches to Prehistoric Populations,* ed. R. Paine, pp. 242–60. Carbondale, Illinois: Occasional Paper 24, Centre for Archaeological Studies. (In press.)

Cohen, M. & Armelagos, G. (Ed.). (1984). *Paleopathology at the Origins of Agriculture.* New York: Academic Press.

Crosby, A. (1972). *The Columbian Exchange.* Westport Connecticut: Greenwood Press.

DuTour, O., Palfi, G., Berato. K. & Brun, J-P. (Eds). (1995). *L'Origine de la Syphilis en Europe avant ou après 1493.* Paris: Editions errance; Centre Archaeologique du Var.

Eaton, S., Konner, M. & Shostak, M. (1988). *The Palaeolithic Prescription.* New York: Harper and Row.

Ember, C. (1983). The relative decline in women's contribution to agriculture with intensification. *American Anthropologist* **85**, 285–304.

Fenner, F. (1970). The effects of changing social organization on the infectious diseases of man. In *The Impact of Civilization on the Biology of Man,* ed. S. Boyden, pp. 48–76. Canberra: Australian National University Press.

Fogel, R. (1984). *Nutrition and the Decline in Mortality Since 1700: Some Preliminary Findings.* Cambridge, Mass.: National Bureau of Economic Research, Working Paper 1402.

Fogel, R., Engerman, S., Floud, R., Friedman, G., Margo, R., Sokoloff, K., Steckel, R., Trussell, T., Villaflor, G. & Wachter, K. (1985). Secular changes in American and British stature and nutrition. In *Hunger in History,* ed. R. Rotberg and T.Rabb. Cambridge: Cambridge University Press.

Garrett, L. (1994). *The Coming Plague.* New York: Farrar, Straus, Girox.

Goodman, A. H., Lallo, J., Armelagos, G. J. & Rose, J. C. (1984). Health changes at Dickson Mounds, Illinois (AD 950–1300). In *Paleopathology at the Origins of Agriculture*, ed. M. N. Cohen & G. J. Armelagos, pp. 271–305. New York: Academic Press.

Harpending, H. (1990). Review of Cohen, *Health and the Rise of Civilization.* In *American Ethnologist* **17**(4), 799.

Harpending, H., Sherry, S., Rogers, A. & Stoneking, M. (1993). The genetic structure of ancient human populations. *Current Anthropology* **34**(4), 483–96.

Hassan, F. (1981). *Demographic Archaeology.* New York: Academic Press.

Hill, K., Hawkes, K., Hurtado, A. & Kaplan, H. (1984). Seasonal variance in the diet of Ache hunter–gatherers in eastern Paraguay. *Human Ecology* **12**, 101–37.

Ichikawa, M. (1981). Ecological and sociological importance of honey to the Mbuti net hunters, Eastern Zaire. *Kyoto University African Studies Monographs* **1**, 55–68.

Jelliffe, D., Woodburn, J., Bennett, F. & Jelliffe, E. (1962). Children of the Hadza hunters. *Tropical Paediatrics* **69**, 907–13.

Jones, R. (1980). Hunters in the Australian coastal savanna. In *Human Ecology in Savanna Environments*, ed. D. Harris, pp. 107–47. New York: Academic Press.

Larsen, C. (1995). Biological changes in human populations with agriculture. *Annual Review of Anthropology* **24**, 185–213.

Larsen, C. & Milner, G. (Eds.) (1994). *In the Wake of Contact.* New York: Wylie Liss.

Lee, R. (1969a) !Kung Bushmen subsistence: an input–output analysis. In *Ecological Studies in Cultural Anthropology*, ed. A. Vayda, pp. 47–79. New York: Natural History Press.

Lee, R. (1969b). Eating Christmas in the Kalahari. *Natural History* **78**(10), 4–22, 60–3.

Martin, D., Armelagos G., Goodman, A. & Van Gerven, D. (1984). The effects of socioeconomic change in prehistoric Africa: Sudanese Nubia as a case study. In *Paleopathology at the Origins of Agriculture*, ed. M. Cohen & G. Armelagos, pp. 193–216. New York: Academic Press.

McNeill, W. (1976). *Plagues and Peoples.* Garden City: Anchor Press.

Mims, C. (1995). Virology research and virulent human pandemics. *Epidemiology of Infection* **115**, 377–86.

O'Brien, J. (1994). Differential high fertility and demographic transitions. In *African population and capitalism*, ed. D. D. Cordell & J. W. Gregory, pp. 173–86. Madison, Wisconsin: University of Wisconsin.

Ortner, D. (1992). Skeletal paleopathology: probabilities, possibilities and impossibilities. In *Disease and Demography in the Americas*, ed. J. Verano & D. Ubelaker, pp. 5–13. Washington DC: Smithsonian Institution.

Pelto, G. & Pelto, I. (1985). Diet and delocalization: dietary change since 1750. In *Hunger in History*, ed. R. Rotberg & T. Rabb, pp. 309–30. Cambridge: Cambridge University Press.

Rafi, A., Spigelman, M., Stanford, J., Donoghue, H. & Zias, J. (1994). DNA of *Mycobacterium leprae* detected by PCR in ancient bone. *International Journal of Osteoarchaeology* **4**, 287–90.

Rose, J. (Ed.) (1985). *Gone to a Better Land.* Archaeological Survey Research

Series 2. Little Rock: Arkansas.

Russell, K. (1988). *After Eden: The Behavioral Ecology of Early Food Production in the Near East and North Africa.* London: British Archaeological Reports, International Series 391.

Sattenspiel, L. & Harpending, H. (1983). Stable populations and skeletal age. *American Antiquity* **48**, 489–98.

Sen, A. (1981). *Poverty and Famines* Oxford: Oxford University Press.

Sen Gupta, P. (1980). Food consumption and nutrition of regional tribes of India. *Ecology of Food and Nutrition* **9**, 93–108.

Simms, S. (1987). *Behavioral Ecology and Hunter–Gatherer Foraging: An Example from the Great Basin.* London: British Archaeological Reports, International Series no. 381.

Speth, J. (1988). Hunter–gatherer diet, resource stress, and the origins of agriculture. Paper presented to *Symposium on Population Growth, Disease, and the Origins of Agriculture. Rutgers University, New Brunswick, New Jersey.*

Spigelman, M. & Lemma, E., (1993). The use of the polymerase chain reaction (PCR) to detect Mycobacterium tuberculosis in ancient skeletons. *International Journal of Osteoarchaeology* **3**, 137–43.

Stuart-Macadam, P. & Kent, S. (Eds.) (1992). *Diet, Demography and Disease.* New York: Aldine.

Truswell, A. & Hansen, J., (1976). Medical research among the !Kung. In *Kalahari Hunter–Gatherers*, ed. R. Lee & I. DeVore, pp. 166–95. Cambridge, Mass: Harvard University Press.

Verano, J. & Ubelaker, D. (Eds.) (1992). *Disease and Demography in the Americas.* Washington DC: Smithsonian Institution Press.

Winterhalder, B. & Smith, E. (Eds.) (1981). *Hunter–Gatherer Foraging Strategies.* Chicago: University of Chicago Press.

Wittfogel, K. (1957). *Oriental Despotism.* New Haven: Yale University Press.

Wood, J., Harpending, H., Weiss, K. & Milner, G. (1992). The osteological paradox: problems in inferring prehistoric health from skeletal samples. *Current Anthropology* **33**(4), 343–70.

15 Eugenics and population policies

A. H. BITTLES AND Y-Y. CHEW

Introduction

Virtually any aspect of the subject of eugenics seems bound to be associated with controversy, and to many individuals in Europe, North America and especially Israel, the very term evokes potent memories of the Nazi era in Germany. Yet, during the early years of the present century, eugenics was seen as holding the key to future human progress, and there was widespread support for studies into the subject from influential supporters on both sides of the Atlantic. In the post-war years, information on the practices conducted in Germany in the name of eugenics attracted near-universal opprobrium which, in Western Europe and the USA, resulted in a rapid decline in support for the ideals espoused by the eugenics movement. More recently, there has been an upsurge of interest and governmental support for the principles of eugenics in several Asian countries, with the twin aims of controlling the numbers of their citizens and ensuring the optimum development of their human potential. To many persons in economically developed countries, any such programme runs the risk of a return to Nazi-style abuses and so must be both regarded with suspicion and resisted. For the purposes of the present volume, the genesis of the eugenics movement in the nineteenth and early twentieth centuries will be briefly reviewed, and two examples of current programmes of population control incorporating eugenic principles assessed against this background.

Eugenics in Western Europe and North America, pre-1945

The origins of the modern eugenics movement can be traced back to the theories espoused by Francis Galton in the mid- to late nineteenth century. In brief, the central tenet of eugenics held that humans should be encouraged to improve their genetic heritage. In a quotation attributed to Galton in 1908 and reproduced on the cover of the now defunct journal *The*

Eugenics Review, this position was summarized as, 'Man is gifted with pity and other kindly feelings; he also has the power of preventing many kinds of suffering. I conceive it to fall well within his province to replace Natural Selection by other processes that are more merciful and not less effective. This is precisely the aim of eugenics'.

It was envisaged that the eugenic ideal could be pursued in two ways: by restricting the reproduction of persons believed to be of inferior genetic potential (negative genetics), and conversely by encouraging individuals thought to possess desirable genetic characteristics to increase their fertility (positive eugenics). With the creation of the Eugenics Education Society in Britain in 1907 and the installation of Galton as its first president, there was a rapid increase of interest in theories propagated on the subject. The Eugenics Record Office was founded by Galton in 1904 and housed in University College London. In 1907 this became the Francis Galton Laboratory for the Study of National Eugenics, and after Galton's death in 1911 monies from his estate were used to fund the Galton Professorship of Eugenics (Blacker, 1952; Keynes, 1993). In the following years the Foundation Professor Karl Pearson, who had been Galton's collaborator and later became his biographer, rapidly expanded the work of the unit, creating databases on variables believed to be inherited as single gene traits and thus amenable to manipulation based on eugenic principles. In this work the eugenics movement attracted favourable attention and support from many persons prominent in British society, including such diverse personalities as Winston Churchill, H.G. Wells and George Bernard Shaw (Kevles, 1986).

Some indication of the level of influence exerted by the Eugenics Education Society and its successor, the Eugenics Society, can be gained from the Departmental Report on Voluntary Sterilization, commissioned by the UK Minister of Health in 1932 (Joint Committee on Voluntary Sterilization, 1934). The Committee, which included among its members the eminent geneticist R. A. Fisher who had succeeded Karl Pearson as Galton Professor of Eugenics at University College London, recommended the legalization of voluntary sterilization for persons deemed to be mentally defective or who suffered from mental disorder; those who suffered from, or were believed to be a carrier of, a grave physical disability which had been shown to be transmissible; or who were believed to be likely to transmit a mental disorder or defect (Joint Committee on Voluntary Sterilization, 1934). Even after the 1939–45 war, eugenic theories continued to be espoused by individuals such as Sir Cyril Burt, who published evidence purporting to show that due to the greater fertility of persons of lower intelligence, the mean IQ of the UK was declining at a rate of 2.0 IQ points per generation (Burt, 1952; see Mascie-Taylor, Chapter 11).

Under the influence of Charles Davenport, and funded by contributions from the Harriman, Kellogg and Carnegie families, eugenic beliefs also became an influential force in the USA during the early decades of the twentieth century, especially after the establishment of the Eugenics Record Office at Cold Spring Harbor in 1910 (Kevles, 1986). In 1912, the IQ testing of immigrants for evidence of feeblemindedness was introduced by the US Public Health Service, and in 1924 quotas were established under the Immigration Restriction Act to govern the number of potential migrants to the USA from differing ethnic and national groups (Beckwith, 1976). As early as 1889, castration had been undertaken with parental approval in the Pennsylvania Training School for Feebleminded Children (Kevles, 1986). Between 1911 and 1930 legislation on compulsory sterilization was introduced in 33 states of the Union, for stipulated offences ranging from criminality, alcoholism, the tendency to commit rape, sodomy or bestiality, and feeblemindedness (Beckwith, 1976). As in the UK, the influence of eugenics declined rapidly after the Second World War. However, in states such as North Carolina, legislation governing sterilization on the grounds of feeblemindedness remained active until 1968, with 1620 persons sterilized for this reason between 1960 and 1968 (Beckwith, 1976; Kevles, 1986). Following a proposal by the geneticist H.J. Muller (1959, 1965), a movement for 'germinal choice' was initiated in the USA, to create sperm banks from which women could opt to be impregnated with samples provided by men believed to be of outstanding genetic potential and intellect. Little evidence is available as to the fate of this programme, although the Nobel Laureate William Shockley was identified as an advocate and donor (Kevles, 1986).

In Germany, the accession to power of the National Socialist Party in 1933 was rapidly followed by enactment of the Sterilization Law which, together with an Amendment passed in the same year, obliged medical practitioners to register patients with defined hereditary disorders, ranging from moral feeblemindedness to physical deformities seriously affecting locomotion, or of a grossly offensive nature (Kevles, 1986; Clay and Leapman, 1995). With the establishment of Hereditary Health Courts in 1934, cases referred for possible sterilization were considered by official tribunals composed of two doctors and a lawyer. Between 1934 and 1939 an estimated 320,000 people were sterilized under the terms of this legislation (Clay and Leapman, 1995). After September 1939, euthanasia also became an offically sanctioned option for mentally retarded people.

The influence of eugenics could be seen in the so-called Nuremberg Laws passed in 1935, which required a compulsory medical examination before marriage. Under these Laws, if one member of a couple was diagnosed with a genetic disease the marriage would be banned, whereas if both were so diagnosed the marriage could proceed after sterilization (Clay and Leapman, 1995). In addition to these negative eugenic measures, in 1935 the Nazi authorities initiated a selective breeding programme to produce children who, it was believed, would ensure the future existence of the Nazi state. The Lebensborn organization entrusted with this task was placed under the personal control of Heinrich Himmler, the head of the SS. In homes specially established in Germany and, after 1939 in German-occupied countries including Austria, Belgium, Denmark, France, Luxembourg, The Netherlands, Norway and Poland, babies fathered by SS men and born to women of approved racial origins were delivered. Some 13,000–14,000 children are believed to have been born as a result of the programme. However, their numbers were greatly increased by an estimated 200,000 children, determined on the basis of a 21 point anthropological examination to be of Aryan stock, who were removed from Poland under the supervision of the Lebensborn organization for fostering by German couples (Clay and Leapman, 1995).

The eugenic legislation passed by the German government had been regarded with both suspicion and mistrust by many geneticists, and the question 'how could the world's population be improved most effectively genetically' was scheduled as a prominent topic of the 7th International Congress of Genetics in Edinburgh, UK. Unfortunately, the dates of the Congress, August 22–30, 1939, proved to be a mere two weeks prior to the onset of hostilities of the Second World War. The Congress President V. I. Vavilov and his accompanying delegation from the USSR were unable to attend, and on the second and third days of the meeting the German, Hungarian, Scandinavian and Swiss delegations all departed. Despite these defections the remaining delegates, who were mainly from the UK and USA, considered a document prepared by a committee which included geneticists of the calibre of J. B. S. Haldane, J. S. Huxley, H. J. Muller, G. Dahlberg and Th. Dobzhansky. Subsequently termed the Geneticists' Manifesto, the end-product was in general balanced in content and tone. For example, it confirmed that environment and heredity comprised 'dominating and inescapable complementary factors in human well-being', besides commenting unfavourably on contemporary race prejudices and unscientific racial doctrines. The Manifesto echoed prevailing eugenic thought by noting that genetic improvement would be dependent on effective means of birth control, both positive and negative. Given such circumstances it was accepted that through conscious selection it would be possible to improve the health, intelligence and temperamental qualities of humans and, as a result, 'everyone might look upon genius, combined of course with stability, as his birthright' (Editorials 1939a,b).

Table 15.1. *Legislation to control population size, Singapore 1966–1974*

1. Establishment of the Singapore Family Planning and Population Board in 1966.
2. Enactment of the Voluntary Sterilization Act in 1969, with an Amendment in 1972, and the introduction of a new Act in 1974 embodying less stringent preconditions for the approval of sterilization.
3. Introduction of the Abortion Act in 1970, and its replacement with a more liberal Act in 1974.
4. A range of associated measures controlling maternity leave, hospital delivery charges, maternal income tax relief, housing allocations, and the enrolment of children in primary education, specifically favouring families with fewer children.

Source: Saw (1990).

Population policies in the Republic of Singapore, 1969–95

With political independence in 1959 there was a strong move by the newly elected Government of the Republic of Singapore to reduce population growth, which was seen as posing a threat to the future economic development of the country. As an interim measure, this reduction was to be achieved by the expansion of family planning facilities through the offices of the Family Planning Association, but in subsequent years a national fertility policy was gradually developed. As detailed by Saw (1990), this policy comprised four main anti-natalist components, which are summarized in Table 15.1.

The apparent degree of success of this legislation in reducing family sizes can be judged from Figure 15.1, which shows total fertility rates (TFR) for the population of Singapore from 1959 to 1988. However, a substantial decline in TFR had already taken place prior to establishment of the Singapore Family Planning and Population Board in 1966, and the subsequent enactment of population control legislation prepared with the assistance of that body (Saw, 1986). Nevertheless, some indication of the extent to which the new laws on sterilization and abortion directly affected population growth can be gauged by reference to official data sources, which reveal that between 1969 and 1988 a total of 112,568 sterilizations were performed in Singapore, and 288,568 induced abortions also were conducted between 1970 and 1988. These figures can be contrasted with the 865,811 live births delivered in Singapore from 1969 to 1988 (Saw, 1988, 1990).

The major decline in TFR from 1970 onwards was in the majority Chinese community, to a rate which was well below replacement population level (Figure 15.2). It would have been apparent from the 1980 Census that, although the TFR in the Chinese community was continuing to

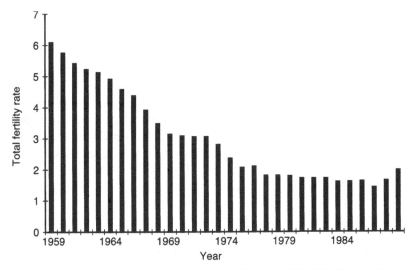

Figure 15.1. Singapore: Annual total fertility rates, 1959–1988. (*Source:* Saw, 1990.)

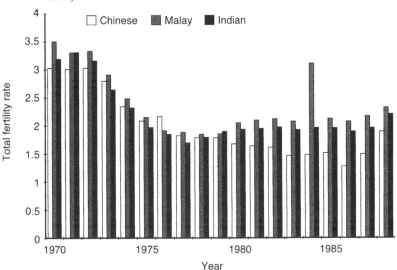

Figure 15.2, Singapore: Annual total fertility rates by ethnic group, 1970–1988. (*Source:* Saw, 1990.)

decline, TFRs for both the Malay and Indian communities had actually reached minimum values in 1978 with a gradual recovery thereafter. These differential patterns of fertility among the three major ethnic communities in Singapore were viewed with growing concern in some quarters (Saw, 1990, appendix C). To add to this worry, by the early 1980s it had become apparent that the very success of governmental anti-natalist policies could have serious effects on future population structure, signalling in particular a significant reduction in the size of the labour force and an increase in the numbers and proportions of dependent persons over 65 years of age.

As indicated by the Prime Minister, Lee Kuan Yew, in the 1983 National Day Rally, there also were concerns as to the intellectual talents of future generations, since data from the 1980 census indicated that family size was inversely related to parental level of completed education, and 13.5% of tertiary-educated women remained unmarried at age 40 (Saw 1990, Appendix A). The Government of Singapore responded to these population predictions by the introduction of a series of measures that sought to increase family sizes, albeit in a selective manner. The new legislation encompassed both positive and negative eugenic components, and aimed simultaneously to promote child-bearing by the better educated and to discourage the large family sizes of couples with lesser educational achievement. The major components of these policy modifications are summarized in Table 15.2.

The scheme for preferential primary school registration of the children of female graduates proved to be electorally unpopular and was rapidly withdrawn. The other measures, including the role of the Social Development Unit, which was perceived as an integral part of governmental strategy in improving the 'quality' of the population, remained in place. A number of these measures were however adjusted in 1987, when the existing Governmental population policy slogan 'Two is enough' was replaced by 'Have three, or more if you can afford it'. With this change, anti-natalist policies in areas such as child relief for personal income tax, and entitlement to public housing purchase, were relaxed to permit the inclusion of a third and occasionally a fourth child in the family (Singh et al., 1991). Costs associated with the hospital delivery of a third child also become allowable under the Medicare scheme introduced in 1984. Other changes included incentives for females working in the public sector to remain in the work-force after the birth of a child. This was to be achieved via unpaid child care leave, the provision of part-time work schedules for women with a child under six years, and additional full-pay leave for mothers with a sick child. The legislation on abortion and sterilization was amended and made more restrictive. Ante- and post-abortion counselling became compulsory for women with secondary or higher education and fewer than three children and, irrespective of their family size, the one week's leave entitlement previously available to public sector employees after they had undergone tubal ligation or vasectomy was withdrawn if they had completed secondary or higher education (Saw, 1990).

Table 15.2. *Legislation to control population size and structure, Singapore, 1984–1985*

1. Phased, preferential primary school registration for the children of multiparous, better-educated women, with priority given to a child whose mother was of graduate or equivalent status and had at least three children.
2. Enhanced regimes of income tax relief for better-educated women bearing children, to encourage their retention in the labour force.
3. A cash incentive offered to poor and/or lesser educated women with two or fewer children to undergo sterilization.
4. Adjustment of birth delivery fees to discourage higher fertility births by poorer, and it was assumed lesser educated, couples.
5. Establishment of the Social Development Unit within the Ministry of Finance, to provide social facilities and computer matchmaking services for single graduate officers in government ministries, statutory bodies and government-owned companies.

Source: Saw (1990).

Recently, the Government of Singapore has introduced legislation to address the problem of the increasing numbers of elderly people. In 1995, the World Health Organization estimated mean life expectancy at birth in Singapore to be 72.4 years for males and 77.4 for females. To reduce the financial dependence of older citizens on the public exchequer the Maintenance of Parents Act (1995) was passed, requiring the progeny of persons over 60 years of age, including illegitimate offspring and step-children, to be responsible for the financial well-being of their parents. Failure to provide maintenance could result in a fine or imprisonment for a term of up to 6 months. In 1995, the Government of Singapore also raised the age of retirement for both males and females from 65 to 67 years, thus reducing the numbers of persons eligible for state old-age pensions.

Population control measures in the People's Republic of China

During the last three decades much has been written on the demographic problems facing mankind, most recently expressed in terms of the Earth's human carrying capacity, i.e., the growing discrepancy between the world's human population and the size of the resource base required to meet the needs of that population (Cohen, 1995; Neel, 1995). As in many developing countries, population data for the People's Republic of China (PRC) have been fragmentary and poor in quality, but the establishment of the National One-per-Thousand Survey of 1982 and the Two-per-Thousand Survey of over 500,000 women in 1988 brought about a marked improvement in this situation (Lavely *et al.*, 1990; Tien *et al.*, 1992). The latest estimates indicate that the population of the PRC is 1237 million, thus comprising over 21% of the global total, and with some 17 million births per annum (PRB, 1997).

The Chinese government demonstrated its intent to restrict population growth by the introduction of the One-child certificate programme in 1979. This programme sought to encourage couples to restrict their family to just one child by offering a series of incentives, including a monetary bonus, which in some areas was increased if the child was female, preferential housing allocation, priority in the allocation of private plots of land, and preferential educational opportunities and subsequent work assignments for their child (Arnold and Zhaoxiang, 1986). Under the terms of the legislation, citizens of the PRC other than members of ethnic minorities could however be specifically and directly penalized for violating the prescribed population policy. Pregnancies to women without birth permits were subject to termination by compulsory abortion and, in cases where a child had been born, financial penalties could be imposed. In the province of Hebei, for a second child these penalties typically ranged from 10% to 50% of the joint income of the husband and wife to be paid annually for periods of 5 to 14 years, with even more severe fines for third or higher order pregnancies (Li, 1995).

Although much has been written on the desirability and the moral rights and wrongs of the programme, rather less information has been forthcoming on its efficacy. Preliminary results showed that, for China as a whole, by 1982 One-child certificates had been issued to 40.3% of couples with a male child only and 34.0% of couples with a single female child (Arnold and Zhaoxiang, 1986). However, the uptake of the programme varied considerably, with the greatest successes recorded in the cities and among those with a university education. This resulted in wide discrepancies across the country with, for example, an 86.2% uptake reported for couples with a male child in Tianjin but only 6.8% for couples with a single female child in Jiangxi. Similarly, the rates of couples renouncing their One-child certificate status ranged from zero for couples with a single son in Beijing to 44.9% for couples with a single female child in Shanxi province (Arnold and Zhaoxiang, 1986).

These figures would suggest strong preference for a male child, which is in keeping with long-established Confucian traditions (Zhaojiang, 1995). Therefore it is not surprising that the values reported for the secondary sex ratio in the PRC differ widely from those reported in Western countries, where a ratio of approximately 106 males per 100 females is regarded as the norm (Bittles *et al.*, 1993). Chinese data collected in 1987 indicate that male child preference strengthens with increasing family size with, for example, secondary sex ratios of 107.7, 117.3, 120.1, 129.1 and 127.0 observed in birth orders 1 to 5 + (Hull, 1990). As discussed by Zhaojiang (1995) this is not a new phenomenon, and information taken from a fertility survey for the years 1936–40 reveals a similar picture, with sex ratios at birth of 1.09, 1.23 and 1.39 for the first, second and third births (Coale and Banister, 1994). What has however changed is the means whereby the reported secondary sex ratio has been achieved, with the former practice of female infanticide being substantially replaced by a combination of under-reporting of female births and sex-selective abortion based on prenatal sex determination (Tien *et al.*, 1992; Zeng *et al.*, 1993; Coale and Banister, 1994).

Data collected country-wide between 1971 and 1986 showed a highly significant increase in induced abortions – from 3.9 million in 1971 to 11.6 million terminations in 1986 (Hull, 1990). More recent information, collected in four counties located in Hebei and Shadung provinces in 1991, indicate a secondary sex ratio for first births of 1.15 yet a reported induced abortion rate of only 0.6%. For second and third pregnancies the sex ratios were 1.35 and 1.40, with 56.1% and 56.4% induced abortions (Smith, 1994). Confirmation that a major reason for this extremely high abortion rate is at least partially driven by the desire for male as opposed to female children is given by the results of a 1989–90 study into the effect of family composition on sex ratio at birth. While the secondary sex ratio for first born children was 105.6, for families which already had 3 + boys and no girls the sex ratio was 64.4, whereas in families with three girls but no boys it was 219.4 (Zeng *et al.*, 1993). Given these trends, it must be assumed that the tertiary sex ratio, which in 1990 was estimated country-wide to be 106.6 (Adlakha and Banister, 1995), must be rising to a significant extent.

It is against this official anti-natalist background, and the unofficial but strong and long-standing male progeny preference, that the recent introduction of a series of measures aimed at improving the overall health of the population of the PRC should be assessed. A further factor to be considered is the number of handicapped persons in the PRC, estimated in 1987 at 51.6 million (Lavely *et al.*, 1990). At the time of its first announcement, the proposed legislation was severely criticized in the Western scientific and medical press on the grounds that it represented a return to the unacceptable practices of the pre-World War II era of eugenics (Editorials, 1994, 1995), and even in its redrafted form this criticism has continued (Bobrow, 1995). Despite these vigorous complaints from economically developed nations, and the intervention of bodies such as the European Society for Human Genetics, the law was duly promulgated at the Eighth National People's Congress and passed into force on June 1, 1995 (Ministry of Health, PRC, 1994).

The 38 Articles of the Law on Maternal and Infant Health Care are sub-divided into seven chapters, and of these it is Articles 7, 8 and 10 of

Chapter 2 on Premarital Health Care that have been particularly subjected to attack. Article 7 states that premarital health care services shall include education in sex, health, reproduction and genetic diseases, and requires a premarital medical examination for both the male and female 'to see whether they suffer from any disease that may have an adverse effect on marriage and child-bearing'. Under the terms of Article 8, the premarital physical check-up should include an examination for 'genetic diseases of a serious nature'. According to Article 10, if the male and female are diagnosed as having 'certain genetic disease of a serious nature which is considered to be inappropriate for child-bearing from a medical point of view', then 'the two may be married only if both sides agree to take long-term contraceptive measures or to take ligation operation for sterility' (Ministry of Health, PRC 1994).

In principle, these stipulations would appear to run counter to part 1 of Article 16 of the United Nations Universal Declaration of Human Rights (1948), which states that 'men and women of full age, without any limitation due to race, nationality or religion, have the right to marry and found a family' (United Nations, 1995). In practice, it is the lack of any precise definition of what is meant by, and included within, the term 'genetic disease of a serious nature' which is particularly worrying to medical geneticists, especially given the gross mis-use of such blanket diagnoses both during the Nazi era and subsequently by other totalitarian regimes.

While the validity of these concerns cannot be denied, it should be noted that equivalent attention has not been paid in Western countries to other puzzling aspects of population health in the PRC. Although the country has made remarkable progress during the last 40 years, as evidenced by present life expectancy figures for males and females of 66.7 and 70.6 years respectively, age-specific death rates remain significantly higher than for other East Asian states such as Hong Kong or Japan. This is especially the case in the first year of life, with rural mortality rates in the PRC of 2413 per 100,000 for males and 2411 per 100,000 for females, and in the one to four age group with rates of 210 per 100,000 and 187 per 100,000 for males and females respectively (World Health Organization, 1995). From the data provided on perinatal and infant deaths, it is not possible to determine the degree to which infanticide might have been a contributory factor, and there is no evidence of an excess of female deaths. Data obtained from other sources including the 1988 Two-per-Thousand Survey do however confirm that female children have increased mortality in the one to four years age group (Choe *et al.*, 1995).

The data from the PRC on fatal accidents in infancy indicate a marked discrepancy between the rates reported for rural and urban areas (Table 15.3). The unique data on the numbers and large proportion of infants

Table 15.3. *China: fatal accidents in infancy per 100,000 population*

	Rural		Urban	
Cause of death	Male	Female	Male	Female
Motor vehicle accidents	2.2	1.3	1.8	1.6
Accidental poisoning	4.7	6.0	3.6	3.1
Accidental falls	5.8	4.1	2.9	3.9
Fire and flames	2.2	2.5	0.7	0
Accidental drowning and suffocation	16.0	11.9	4.4	2.0
Accidental mechanical suffocation	160.7	192.6	32.0	33.0
Struck by falling object	0	0.3	0	0.8
Other accidents	21.8	31.7	12.0	7.5
Homicide	1.1	1.3	1.1	1.6
All external causes of injury and poisoning	217.8	259.7	60.8	55.4

Source: World Health Organization (1995).

who are recorded as dying following accidental mechanical suffocation are especially obvious, with disproportionately higher rates reported in rural as opposed to urban areas for males ($\times 5.0$) and females ($\times 5.8$). The excess of female newborns dying from this cause is also noteworthy, once again with the greatest effect observed in rural areas.

Certain aspects of adult mortality in the PRC differ significantly from the patterns recorded in neighbouring countries of East Asia and elsewhere. For example, suicide rates are elevated in the rural population at all age intervals (Table 15.4). The marked excess of deaths by suicide among females in the age intervals 15–24, 25–34 and 35–44 years is a very significant finding, which runs counter to reports from other populations in which male suicide consistently predominates at these ages (Desjarlais *et al.*, 1995). No ready explanation for this anomaly is forthcoming, other than a presumption that some facet or facets of the lives experienced by these young women in the PRC is specifically and uniquely stressful. At the same time, it has been noted that suicide, and the threat of suicide, is one of the few traditional avenues of domestic power and forms of protest which are open to Chinese women (Desjarlais *et al.*, 1995).

Discussion

As discussed above, the re-emergence of eugenic doctrines in East and South-East Asia justifiably has raised strong ethical worries, especially among medical geneticists who are concerned that technologies developed to assist patients in making informed decisions on their reproduction may be mis-used by governmental agencies. More generally, there is the worry

Table 15.4. *China: suicide rates per 100,000 population*

Age in years	Rural		Urban	
	Male	Female	Male	Female
5–14	0.7	0.9	0.6	0.6
15–24	17.4	36.7	5.6	10.6
25–34	21.5	38.5	7.0	8.5
35–44	22.0	29.7	8.0	8.3
45–54	30.2	33.3	10.0	9.2
55–64	49.8	42.7	12.4	14.0
65–74	97.5	74.3	23.8	22.2
75 +	145.7	102.6	41.5	41.9

Source: World Health Organization (1995).

that the entire subject area of medical genetics may suffer from a loss of public confidence. On the basis of the information to hand the first of these concerns may be exaggerated, as there is little evidence to show that policies enacted by national governments to control their populations actually work. For example, as discussed above, even in Singapore which is a compact and tightly governed city-state with only 3.5 million citizens, the introduction of parliamentary legislation to control population size, structure and equality has necessitated considerable effort, resulted in significant policy shifts and reversals, and been of limited relevance or value.

The political system in the PRC is obviously very different to Singaporean parliamentary democracy, and it might be expected that under a non-democratic form of government it would be easier to enforce legislation. Yet the data now available on the workings of the One-child policy suggest that it has not been particularly successful in meeting its aim, with the majority of Chinese women continuing to pursue the traditional goal of producing a male child to carry on the family line and to provide security in old age (Li, 1995), in some cases unintentionally assisted by the unreliability of the available contraceptive options (Wang and Diamond, 1995). Therefore the probability that the 1995 Law on Maternal and Infant Health Care can or actually will be enforced appears remote, especially in more traditional rural areas and with an apparent, gradual lessening of centralized government control.

It must however be accepted that the size and changing structure of the Chinese population means that the government of the PRC is faced with a number of major problems, several of which are predominantly social in nature. For example, there is evidence that during the period 1982 to 1986 there was a marriage squeeze, exacerbated by the restricted geographical mobility of the population, which could only be resolved by men marrying much younger women (Tien *et al.*, 1992). Given the marked fertility reductions during the second half of the 1970s and the increasing imbalance in the secondary sex ratio in recent years, it is predicted that a new marriage squeeze is about to commence (Tien *et al.*, 1992), with an estimated one million excess males per year in the first-marriage market after the year 2010 (Tuljapurkar *et al.*, 1995). A similar trend is expected in other East Asian countries including Vietnam, where an official one-or-two-child family policy has been in operation since 1988 (Goodkind, 1995). In South Korea, during the last decade there has been a rapidly rising secondary sex ratio with increasing birth order, resulting in a sex ratio for last-born children of 228.6 (Park and Cho, 1995). This factor has led to speculation that a shortage of brides in rural areas could in time lead to increased homosexuality or even polyandry (Park and Cho, 1995), which echoes similar predictions made for Western society a generation earlier (Etzioni, 1968).

Chinese records from the seventeenth century onwards indicate that infant abandonment was a relatively common occurrence in certain provinces, and this problem appears to be re-emerging in the PRC (Johnson, 1996). In a survey reported on over 16,000 abandoned children in Hunan province between 1986 and 1990, most were from rural areas, 92% were female and usually very young, and 25% had some form of handicap. Abandonment effectively is correlated with reduced survival, since the mortality rates in the welfare centres for foundlings exceed 40% in some of the major, state-run orphanages, and can reach as high as 80% in smaller, more remote or poorly equipped centres, possibly also reflecting a high proportion of handicapped, abandoned infants (Johnson, 1996).

With statistics of this type, and the earlier 1987 estimate of 51.6 million handicapped persons throughout the country (Lavely *et al.*, 1990), it could be argued that the Law on Maternal and Infant Health Care is a sensible and desirable piece of legislation. In addition, it should be noted that under the terms of Chapter 6, Article 37 of the Law, investigations to determine the sex of a foetus are prohibited, which should act to curb current excesses in that area and in time might alleviate the problems associated with the increasingly imbalanced tertiary sex ratio. Unfortunately, the legacy of Nazi Germany and other examples of pre-World War II eugenic legislation in Western Europe and the USA cannot easily be forgotten, and the maladroit population screening programmes for inherited haemoglobin disorders in the USA (Motulsky, 1973, 1974) also should serve as a lesson of how well-intentioned efforts can turn out to be socially damaging and counter-productive. Irrespective of such precedents, it seems improbable that the Government of the PRC will pay undue attention to Western concerns. Therefore, whether or not the provisions of the Maternal and

Infant Care legislation are fully brought to bear, ultimately will depend on its practicality and the degree of its acceptability to the Chinese people.

References

Adlakha, A. & Banister, J. (1995). Demographic perspective on China and India. *Journal of Biosocial Science* **27**, 163–78.
Arnold, F. & Zhaoxiang, L. (1986). Sex preference, fertility, and family planning in China. *Population and Development Review* **12**, 221–46.
Beckwith, J. (1976). Social and political uses of genetics in the United States: past and present. *Annals of the New York Academy of Sciences* **265**, 46–58.
Bittles, A. H., Mason, W. M., Singarayer, N., Shreeniwas, S. & Spinar, M. (1993). Determinants of the sex ratio in India: studies at national, state, and local levels. In *Urban Ecology and Health in the Third World*, ed. L. M. Schell, M. T. Smith & A. Bilsborough, pp. 153–67. Cambridge: Cambridge University Press.
Blacker, C. P. (1952). *Eugenics: Galton and After*. London: Duckworth.
Bobrow, M. (1995). Redrafted Chinese law remains eugenic. *Journal of Medical Genetics* **32**, 409.
Burt, C. (1952). *Intelligence and Fertility: the Effect of the Differential Birthrate on Inborn Mental Characteristics. Occasional Papers on Eugenics, 2*. London: Cassell.
Choe, M. J., Hongsheng, H. & Feng, W. (1995). Effects of gender, birth order, and other correlates on childhood mortality in China. *Social Biology* **42**, 50–64.
Clay, C. & Leayman, M. (1995). *Master Race – the Lebensborn Experiment in Nazi Germany*. London: Hodder and Stoughton.
Coale, A. J. & Banister, J. (1994). Five decades of missing females in China. *Demography* **31**, 459–79.
Cohen, J. E. (1995). Population growth and Earth's human carrying capacity. *Science* **269**, 341–6.
Desjarlais, R., Eisenberg, L., Good, B. & Kleinman, A. (1995). *World Mental Health*, pp. 69–71. New York: Oxford University Press.
Editorial (1939a). Men and mice at Edinburgh. *Journal of Heredity* **30**, 371–3.
Editorial (1939b). Social biology and population improvement. *Nature* **144**, 521–2.
Editorial (1994). China's misconception of eugenics. *Nature* **367**, 1–2.
Editorial (1995). Western eyes on China's eugenics law. *The Lancet* **346**, 131.
Etzioni, A. (1968). Sex control, science and society. *Science* **161**, 1107–12.
Goodkind, D. M. (1995). Vietnam's one-or-two-child policy in action. *Population and Development Review* **21**, 85–111.
Hull, T. H. (1990). Recent trends in sex ratios at birth in China. *Population and Development Review* **16**, 63–83.
Johnson, K. (1996). The politics of infant abandonment in China. *Population and Development Review* **22**, 77–98.
Joint Committee on Voluntary Sterilization (1934). *Report of the Departmental Committee on Sterilization*. London: HMSO.
Kevles, D. J. (1986). *In the Name of Eugenics. Genetics and the Uses of Human Heredity*. London: Pelican.
Keynes, M. (1993). Sir Francis Galton: a man with universal scientific curiousity. In *Sir Francis Galton: The Legacy of His Ideas*, ed. M. Keynes, pp. 1–32.

London: Macmillan.

Lavely, W., Lee, J. & Feng, W. (1990). Chinese demography: the state of the field. *Journal of Asian Studies* **49**, 807–34.

Li, J. (1995). China's one-child policy: how and how well has it worked? A case study of Hebei Province, 1979–88. *Population and Development Review* **21**, 563–85.

Ministry of Health, PCR (1994). *Law of the People's Republic of China on Maternal and Infant Health Care*. Beijing: Legislative Affairs Commission of the Standing Committee of the National People's Congress of the People's Republic of China.

Motulsky, A. G. (1973). Screening for sickle cell hemoglobinopathy and thalassemia. *Israeli Journal of Medical Science* **9**, 1341–9.

Motulsky, A. G. (1974). Brave new world? *Science* **185**, 653–63.

Muller, H. J. (1959). The guidance of human evolution. *Perspectives in Biology and Medicine* **3**, 1–43.

Muller, H.J. (1965). Letter. *Eugenics Review* **57**, 101–4.

Neel, J. V. (1995). The neglected genetic issue – the why and how of curbing population growth. *American Journal of Human Genetics* **56**, 538–42.

Park, C. B. & Cho, N.-H. (1995). Consequences of son preference in a low-fertility society: imbalance of the sex ratio at birth in Korea. *Population and Development Review* **21**, 59–84.

PRB (1997). *World Population Data Sheet*. Washington, DC: Population Reference Bureau.

Saw, S-H. (1986). A decade of fertility below replacement level in Singapore. *Journal of Biosocial Science* **18**, 395–403.

Saw, S-H. (1988). Seventeen years of legalized abortion in Singapore. *Biology and Society* **5**, 63–72.

Saw, S-H. (1990). *Changes in the Fertility Policy of Singapore*. Singapore: The Institute of Policy Studies.

Singh, K., Fong, Y. F. & Ratnam, S. S. (1991). A reversal of fertility trends in Singapore. *Journal of Biosocial Science* **23**, 73–8.

Smith, H. L. (1994). Nonreporting of births or nonreporting of pregnancies? Some evidence from four rural counties in North China. *Demography* **31**, 481–6.

Tien, H. Y., Tianlu, Z., Yu, P., Jingneng, L. & Zhongtang, L. (1992). China's demographic dilemmas. *Population Bulletin* **47**, 1. Washington, D.C.: Population Reference Bureau.

Tuljapurkar, S., Li, N. & Feldman, M. W. (1995). High sex ratios in China's future. *Science* **267**; 874–6.

United Nations (1995). *The Universal Declaration of Human Rights* (General Assembly of the United Nations, October 10, 1948). New York: Information and External Relations Division of the United Nations Population Fund.

Wang, D. & Diamond, I. (1995). The impact on fertility of contraceptive failure in China in the 1980s. *Journal of Biosocial Science* **27**, 277–84.

World Health Organization (1995). *1994 World Health Statistics Annual*. Geneva: World Health Organization.

Zeng, Y., Tu, P., Gu, B., Xu, Y., Li, B. & Li, Y. (1993). Causes and implications of the recent increase in the reported sex ratio at birth in China. *Population and Development Review* **19**, 283–302.

Zhaojiang, G. (1995). Chinese Confucian culture and the medical ethical tradition. *Journal of Medical Ethics* **21**, 239–46.

16 The policy implications of health inequalities in developing countries

CAROLYN STEPHENS

Introduction

No man is an *Iland*, intire of it selfe; every man is a peece of the *Continent*, a part of the *maine*; if a *Clod* be washed away by the *Sea, Europe* is the lesse, as well as if a *Promontorie* were, as well as if a *Mannor* of thy *friends* or of *thine owne* were; any mans *death* diminishes *me*, because I am involved with *Mankinde*; And therefore never send to know for whom the *bell* tolls; It tolls for *thee*.

John Donne, Meditations XVII (1609)

As John Donne suggested in 1609, the life and death of individuals are linked inextricably to those of others in their society. Donne's was a philosophical argument about the commonality of human experience and the need to recognize that each of us is connected in social and physical terms. More than 300 years later, Donne's words ring true as analysts re-appraise our understanding of the relationships between individuals and their health, and the global or societal forces that affect them.

Traditional notions of health and its determinants are challenged by the current evidence that diffuse but profound forces in the global society can have dramatic implications for the health of populations and individuals. It is gradually becoming clear that the actions of individuals and groups in one society can have negative and relatively direct effects on other people and societies across the globe. Global environmental changes, in part attributable to the actions of minority northern populations with unsustainable consumption patterns, threaten health for the population of the whole planet (Korten, 1996). Similarly, global economic changes, rooted in the systems of a small proportion of the global population, have had massive effects on economic conditions for the majority of the world's population (Hutchful, 1987; Jackson, 1990; Kanji, 1995, unpub. data). A combination of global pressures – macro-economic, ideological and social

– are exacerbating the degree of polarization between and within societies (Korten, 1996).

It is in this overall context one needs to understand health inequalities in 'developing' countries. It is important to be aware that terms which hitherto describe non-North American and European regions of the world are changing rapidly, as economic 'tigers' in Asia overtake traditional 'developed' nations, and as one recognizes the depth of economic deprivation in parts of eastern Europe. Studies in this field use a range of terms that are unsatisfactory. 'Developing' or less developed countries is used as a summary expression to mean Africa, South East Asia and Latin America. The terms North and South summarize the division of the economically powerful G7 group versus the mostly southern countries of the G77. However, the greatest degree of health inequality remains between the North and the South (United Nations Development Programme, 1992; UNRISD, 1995) and this divide appears to be increasing (World Health Organization, 1995). There is also evidence that alongside the economic and social polarization between nations internationally, there are trends of polarization in the health experience of individuals and groups within and between national and regional societies (Elo and Preston, 1996).

What do these macro-changes mean for an understanding of policies necessary to tackle inequalities in health in the so-called developing countries of Africa, Asia and Latin America? Firstly, the evidence of the effects of global processes challenges conceptual understanding of the influences of social and economic conditions on health. Understanding the impact of global forces will not be achieved through the intuitive or learned approach of either those trained in human biology or those working in public health epidemiology. Whilst we are trained to explore in detail the interplay of biological, environmental and genetic factors in the creation of health and disease, we are less equipped to grapple with the idea that diffuse macro-political and social processes operating internationally can affect health in relatively direct ways.

Secondly, the rapidly changing international context challenges traditional policy responses to poverty and development. As Phillips and Verhasselt (1994) suggest, the health of individuals and populations is inextricably linked with 'development'. Development is an elusive term, but has in the past meant a form of sustained economic growth that provides a population with the means to survive and thrive through provision of basic conditions of food security, shelter, work and health care. Under-development has been characterized as limited economic growth, with the consequence that large proportions of the population are exposed to physical and biological hazard. Large-scale exposure to these hazards in the South predicts the levels of ill health and disease experienced. Development theorists have debated for centuries over the means to a form of development that ensures basic quality of life and survival for populations. There has been relatively little discussion over the goal of development. Yet, recent evidence from both North and South challenges basic tenets of the development theory. It suggests that the elusive model of 'sustained economic growth' may destroy the possibility of health for future generations. In addition, health inequalities do not go away with the achievement of survival for all. This fact is of great importance for policy development in countries still struggling to stop their populations dying below the age of five.

This chapter discusses a small part of this overall conundrum. It examines policies aimed at tackling health inequalities in developing countries. It discusses two substantial influences on the development of policies: the focus of health research in developing countries to date; and the focus of development discussions on poverty and inequality. This is followed by a review of strategies addressing health inequalities in developing countries, and explores the urban environment as a means to highlight more complex dynamics of inequality in the cities of the South. The chapter concludes by discussing the wider implications of health inequalities in such societies.

Basic concepts – how inequality has been understood

To act on a problem demands that the problem is recognized to exist. This section first assesses the research basis for defining public health priorities in developing countries. It then looks at the literature on policy approaches towards major health priorities.

The epidemiology of inequality

Policy on health inequalities in developing countries can be related in part to the historical focus of health (epidemiological) research in developing countries generally. This itself is linked to basic conceptual notions of epidemiology that health (or rather disease) is determined by physical and biological environments, modulated by social environments, but with a focus on the physical and biological factors closest (proximal) to the outcome. Identifying the key agents of disease has been a central tenet of health research for over a century (White, 1991). Thus, there has been over 100 years of epidemiological discussion about the relative importance of individual biological and physical risks to health. Schaeffer (1993) argues that 'because they control how resources are used and how goods and risk exposures are created, social environments are the dominant determinants of human well-being'. The existence of health inequalities, between individuals, social groups and nations, is thus immmediately a result of the distribution of risk exposures between individuals and groups. Indirectly, exposure inequalities are a result of the social distribution of power at local, national and international levels, which determines the distribution of the basic means to survive. Epidemiology has not concentrated on the causes of the social distribution of determinants of health, but has focused on the relationships between different exposures and health outcomes. Thus, a list has evolved of the 'basic' human needs necessary for survival – food, potable and plentiful water, shelter, disposal of human and other wastes and protection from vectors of disease (Basta, 1977; Berman, 1991; Aaby, 1992; Bradley *et al.* 1992; Bradley, 1993). Whilst many public health specialists share Schaeffer's concern for the broader determinants of the direct risk factors, research attention has focused on individual risks and outcomes. Where the direct risks for survival of specific groups are very evident, understanding of health *inequalities* suffers conceptually – the problem is isolated artificially (see Porter and Ogden, Chapter 6).

Indeed, it is important to note that conceptually, very few analysts draw the important distinction between health inequalities and health inequity. This is a distinction between distribution of risk and distribution of control – in other words, it is the difference between a description, i.e. *inequality in distribution of a risk or a health effect*, and an analytic term ascribing responsibility, i.e. *inequity in the distribution of benefits and disbenefits* (see also Strickland and Shetty, Chapter 1).

It is rare to see evidence of health inequalities between social groups in developing countries, but even more rare to see an analysis of health inequities. The distinction is not academic. Health inequalities reflect individual or group differences in exposures to risks, in personal susceptibility to illness and in differential access to and use of treatment facilities. Indirectly, health inequalities reflect the extent to which individuals and groups have control over their own and others' exposure to risk and management of health effects. If one social group avoids death or illness consistently, while another does not, health *inequalities* may be the result of differences between groups in exposures to environmental risks and abilities to treat their health effects. There are inequalities, for example, in access to water and sanitation facilities or education opportunities between groups within developing countries – these may lead to health *inequalities*, often in incidence and prevalence of infectious diseases. However, if one group benefits to the disbenefit of another group, this is a *health inequity*. Using the same example of water, the poor often have least access to minimum quantities of potable water. The poor in cities of developing countries also pay higher prices than the wealthy for limited quantities and for poor quality water from vendors. This has been described as a doubly regressive taxation in which one group are doubly disbenefitted (in health and economic terms) while another doubly gains (Benneh *et al.*, 1992).

A second trend of health research in developing countries which has influenced policies related to health inequality is the focus on particular age-groups or diseases. Murray *et al.* (1992), discussing why adult health in developing countries is still so little understood, attribute lack of data on health of any age groups over 15 years to the intellectual and research focus of international public health to date. Attention in the past and to a substantial degree even now is concentrated on two major public health areas – those of tropical diseases and the health of children. The term 'tropical diseases' is not necessarily limited in definition, but it has tended to mean the analysis of major infectious diseases, for example, diarrhoeal and gastro-enteric diseases, malaria and schistosomiasis. The burden of these diseases was demonstrated to be large, especially for rural populations and particularly in some tropical regions of the world (World Bank, 1993). Research emphasis on child health was a natural corollary to that on infection, since children are a particularly vulnerable group for the ill health and deaths caused by infectious tropical diseases. The roots of interest in child health stem also from the overwhelming evidence of high mortality burdens for children in developing countries, linked to epidemiological data which showed that the majority of these deaths 'were attributable to a short list of preventable communicable diseases' (Murray *et al.*, 1992).

These analytic trends have influenced substantially which data on health have been available to policy-makers and analysts in developing countries. They have determined which public health problems have been highlighted and which have been lost to view, and have also influenced data on health inequalities in developing countries. For example, large data gaps have existed for particular populations in developing countries: adult health has been ignored (Feachem *et al.*, 1990); health of the elderly has only recently appeared on the agenda (Kalache and Aboderin, 1995); occupational diseases are still under-discussed; urban health has, until recently, been assumed to be good relative to rural health (Harpham and Stephens, 1991); and chronic and psycho-social diseases have been assumed to be a problem of industrialized, not developing countries (Harpham *et al.*, 1990; Harpham and Stephens, 1991; Rossi-Espagnet, 1991; Williams, 1991; Stephens, 1995).

The combined influence of these analytic trends has reinforced an impression that the major policy question facing decision-makers in developing countries is the resolution of problems related to the physical deprivation of 'absolute poverty'. This poverty has been characterized by lack of basic services or 'basic needs' and evinced by lack of survival of the population. These problems have been construed to exist most severely in rural areas of developing countries (Lipton, 1988; Hardoy and Satter- thwaite, 1991). To explain what this means one needs to turn to the development literature.

The development debate on inequalities

Understanding past policies to alleviate health inequalities in developing countries involves understanding the influence of the development theory. Generalizing misleads, but it is necessary to have some idea of these debates in order to understand how and what policies have developed. From the 1950s to the 1970s, discussion focused on the *process* of econ- omic development – it centred essentially around the eternal poles of economic growth or (re-)distribution. Growth theorists argued that through 'modernization' and 'trickle-down' of economic benefits, an im- proved quality of life would eventually reach all human beings. Distribu- tion theorists (see Abel-Smith and Lieserson, 1978; Walt and Vaughan, 1981; Philips and Verhasselt, 1994 for discussions of this in relation to health) argued that it was more important to look at unemployment and unequal distribution of assets, no matter what the rate of growth or level of economic wealth. If distribution worsened, despite growth, then develop- ment could not be said to take place. In the face of massive levels of absolute poverty, measured by perceptible lack of the means to survive, the first priority seemed to be equitable distribution of access to 'basic needs', facilitated by increased 'community participation'. As Abel-Smith and Lieserson (1978) put it, 'What is important is how many people are lacking in basic nutrition, clothing and shelter, access to work, education and health services, and opportunities to participate in community life'. Fried- man (1992) calls this 'essential needs'. In terms of the distinction between inequality and inequity, the basic needs approach, at best, focuses on equal distribution of basic assets.

This discussion reflects nineteenth century thought and persists today. In 1990, the *World Development Report* (World Bank, 1991) defined pov- erty as the 'inability to attain a minimal standard of living', measured by household incomes and expenditure. Such data are often supplemented by welfare measures of health, literacy and access to public goods (World Bank, 1991). The same report distinguishes poverty from inequality, 'Pov- erty is not the same as inequality. The distinction needs to be stressed. Whereas poverty is concerned with the absolute standard of living of a part of society – the poor – inequality refers to relative living standards across the whole society' (World Bank, 1991). This is a denial of equality – it does not even broach inequity. Many authors disagree, arguing that absolute measures of 'poverty' (income, facilities, occupation, etc.) ignore the complex ways in which poor people actually obtain, or are prevented from obtaining, resources (Sen, 1981; Chambers, 1992; Townsend, 1993; Amis and Rakodi, 1994; Moser, 1994 unpub. data). Sen's (1981) and Chambers' (1982) rural poverty work, and that of Amis and Rakodi (1984) in cities, argue that analysis 'must include an understanding of vulnerability, which emphasizes the importance of assets and debt, as well as access to public resources and political process'. The last part of this emphasis, on political process and power, introduces equity.

Recent theoretical discussion of 'vulnerability' in developing countries has not substantially diminished the tendency to analyze the problems of those in the least powerful positions as simply a question of 'poverty' measured in absolute series of inadequacies in physical standards of living, mediated by their economic position in society. Ownership or access to physical goods in the environment (or rather lack of them) is still commonly used as a proxy for poverty, (Bisharat and Zagha, 1986; Hardoy and Satterthwaite, 1989, 1991; Benneh et al., 1990; Harpham et al., 1990; Bateman and Smith, 1991 unpub. data; Bruce Tagoe, 1992). The notions of human agency and power implied by concepts such as vulnerability have permeated conceptually, but not much further.

The implications of the analytic trends for policies to address health inequality

This discussion highlights two analytic trends that have influenced significantly health policy in developing countries. Firstly, focusing on immediate exposures and specific health outcomes for specific vulnerable age groups has a natural corollary in a focus on the absolute material symptoms of 'poverty'. Major health problems seem linked to the immediate symptoms of absolute poverty, measured by access to 'basic needs'. The parallel development debate has revolved, until recently, around the distribution of these material symptoms of absolute poverty – equality in distribution of basic needs. At the macro-level, the means to distribute these basic needs equally has been managed by governments which also believed in distribution of assets such as education, land and political process (Jackson, 1990). As the following sections will describe briefly, the distribution of power has not been achieved, while the distribution of basic needs has achieved partial success.

Tackling inequalities – evidence from past policies

This is the partial success story of survival. During the last 50 years, average life expectancy at birth in developing countries increased from 40 to 63 years and child mortality fell from 280 to 106 per 1,000. Comparing progress in developing countries against developed countries, the United Nations Development Programme (1992) reports:

- Average life expectancy increased between 1960 and 1990 from 67% to 84% of the level in the North.
- Daily per capita calorie supply increased between 1964/66 and 1984/86 from 72% to 80% of the level in the North.
- Infant and child mortality rates were halved between 1960 and 1990 (an achievement which took more than a century in industrialized countries).

What explains these apparent successes?

The overall story – distributing the means to survival

Approaches underpinning the 'survival' of children in developing countries have been diverse. They have not necessarily been aimed at reducing health inequalities within societies of the developing countries. Some of the major successes have stemmed from population-based approaches aimed at individual diseases in children. Immunization against communicable diseases is one example. The eradication of smallpox is notable – 80% of the world's children are now immunized against the basic vaccine-preventable diseases. The World Health Organization (1995) reports that through such means three million people in 1993 were saved from early death.

In broader developmental terms, discussion of health policy in developing countries was, from the mid-1950s to the early 1980s, often made in the context of macro-level political commitment to equality in access to basic needs, often allied to development policies aimed at distribution of power (Abel-Smith and Lieserson, 1978). The Alma Ata declaration of 1979 on comprehensive primary health care fitted into this general development commitment. Equality in health was a development goal pursued at a societal level (Abel-Smith and Lieserson, 1978). During the 1970s and 1980s considerable attention was also devoted to the success in terms of survival of a selected group of countries and states which had achieved considerable reductions in indicators of child health, such as mortality rates for infants and under 5-year-olds. This was achieved through a package of micro- and macro-reforms to policy at a national level, based on principles of political equity (Post and Wright, 1989). The impetus for such reforms came from macro-development policies, typified by post-colonial governments in Africa, Asia and Latin America, which, in many cases, moved towards redistribution of political power and basic services to all groups after decades of selective distribution of power and assets. These reforms were substantial but they were put forward by their advocates as 'low cost'. The notion of 'good health at low cost' emerged (Philips and Verhasselt, 1994), based on the experience of a few countries and states, notably China, Kerala in India, Sri Lanka and Costa Rica. The success of these nations and states in terms of basic health indicators, such as infant mortality, suggested the importance of several key elements: education for *all* with an emphasis on primary and secondary schooling; political will; equitable distribution throughout rural and urban populations of public health measures and primary health care; and assurance of adequate food for all.

Health transitions and economic dilemmas

By the late 1980s, two themes had emerged – one political and economic, the other health and demographic. The 'health transition' was being discussed widely, particularly in Latin America (McKeown, 1988; Bobadilla *et al.*, 1990; Caldwell, 1991; Aaby, 1992). The health transition is the net result of three components: the demographic component – i.e. the ageing of population related to declining fertility and mortality rates; the risk factor component – i.e. changes in exposure to the underlying causes of specific diseases through policies such as vaccination and environmental sanitation; and the therapeutic component – i.e. the changes in the probability that an ill or infected individual will die as a result of changes in access to, use of and effectiveness of curative health services. This transition seemed to suggest that there was a natural development trajectory which was being achieved by some countries through pursuit of broad 'equality in basic needs' policies.

Alongside this trend, was the shift in development policy during the 1980s. Internationally, the mood changed to suggest that policies aimed at equity and equality were not 'low cost'. International debts accrued by developing countries were attributed to mismanagement and over-ambitious policies of distribution – ignoring the larger dynamics of international trade and economic and political power (Jackson,1990). These debts left many countries extremely vulnerable to the macro-economic re-emphasis on 'trickle-down' policies, promoted vigorously by international agencies such as the World Bank and the International Monetary Fund (Jackson, 1990). Equality of distribution of assets at the macro-level was explicitly off the agenda. Although poverty was still on the agenda, the emphasis was on absolute poverty, suggested or imposed macro-economic restructuring of economies, and promoted targeting and 'basic packages' of care and services to the targeted (World Bank, 1993). While academics criticized this retrenchment of policies of equality and equity, and began to suggest a crisis of sustainable development (Redclift, 1984, 1987; Jackson, 1990), international agencies continued to dominate the policy agenda of developing countries in terms of macro-economics and also of health. However, there are changes occurring now. To understand why this might be so and to conclude whether survival is the extent of the debate on health inequalities in developing countries one needs to look at the group conceived as the potential villain in the South – the urban populations and their health.

Urban health and inequalities in developing countries

Academics interested in health and equality in the South over the last few decades have focused often on the urban–rural divide. This is perhaps most notoriously encapsulated in Michael Lipton's work 'Why poor people stay poor: urban bias in world development?' (Lipton, 1988). Lipton argued that 'the most important class conflict in the poor countries of the world today is not between labour and capital but between the rural classes and the urban classes'. Urban dwellers were seen as primarily parasitic on the rural population. Overall, a consequence of this theory was to concentrate research support and policy focus on resolving rural–urban differences (Hardoy and Satterthwaite, 1991). For decades, aggregate urban health data aided the perspective that the rural populations were disadvantaged. Substantial research now refutes this perspective (Hardoy and Satterthwaite 1991; Harris, 1992; Stren *et al.*, 1992; Kasarda and Parnell, 1993), including from the point of view of health, and there is recognition of the fact that disadvantaged groups in cities are sometimes worse off than their rural counterparts (Harpham *et al.*, 1990; World Health Organization, 1992, 1993). More importantly, it is now clear that the notion of policy based on rural–urban divides misunderstands profoundly the degree of connection between rural and urban populations. This brings us to a brief discussion of health inequalities in the cities of developing countries.

Urban health inequalities

By 2025, it is predicted that four out of five urban people will live in towns and cities of developing countries (Kasarda and Parnell, 1993). Living in urban areas does not produce benefits for many people – between 30% and 70% of people in cities and towns of the South live in 'poverty', characterized by household and neighbourhood environmental deprivation and in circumstances of extreme social and economic stress. Urban poverty is the 'critical theme of the 1990s' for the South (Cheema, 1992) as it was at the turn of the nineteenth century for the North.

Urban inequality is a comparatively new emphasis of research (Stephens *et al.*, 1994, Stephens, 1995). Inequity is under-explored. Tangible evidence of the large-scale lack of 'basic needs' in cities has led to a research agenda in urban areas focused on child health and infectious diseases (Stephens *et al.*, 1990; Stephens, 1995). If health inequalities are discussed it is in these terms. Thus, there is evidence of severe inequalities in health related to access to basic needs, both between cities and within them (Hardoy *et al.*, 1990, 1992; Hardoy and Satterthwaite, 1991). Within countries, there is some evidence that inequalities in child mortality and morbidity exist between larger and smaller cities (Timaeus and Lush, 1995), with poor children in small urban areas disadvantaged compared to their counterparts in big cities (Brockerhoff, 1993). Looking within cities, there are now hundreds of studies of low-income and squatter communities in the South, that suggest high death and disease rates for infants (under 1-year-old) and children (under 5 years old) – between two and ten times higher in deprived than in non-deprived areas of cities (Bradley *et al.*, 1992).

Explanations for these health inequalities still focus on individual and group access to basic living conditions. One writer (Brockerhoff, 1995) concludes that 'overwhelming empirical evidence from all developing regions now links poor housing conditions in urban areas to childhood diseases and injuries'. Many other reviews come to similar conclusions (Tabibzadeh *et al.*, 1989; Harpham and Stephens, 1991; Bradley *et al.*, 1992; Hardoy *et al.*, 1992). The evidence suggests a large-scale need for better basic living conditions. The lack of 'basic needs' for large groups of urban people affects survival, particularly of children, and this means continued impact of infectious diseases for other vulnerable groups (particularly pregnant women and the elderly) who live in areas without 'basic needs' (Bradley *et al.*, 1992; Stephens *et al.*, 1994). However, most studies do not describe inequalities between groups within the city as a whole, nor do they look at the whole population, or look at inequities in distribution of control over risks. The picture is further complicated when one does.

The urban health transition? The picture gets more complicated

In some regions of the world, policy-makers have pursued the alleviation of urban health inequalities through the distribution of basic needs, just as Europe did. This has improved survival of urban poor children, as well as improving substantially basic living conditions for the poor and overall population (Monteiro, 1982; Bisharat and Zagha, 1986; Bobadilla *et al.*, 1990). There is now evidence that in some areas of the world, urban populations are moving rapidly through the health transition. This is seen in the shift in patterns of urban causes of death from infectious to chronic diseases. Heart diseases and neoplasms, the illnesses of adults and the elderly, now emerge as important urban problems. This is the case in urban areas as diverse as São Paulo in Brazil (FUNDACAO SEADE, 1990), Cape Town in South Africa (Medical Officer of Health, 1992), Accra in Ghana (Stephens *et al.*, 1994) and even in urban areas of Maharashtra in India (Mutatkar, 1994, unpub. data).

The emergence of changes in the patterns of death and disease in cities as a whole implies that, in some cities, a broad set of policies bringing access to basic needs to most or all people has succeeded in assisting the survival of urban populations past the risks of the infectious and parasitic diseases associated directly with unhygienic living conditions. This, combined with access to basic health care and vaccination has addressed substantially the *survival* of the urban poor children. This is not surprising. A combination of improved nutrition, access to potable water and sanitation and improved housing and labour conditions were the recipe for improved survival of the urban poor in the North (McKeown, 1976, 1979). When access to these is combined with basic health services, which include vaccination against major killers such as measles, diphtheria, tuberculosis, tetanus and polio, the impact of urban deprivation on health reduces substantially. There is evidence that this 'basic needs' package has achieved substantial gains for some countries, states and cities. Internationally, infant and child mortality rates were more than halved between 1960 and 1990 – an achievement which took more than a century in industrialized countries – (United Nations Development Programme, 1992). Immunization against communicable diseases is now extensive, with notable successes in urban areas (Wills, 1995).

Obviously, for many cities, this health transition has not been achieved. However, examining health in cities that have passed the 'Sanitary crisis' raises important questions in relation to health inequalities and equity. The first is whether the 'health transition' brings health equality, and secondly, whether dealing with the physical symptoms of urban poverty in isolation is adequate to alleviate the broader health implications of structural inequities. The final section discusses this briefly.

Urban systems and the new health inequalities

Recently, concerns have been expressed over the lack of logic in tackling urban poverty in isolation of the whole urban system. This tends to be an argument put forward by urban planners, macro-economists, and social policy experts to deal with the scale and complexity of urban poverty, mostly in its 'basic needs' sense. Evidence on health inequalities is also compelling, but rare. As suggested earlier, until recently, most health analysis has been focused on the poor only, or on children and infectious or parasitic diseases. Thus, attention has focused on isolated living standards (or at best, levels of vulnerability), single diseases and particular age-groups, but rarely adults. Very little information has been available to express the health impacts of urban conditions in terms of the 'extent of the problem' within cities as a whole (Tabibzadeh et al., 1989). If an extensive analysis is done of overall inequalities, it becomes clear that an isolated focus on the lack of basic needs for particular groups and their immediate consequences in terms of infectious disease in part obscures a much bigger inequality picture for adults and for other health impacts (Stephens et al., 1994, 1996).

In reality this not a new story. That health inequalities would not go away after basic needs were met could have been predicted from the literature from developed countries (see Macintyre, Chapter 2; Power and Matthews, Chapter 3; Ben-Shlomo and Marmot, Chapter 17). In addition, it has been argued for some years, on the basis of little actual evidence, that a 'double burden' of potential health impacts may exist for poor groups in cities (Tabibzadeh et al., 1989; Harpham et al., 1990; Stephens et al., 1990; Harpham and Stephens, 1991; Stephens et al., 1994); in other words, the poor experience higher rates of both infectious and chronic diseases. There is increasing evidence to support this. Alongside or replacing child health problems in cities, there are large-scale inequalities in adult health between groups in urban areas of the South. In some cities of Latin America (for example São Paulo and Santiago), the health transition has occurred. However, patterns of health inequalities continue, particularly in areas such as violence, heart disease and adult illness in general (Stephens et al., 1994; Moser and Holland, 1995; Stephens, 1995). This suggests the complex impacts of disadvantage – even if a poor individual survives the infectious disease risks of childhood, the early health impacts of poverty appear to be succeeded by inflated risks of non-communicable diseases, particularly violence and, increasingly, circulatory diseases, in adulthood (Frenk et al., 1989; Stephens et al., 1990; Stephens et al., 1994).

Perhaps the most compelling argument on the complex policy implications of health inequalities in developing countries comes from the data on violence in cities. This has been described as 'epidemic' for many regions including many urban areas of Latin and North America, along with African and increasingly, Asian cities (Pan American Health Organization, 1990; de Noronha, 1993; Guerrero, 1993; Hasan, 1993; Pinheiro,

1993; Ndiaye, 1993; Zwi, 1993). In public health terms, deaths from violence may now overshadow infectious diseases as adolescent killers in some urban environments (Laurenti, 1972; Pan American Health Organization, 1990; SEMPLA, 1992). For example, violence (mostly homicides) accounted for 86% of all deaths in boys aged 15–19 in São Paulo in 1992 and over half of all deaths in 5 to 14-year-olds (SEMPLA, 1992; Stephens *et al.*, 1994). This health impact does not fall evenly – in São Paulo in 1992, death rates from homicides were 11 times higher in adolescent boys in deprived areas than for adolescents in wealthier areas (Stephens *et al.*, 1994). Evidence of similar health inequalities in violence is emerging in cities across the developing world. Explanations for these post-health transition urban inequalities are in their infancy. The 1996 UN Summit on Human Settlements continued to focus on child health and the sanitary environment, adding only the notion of 'criminality' and the need to control it through 'security' as a means of resolving the new health inequalities.

Conclusions

How does one conclude this puzzle? Policies on health inequalities in developing countries have been treated as a question of distribution of basic needs and basic survival. Distribution of the political process and power was part of this package intitially, but it has been the most controversial and most difficult to sustain. Data on the extent of health inequalities in the growing cities of the developing world beg a more sophisticated approach.

At the very least, a thorough equity slant could be placed on descriptions of child health inequalities and the social control of risk factors linked to them. In cities it is possible to do this – it becomes clear that poor children not only experience higher rates of illness than their wealthier neighbours, but also that their parents often pay more for the contaminated water which they drink (Cairncross and Kinnear, 1992; Benneh *et al.*, 1993) and the limited services they have access to. Putting this puzzle together suggests, simply, that the poor often pay higher prices for their ill health and discomfort, than their wealthier neighbours, who control access to resources and pay less for good health, while maintaining personal access to 'luxuries' such as private swimming pools and plentiful water supplies to maintain spacious gardens. In this context, it is not just the lack of basic needs of one group which could be addressed, but the inequities in control and use of the basic resources within cities.

More fundamentally, the new health inequalities in developing countries suggest the validity of John Donne's belief that we are all connected. The overall health experience of the North and South seems very different, with the North struggling to control obesity and chronic diseases in an ageing population and the South continues to grapple with basic food security and continued burdens of infectious diseases hitting a youthful population. Yet at the same time, there is increasing evidence that the 'worlds' which have been regionally and economically distinguished (North/South; developed/less developed) are converging. This is true in an economic and, increasingly, a health sense. At the bottom of society are those who experience the double burden of health inequalities. At the top are those who now share a common international position and exhibit similar traits of high material consumption. Those at the top are in the minority, but they are a powerful international elite at home in Rio, Abidjan, Cape Town, Paris or London. They share another commonality – their highly advantaged health experience.

Debates have raged over the last decades about the means to development, but the questions of globalization and over-development have only recently surfaced. Inequality within societies is returning to the analytic agenda after a decade of retrenchment. The 1994 Human Development Report of the United Nations Development Programme argued that, 'the purpose of development is to create an environment in which all people can expand their capabilities, and opportunities can be enlarged for both present *and future* generations'. The report goes on to suggest that the real foundation of human development is universalism in acknowledging the life claims of 'everyone'. Universalism as a development goal is not yet discussed in terms of the challenges posed by the health inequalities associated with complex outcomes like violence. At present, it is more common to attribute complex post-transition health inequalities in cities of developing countries to the behavioural traits of the large group of people who die prematurely from violence or heart disease. It is not recognized that such outcomes are likely to be associated to structural inequities in distribution of control over resources and opportunities to gain control. There is still minimal recognition that inequalities will not be resolved without tackling inequities in the distribution of *control* over resources within cities and nations, but most importantly internationally.

Finally, the concept of future generations (that 'everyone' includes future generations) is a theme yet to be incorporated into national and international development plans. Global forces, both economic and social, have always been the ultimate determinants and predictors of the distribution of health and disease for people. However, this is perhaps the moment in history when we are most acutely aware of the phenomenon, to the extent that we now have an expression – globalization – which tries to summarize the concept of international connection. The 1995 United Nations Social Summit in Copenhagen concluded that, 'No society has ever become "developed" by intensifying exclusion and polarization between those who have resources and those who do not' (UNRISD, 1995). This society is now international.

If we go back to the beginning of this chapter, we can now recognize the kind of connectivity that John Donne emphasized in the seventeenth century. The evidence implies most thoroughly that, just as health inequalities in the North are explicable as part of a complex process of social exclusion and cohesion, so inequalities in the South are not simply about mechanisms to distribute basic needs within those nations. The poor in developing countries will not achieve health equality simply through access to basic needs. Few seem yet to recognize that we may have come finally to a potential impasse in our development trajectories and in the ways in which we understand how to achieve health for all people. If the analysts of sustainability and global change are correct, the trajectory must be shifted at the top in terms of models of development that encourage over-consumption, polarization of power and increasing social exclusion and anomie – a notion which has yet to be addressed internationally.

References

Aaby, P. (1992). Lessons for the past: Third World evidence and the reinterpretation of developed world mortality declines. *Health Transition Review* 2 (supp. issue), 155–83.

Abel-Smith, B. & Lieserson, A. (1978). *Poverty, Health and Development*. Geneva: World Health Organization.

Amis, P. & Rakodi, C. (1994). Urban poverty: issues for research and policy. *Journal of International Development* 6 (5), 627–34.

Basta, S.S. (1977). Nutrition and Health in low income urban areas of the third world. *Ecology of Food and Nutrition* 6, 113–24.

Benneh, G., Songsore, J., Nabila, J. S., Amuzu, A. T., Tutu, K. A., Yangyuoru, Y. & McGranahan, G. (1993). *Environmental problems and the Urban Household in the Greater Accra Metropolitan Area (GAMA) – Ghana*. Stockholm: Stockholm Environment Institute.

Benneh, G., Nabila, J. S., Songsore, J., Yankson, P. & Teklu, T. (1990). *Demographic Studies and Final Projections for Accra Metropolitan Area (AMA)*, pp. 1–90. Accra: UNDP/UNCHS/Ministry of Local Government.

Berman, S. (1991). Epidemiology of acute respiratory infections in children of developing countries. *Review of Infectious Diseases* 13(6), S454–62.

Bisharat, L. & Zagha, H. (1986). *Health and Population in Squatter Areas of Amman: A Reassessment After Four Years of Upgrading*. Amman: UNICEF Monograph.

Bobadilla, J. L., Frenk, J., Frejka, T., Lozano, R. & Stern, C. (1990). *Health Sector Priorities Review. The Epidemiological Transition and Health Priorities,* pp. 1–26. Washington, DC: Population & Human Resources Dept., World Bank.

Bradley, D., Stephens, C., Harpham, T. & Cairncross, S. (1992). *Urban Management Program. Discussion Paper 6. A Review of Environmental Health Impacts in Developing Country Cities*. Washington, D.C: The World Bank.

Bradley, D. J. (1993). Environmental and health problems of developing countries. In *Environmental Change and Human Health*, pp. 234–46. Chichester: Wiley Publications.

Brockerhoff, M. (1993). *Child Survival in Big Cities: Are the Poor Disadvantaged?* New York: Population Council Research Division Working Papers.

Brockerhoff, M. (1995). Child survival in big cities: the disadvantages of migrants. *Social Science and Medicine* **40**(10), 1371–83.

Cairncross, S. & Kinnear, J. (1992). Elasticity of demand for water in Khartoum, Sudan. *Social Science and Medicine* **43**(2), 183–9.

Caldwell, J. C. (1991). Major new evidence on health transition and its interpretation. *Health Transition Review* **1**(2), 221–4.

Chambers, R. (1992). Poverty in India: concepts, research and reality. In *Poverty in India; Research and Policy*, ed. B. Harriss, S. Guahn & R. H. Cassen, pp. 301–32. Bombay: Oxford University Press.

Cheema, S. (1992). The Challenge of Urbanization. In *Cities in the 1990s. The Challenge for Developing Countries*, ed. M. Harris, pp. 26–33. London: UCL Press.

de Noronha, J. C. (1993). Drug markets and urban violence in Rio de Janeiro: a call for action. *The Urban Age* **1**(4), 9.

Elo, I. T. & Preston, S. H. (1996). Educational differentials in mortality: United States, 1979–85. *Social Science and Medicine* **42**(1), 47–57.

Frenk, J., Bobadilla, J. L., Sepulveda, J. & Cervantes, M. L. (1989). Health transition in middle-income countries: new challenges for health care. *Health Policy and Planning* **4**(1), 29–39.

Friedman, J. (1992). Rethinking poverty: the (dis) empowerment model. In *Empowerment. The Politics of Alternative Development*, ed. J. Friedman, pp. 65–71. London: Blackwell.

FUNDACAO SEADE, (1990). *Informe Demografico. Migracao no interior do Estado de São Paulo. Fundacao SEADE* **23**, 1–207.

Guerrero, R. (1993). Cali's innovative approach to urban violence. *The Urban Age* **1**(4), 12–13.

Hardoy, J. E., Cairncross, S. & Satterthwaite, D. (1990). *The Poor Die Young: Housing and Health in Third World Cities*. London: Earthscan Publications.

Hardoy, J. E., Mitlin, D. & Satterthwaite, D. (1992). *Environmental Problems in Third World Cities*. London: Earthscan Publications.

Hardoy, J. E. & Satterthwaite, D. (1989). *Squatter Citizen: Life in the Urban Third World*. London: Earthscan Publications

Hardoy, J. E. & Satterthwaite, D. (1991). Environmental problems of Third World cities: a global issue ignored? *Public Administration and Development* **11**, 341–61.

Harpham, T. & Stephens, C. (1991). Urbanization and health in developing countries from the shadows into the sportlight. *Tropical Diseases Bulletin* **88**(8), 1–35.

Harpham, T., Vaughan, P. & Lusty, T. (1990). *In the Shadow of the City*

Community Health & the Urban Poor. Oxford: Oxford University Press.
Harris, M. (1992). Introduction. In *Cities in the 1990s. The Challenge for Developing Countries*, ed. M. Harris, pp. ix–1. London: UCL Press.
Hasan, A. (1993). Karachi and the global nature of urban violence. *The Urban Age* 1(4), 1–4.
Hutchful, E. (Ed.) (1987). *The IMF and Ghana. The Confidential Record.* London: Zed Books Ltd.
Jackson, B. (1990). *Poverty & the Planet. A Question of Survival.* London: Penguin Books.
Kalache, A. & Aboderin, A. (1995). Stroke: the global burden. *Health Policy and Planning* 10(1), 1–21.
Kasarda, J. D. & Parnell, A. M. (1993). Introduction: Third World Urban Development Issues. In *Third World Cities: Problems, Policies and Prospects*, ed. J. D. Karsada & A. M. Parnell, pp. ix–1. London: Sage Publications.
Korten, D. (1996). Sustainability and the global economy: beyond Bretton Woods. *Forests, Trees and People.* June.
Laurenti, R. (1972). Alguns aspectos epidemiologicos da mortalidade por acidentes de transito de veiculos a motor na cidade de São Paulo, Brasil. *Revista de Saude Publica* 6(4), 339–41.
Lipton, M. (1988). Why poor people stay poor: urban bias in world development. In *The Urbanization of the Third World*, ed. J. Gugler, pp. 41–51. New York: Oxford University Press.
McKeown, T. (1976). *The Modern Rise of Population.* London: Edward Arnold.
McKeown, T. (1979). *The Role of Medicine. Dream, Mirage or Nemesis?* Oxford: Basil Blackwell.
McKeown, T. (1988). *The Origins of Human Disease.* Oxford: Basil Blackwell.
Medical Officer of Health (1992). *Annual Report: Cape Town, 1992.* Cape Town: Ministry of Health Report.
Monteiro, C. A. (1982). Contribuicao para o estudo do significado da evolucao do coeficinete da mortalidade infantil no Municipio de São Paulo (SP), Brasil nas tres ultimas decadas (1950–79). *Revista de Saude Publica* 16(1), 7–18.
Moser, C. & Holland, J. (1995). *A Participatory Study of Urban Poverty and Violence in Jamaica: Summary Findings.* Washington DC: Urban Development Division, World Bank.
Murray, C., Yang, G. & Qiao, X. (1992). Adult mortality: levels, patterns and causes. In *The Health of Adults in the Developing World*, ed. R. Feachem, T. Kjellstrom, C. Murray, M. Over & M. Phillips, pp. 23–111. Washington DC: World Bank.
Ndiaye, M. (1993). Dakar: youth groups and the slide towards violence. *The Urban Age* 1(4), 7.
Pan Américan Health Organization (1990). Violence: a growing public health problem in the region. *Epidemiological Bulletin* 11(2), 1–7.
Philips, D. R. & Verhasselt, Y. (1994). Introduction: health and development. In *Health and Development*, ed. D. R. Philips, & Y. Verhasselt, pp. 3–33. London: Routledge.
Pinheiro, P. S. (1993). Reflections on urban violence. *The Urban Age* 1(4), 3.
Post, K. & Wright, P. (1989). *Socialism and Underdevelopment.* London: Routledge.
Redclift, M. (1984). *Development and the Environmental Crisis. Red or Green*

Alternatives. London: Routledge.
Redclift, M. (1987). *Sustainable Development: Exploring the contradictions*. London: Routledge.
Rossi-Espagnet, A., Goldstein, G. B. & Tabibzadeh, I. (1991). Urbanization and health in developing countries: a challenge for health for all. *World Health Statistics Quarterly* **44**(4), 186–245.
Schaefer, M. (1993). *Health, Environment and Development: Approaches to Drafting Country-Level Strategies for Human Well-Being Under Agenda 21*. Geneva: World Health Organization.
SEMPLA (1992). *Base de Dados para o Planejamento*. São Paulo: State Publication of SEMPLA.
Sen, A. (1981). *Poverty and Famines: An Essay on Entitlement and Deprivation*. Oxford: Clarendon Press.
Smith, R. (1995). The WHO: change or die. *British Medical Journal* **310**, 543–4.
Stephens, C. (1995). The urban environment, poverty and health in developing countries. *Health Policy and Planning* **10**(2), 109–21.
Stephens, C., Harpham, T., Bradley, D. & Cairncross, S. (1990). *Health Impacts of Environmental Problems in Urban Areas of Developing Countries. An Analysis of the Epidemiological Evidence for the WHO Expert Committee on Urbanization*. Geneva: World Health Organization.
Stephens, C., Timaeus, I., Akerman, M., Avle, S., Borlina-Maia, P., Campanario, P., Doe, B., Lush, L., Tetteh, D. & Harpham, T. (1994). *Environment & Health in Developing Countries: An Analysis of Intra-Urban Differentials Using Existing Data. Collaborative Studies in Accra and Sao Paulo and Analysis of Urban Data of Four Demographic and Health Surveys*. London: London School of Hygiene and Tropical Medicine.
Stephens, C., McGranahan, G., Bobak, M., Fletcher, A. & Leonardi, G. (1996). Urban environmental impacts. In *World Resources Report*, pp. 31–55. Washington, DC: World Resources Institute.
Stren, R., Bhatt, V., Bourne, L. & Hardoy, J. E. (1992). *An Urban Problematique: The Challenge of Urbanization for Development Assistance*. Toronto: CIDA.
Tabibzadeh, I., Rossi-Espagnet, A. & Maxwell, R. (1989). *Spotlight on the Cities Improving Urban Health in Developing Countries*. Geneva: World Health Organization.
Timaeus, I. M. & Lush, L. (1995). Intra-urban differentials in child health. *Health Transition Review* **5**(2), 163–90.
Townsend, P. (1993). *The International Analysis of Poverty*. Hemel Hempstead: Harvester Wheatsheaf.
United Nations Development Programme, (1992). *Human Development Report 1992*. Oxford: Oxford University Press.
United Nations Development Programme, (1994). *Human Development Report 1994.*, Oxford: Oxford University Press.
UNRISD (1995). *UNRISD Social Development News, No. 12*. Geneva: UNRISD Social Development Division.
UNRISD (1995). *UNRISD Social Development News, No. 13*. Geneva: UNRISD Social Development Division.
Walt, G. & Vaughan, P. (1981). *An Introduction to the Primary Health Care Approach*. London: Ross Institute, London School of Hygiene and Tropical Medicine.

White, K. L. (1991). *Healing the Schism. Epidemiology, Medicine and the Public's Health*. New York: Springer-Verlag.

Williams, B. T. (1991). *Health Burden of Urbanizaton*. Background Technical Paper for The World Health Organization Technical Discussions on Urbanization. Geneva: World Health Organization.

Wills, C. (1995). A journal of the plague weeks. *Independent on Sunday* **23 July**.

World Bank (1991). *World Development Report 1991 Poverty*. Oxford: Oxford University Press.

World Bank (1993). *World Development Report 1993 Investing in Health*. Oxford: Oxford University Press.

World Health Organization (1992). *WHO Commission on Health and Environment Report of the Panel on Urbanization*, pp. 1–160. Geneva: World Health Organization.

World Health Organization (1993). *The Urban Health Crisis: Strategies for Health for All in the Face of Rapid Urbanization*. Geneva: World Health Organization.

World Health Organization (1995). *The World Health Report 1995: Bridging the Gaps*. Geneva: World Health Organization.

Zwi, A. (1993). *Public Health and the Study of Violence: Are Closer Ties Desirable?* Monograph. London: London School of Hygiene and Tropical Medicine.

17 Policy options for managing health inequalities in industrial and post-industrial countries

YOAV BEN-SHLOMO AND MICHAEL G. MARMOT

Introduction

There are three assumptions behind any policy intervention that aims to reduce inequalities in health. Firstly, that this is an important area. Secondly, that any observed inequalities are amenable to manipulation or intervention, and finally that either health or more broadly, social policy can actually have a direct effect in reducing the degree and level of current inequalities in health. This chapter will present evidence that the links between inequalities and health (see Macintyre, Chapter 2) are potentially amenable to intervention, and that policy may therefore have substantial positive effects on the health of populations. What sort of policy and how it should be implemented is less clear and beyond the scope of this chapter in which we have aimed at covering a wide variety of different areas. We have, therefore, by necessity, only scratched the surface of what are complex political issues.

It may seem self-evident that it is important to try and reduce inequalities in health on moral, pragmatic and economic grounds. Many countries explicitly support the first 'Health For All 2000' target, which aims for equity in health by improving the level of health of the disadvantaged and thereby resulting in a 25% reduction in health status differences between social groups. On a more pragmatic level, setting 'targets' for improving population health indirectly forces one to address the issue of inequalities. *The Health of the Nation* report (Department of Health, 1992) states that ischaemic heart disease mortality should be reduced by 40% by the year 2000 for people under 65 years and 30% for those between 65 and 74 years although no mention is made of reductions at a sub-group level for different social groups. It is fairly obvious that differential rates of decline across social groups make it progressively harder to achieve these targets and poorer populations, who have higher rates of disease, have more potential opportunity to reduce disease risk. From an economic perspec-

308

tive it is clear that the burden of premature morbidity and mortality is an economic drain both in terms of health care expenditure as well as lost productivity and social benefits. Whether reducing health inequalities will be overall cost-effective is less clear as improvements in life expectancy will be counter-balanced with increasing costs of care for the elderly (Evandrou *et al.*, 1992). Poorer sections of society are more likely to be dependent on state support for providing long-term care, rather than having private insurance or other financial means.

Are inequalities in health inevitable ?

Inequalities in health have been well documented from the industrial revolution onwards (Woods and Williams, 1995). Since the 1860s, when mortality data became available, overall population mortality rates in England and Wales have been declining and have continued to progressively decline over this century (Davey-Smith and Marmot, 1991). Indirect evidence suggests social differences in life expectancy have existed for over the last two centuries (Davey-Smith *et al.*, 1992). Routine mortality statistics on social class in England and Wales were first published for the 1921 census and continue on a decennial basis. They demonstrate clear social gradients in mortality risk over this period (Pamuk, 1985). More recently, mortality differentials have appeared to increase between social groups (Marmot and McDowall, 1986; Najman, 1993; Phillimore *et al.*, 1994) suggesting that poorer sections of society are not experiencing the same degree of health benefits.

Comparison of social differentials between European countries is complicated by problems of classification and socio-cultural differences in occupational status. Despite these methodological limitations, there is clear evidence that some countries appear to have far wider social class gradients than others as measured by educational status (Kunst and Mackenbach, 1994). Countries such as Sweden and Denmark have narrower mortality differences whilst France and Italy have more marked differences. Similarly within England and Wales, regional differences are clearly seen in the strength of the relationship between area deprivation and mortality (Eames *et al.*, 1993) (see Figure 17.1). A similar increase in deprivation is associated with a far greater increase in mortality for regions in the North of the country compared to the South. These temporal and geographical data provide evidence that inequalities are widespread and have existed for as long as we have documented them. However, the variation in the degree of inequalities suggests that they are not inevitable and can at least be attenuated if not abolished.

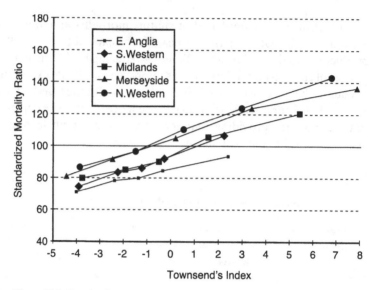

Figure 17.1. Standardized Mortality Ratio by Townsend's Index.

Why should policy have an impact on inequalities?

Criteria are useful when assessing the potential role of policy to manage inequalities (see Table 17.1). It is sensible to focus on areas that are of considerable public health importance in terms of mortality, morbidity and quality of life, and have marked social inequalities as any degree of success in these areas will have the greatest impact. Research evidence must also exist as to what factors might cause the disease and preferably intervention studies should demonstrate benefits, although these may not always be available in the form of randomized controlled trials, and both observational and qualitative data may be required to guide interventions. Policy makers do not necessarily rely on the quantity or quality of research evidence when formulating policy (Walt, 1994). Studies should also provide evidence that intervention actually reduces inequalities, as either the uptake or effectiveness of the intervention may differ by socio-economic status (Mackenbach, 1995). For example, health promotion activities to decrease smoking have probably increased differentials in smoking status as individuals of lower socio-economic status have faced greater difficulties in giving up smoking (Graham, 1988; Pierce *et al.*, 1989; Novotny *et al.*, 1990; Ben-Shlomo *et al.*, 1991). Any policy suggestion must be viewed in its wider political context and its impact on non-health related areas, as it will certainly have an impact beyond the health sector.

We have adopted the model proposed by Dahlgren and Whitehead

Table 17.1. *Criteria to assess potential of policy interventions*

- Identify an area of public health importance
- Examine whether there are mortality and or morbidity inequalities
- Determine whether there is evidence as regards aetiology
- Determine whether there is evidence on effective interventions
- Ensure that interventions actually reduce socio-economic differentials
- Review barriers to implement policy.

(1991) which conceptualizes layers of influence. At the centre are characteristics of individuals, such as sex, age and genetic factors, which cannot be altered. Over this come the following layers: individual life style; living and working conditions; social and community factors; and general socio-economic, cultural and environmental circumstances. This model resembles recent developments in epidemiological theory (Krieger, 1994; Susser and Susser, 1996), which emphasize the importance of macro- as well as micro-level determinants for behaviour and disease risk (see Porter and Ogden, Chapter 3). We have chosen to illustrate this approach with two examples, mortality and morbidity from ischaemic heart disease (IHD) and accidents or violence, since both of these contribute heavily to the social class differences in years of potential life lost (Blane *et al.*, 1990; Wilkinson, Chapter 4) (see Figure 17.2).

Individual lifestyle factors

We do not intend to discuss interventions for every individual risk factor. Some of these, such as hypertension and raised cholesterol, are usually approached from a conventional 'high risk' medical model, and are discussed further in relation to ensuring equity of access to health care (for a recent review of interventions in the elderly for the prevention of IHD and stroke see Ebrahim and Davey-Smith, 1996).

Systematic reviews of reducing inequalities

A recent review, commissioned by the Department of Health, examined all intervention studies with an experimental design that was intended to reduce health inequalities by targeting poorer sections of the population (Arblaster *et al.*, 1995). Only 94 studies could be identified and many were of dubious methodological quality. The characteristics found to be associated with greater intervention success are summarized in Table 17.2. The authors concluded that, 'it is important that strategies developed to reduce

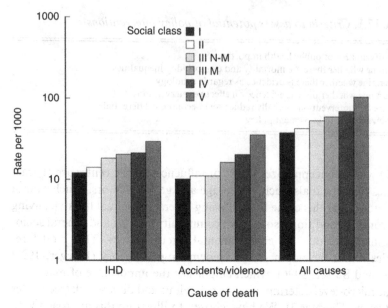

Figure 17.2. Years of potential life lost. Age adjusted rates per 1000 population (men 16–64 years, 1981). IHD: ischaemic heart disease. (Taken from Blane *et al.*, 1990.)

inequalities are not assumed to be having a positive impact simply because the aim is "progressive" and so rigorous evaluations of promising interventions are important.' The paucity of evidence in support of interventions to reduce inequalities has led some to take a nihilistic view of health service interventions (Foster, 1996). However, there are many effective medical interventions, which may not be provided equitably to all sections of society.

Smoking

Smoking is an important risk factor for a wide variety of diseases including IHD (Doll *et al.*, 1994). Smoking is far more common in the lower socio-economic groups, the unemployed, those not married, lone mothers, and people living in council estates (Townsend, 1995). General practitioner advice can produce significant results in cessation rates although these are limited to only a small proportion (3%) of patients (cited in Arblaster *et al.*, 1995). An enhanced midwifery care at home project also found that twice as many mothers gave up smoking or reduced their consumption of cigarettes than in the control group (cited in Townsend,

Table 17.2. *General characteristics associated with successful interventions*

Prior to intervention
- Needs assessment
- Community commitment

Type of interventions
- Intensive approaches
- Multidisciplinary approaches
- Multifaceted interventions
- Variety of settings

Delivering intervention
- Face-to-face interactions
- Culturally appropriate
- Appropriate agent delivering intervention
- Sufficient training of those delivering intervention

Additional factors
- Support materials and resources
- Prompts and reminders
- Developing skills of participants

1995)). Nicotine patches have also been found to facilitate quitting, but as yet are not available on the National Health Service (NHS) and so are less likely to be used by poorer patients (Ling Tang *et al.*, 1994). Free provision of patches to people on unemployment benefit or income support would seem a simple way of targeting care to this section of the population (Townsend, 1995). Townsend (1995) has examined the impact of advertising, restricting availability and pricing policy. She concludes that restricting advertisements and other forms of promotion is likely to reduce smoking by around 7.5%. Although there is no evidence on whether this impact would be similar across all social groups, advertising is more common in poorer areas as well as the tabloid press. Increasing the legal age for buying cigarettes from 16 to 18 years is also likely to be effective, although this is unlikely to stop adolescent experimentation with cigarettes. Cigarette taxation is a powerful method of reducing consumption. Temporal data show that consumption mirrors real price values of cigarettes; a 1% increase in the price results in a 0.55% fall in consumption. Such price elasticity is even more marked for the poor, and for teenagers. Between 1980 and 1988, the prevalence of smoking amongst Canadian teenagers fell from 45% to 22% in line with marked price increases. Townsend *et al.* (1994) have suggested that price increases should be instituted in tandem with increased financial support for the poorest, as they spend a disproportionate amount of their income on cigarettes and some will fail to give up despite increased cost.

Alcohol

Alcohol, like smoking, is an individual risk factor that plays an important role in causing diseases such as cirrhosis, but indirectly contributes to accidents and violence. As accidents are more common in young persons, it has been estimated that alcohol contributes to 40% of the relevant years of life lost. Alcohol is also associated with both suicide and violent crimes towards others. Time series analyses demonstrate an association between measures of alcohol consumption and deaths from road traffic accidents, suicide and homicide (Edwards *et al.*, 1994). The relationship between alcohol consumption and socio-economic status is far less clear than that with smoking. Some studies show no social gradient for men while for women an inverse relationship is seen which suggests that women of higher socio-economic status are more likely to consume more alcohol (Braddon *et al.*, 1988). Data from the General Household Survey in the UK demonstrate that the proportion of heavy drinkers amongst men is relatively equal across all occupational groups (OPCS, 1989). Young men (18–24 years) are more likely to be heavy drinkers and show more of a social gradient (Figure 17.3). These data do not, however, distinguish between binge as opposed to controlled drinking and the settings where consumption occurs, which are relevant to the risk of an accident. An intoxicated manual employee is more likely to risk some serious work-related accident than his non-manual counterpart.

Lederman was probably the first to postulate a distribution theory of alcohol consumption, so that changes in the mean alcohol consumption would be accompanied by changes across the whole distribution including the heavy drinkers (Edwards *et al.*, 1994). Hence the proportion of heavy drinkers is related to the 'wetness' of the society in which they live. This has been empirically demonstrated by data from the Intersalt study (Intersalt Cooperative Research Group, 1988) which measured alcohol consumption across 52 centres in 32 countries worldwide. A correlation of 0.97 was found between the mean level of consumption and the percentage of heavy drinkers (Rose and Day, 1990). Interestingly, the mean consumption is only weakly related to the percentage of abstainers suggesting that socio-cultural factors are more powerful determinants of the amount of alcohol consumed by current drinkers rather than drinking status *per se* (Colhoun *et al.*, 1997).

Several studies have found that medical advice in primary care settings can reduce alcohol consumption (Wallace *et al.*, 1988; Anderson and Scott, 1992). This approach is limited as only a small proportion of participants appear to benefit, and those subjects who are willing to take part in such research studies are non-representative of the heavy drinking

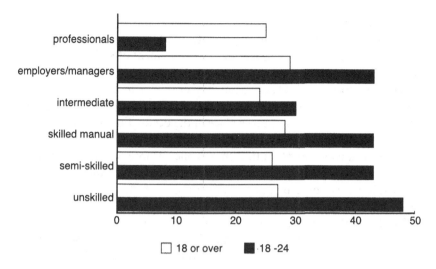

Figure 17.3. Heavy alcohol consumption for men by socio-economic status.
(Data taken from the General Household Survey 1986: OPCS, 1989.)

population. As Griffith Edwards and colleagues (1994) conclude, preventive measures that influence the generality of drinkers will often impact on heavy or problematic drinkers. They recommend the following measures: (a) taxation policies to increase cost of alcohol; (b) reducing physical access to alcohol by influencing minimum legal drinking age, restrictions on hours and days of sale, number of outlets, etc.; (c) drink-driving counter measures; and (d) increased education in schools, mass media, warning labels and advertising restrictions on the hazards of alcohol consumption.

Working conditions

Work characteristics

Traditionally, occupational health has focused on physical and environmental exposures in the workplace. Low status workers are, in general, more exposed to such hazards than high status workers. These same hazards may cause more general environmental exposure, for example through industrial wastes. Although such physical and chemical hazards are, rightly, a cause for concern it is arguable that, among white collar workers in particular, psycho-social characteristics of work are more potent causes of ill health.

Siegrist (1995) has suggested four important reasons for the central role of work in industrialized societies. Firstly, work is a major determinant of individual income levels, which, of course, have a major impact on living conditions, opportunities, lifestyle, and social and psychological well-being. Secondly, work may both promote and limit personal growth and development. It shapes life goals and the assessment of self in relation to others. Thirdly, occupation is a measure of social status. Not only is it a criterion of social stratification, but it also relates to esteem and social approval. Fourthly, much of social and psychological experiences, as well as environmental exposure, takes place in the occupational setting.

A comprehensive literature review on stressors related to the organization of work concluded that they shared one or more common characteristics (Kristensen, 1989): (1) lack of control; (2) lack of meaning; (3) lack of predictability; (4) over or under-stimulation; and (5) conflict. This provided general support for the Karasek model (Karasek and Theorell, 1990). Karasek had postulated that a combination of high demands and low control at work were associated with increased cardiovascular risk. Most subsequent studies have found support for the association between low control and increased risk, but not all support the two factor demand–control model (Schnall et al., 1994).

Siegrist (1995) has elaborated a somewhat different two-factor model. His model takes into account the role of personal coping mechanisms and adaptation to work demands and is based on an effort–reward imbalance. He emphasizes that effort at work is a function of both extrinsic demands and intrinsic motivation. Rather than focus on control, he focuses on rewards of three types: money; esteem and status; and job security. Like the demand–control model, this approach posits that high effort and low reward produced sustained distress which results in increased cardiovascular risk. There is support for this hypothesis in studies of both white-collar and blue-collar occupations (Siegrist and Matschinger, 1989; Siegrist et al., 1990, 1991). Low control in the workplace (Marmot et al., 1991) and, to a lesser extent, effort–reward imbalance tend to be more common at the lower end of the occupational hierarchy. They may, therefore, be important contributors to socio-economic differences in health.

While the evidence for these associations is strong in observational studies, it is not straightforward to organize intervention studies (Theorell, 1995). The appropriate point of intervention depends on the extent to which one views work stress either as a property of individual perception or as a characteristic of the work environment. If it is a characteristic of the individual, the appropriate intervention might be stress counselling or relaxation techniques. If it is related to the work environment the appropriate intervention would be the job design. Although there are clear individual differences, certain jobs do seem to be characterized by high risk. If this risk is related to psycho-social characteristics, it does suggest that the appropriate level of intervention should be the workplace.

Job insecurity

There is a feature of the Siegrist model (1995) which may be important in this context. Among the rewards in the Siegrist formulation are job security and social status. Job insecurity is the other side of labour market flexibility and it has become a feature of some economies (Uchitelle and Kleinfield, 1996). The effect of job insecurity on health has not been the focus of much investigation. There is evidence from the Whitehall II study that anticipation of job loss is associated with deterioration in health (Ferrie *et al.*, 1995). There is evidence from the 1958 birth cohort that people who have experienced unemployment have worse health even when they return to employment (Bartley *et al.*, 1996). Evidence from the USA suggests that people who have experienced unemployment tend to be re-employed in lower status jobs. High rates of unemployment and high turnover in the labour market may therefore induce insecurity even among those who are in relatively stable jobs.

Unemployment

There was, at one point, a vigorous debate as to whether unemployment was associated with deterioration in health. One part of the debate relates to the question of health selection (Power and Matthews, Chapter 3). This argument suggests that if people who are sick are those most likely to become unemployed, causation would be 'the wrong way' of looking at the question. The balance of evidence suggests that higher mortality risks among the unemployed cannot be explained on this basis (Bartley, 1991). A second issue relates to whether the unemployed have apparently higher risk because of their lower socio-economic status or for some other reason. This argument can be separated into two parts. Unemployed people have a higher mortality than those in the same social class who remain employed (Moser *et al.*, 1984). However, part of the reason for the increase in health risk of the unemployed may be worsening economic circumstances. Their increased health risk may also relate to those characteristics reviewed above and found to be associated with adverse working conditions, i.e. lack of control, lack of meaning, lack of predictability, understimulation and conflict.

  The factors that influence job insecurity and levels of unemployment are outside the realms of influence traditionally associated with health. Nevertheless, these may be powerful influences on the health of populations and in part contribute to the socio-economic differences in health.

Social and community influences

The concept of a community varies by discipline or theorist (for a detailed review see Patrick and Wickizer, 1995). Patrick and Wickizer (1995) usefully group definitions of community into three broad categories: community as *place*; community as *social interaction*; and community as *social and political responsibility*. These three categories will obviously have some relationship to geographical proximity, but geography may be unimportant when describing 'personal communities'. Obviously the physical environment of a community may affect health directly through toxic exposures (see Schell and Czerwinski Chapter 7). The need to monitor (Elliott *et al.*, 1992) and legislate against such hazards is clearcut. More intriguing is the potential role of the community in both providing psycho-social support and empowering individuals to alter their behaviour (see Wilkinson, Chapter 4).

Area effects

Many previous ecological studies have shown the strong relationships between area deprivation and mortality from IHD and accidents and violence (Centerwell, 1984; Carstairs and Morris, 1991). These relationships have been interpreted as simply reflecting the increased risk of individuals of lower socio-economic status who reside in such areas (Sloggett and Joshi, 1994). But evidence exists to show that poor areas may have worse health due to factors over and above the characteristics of their residents. The Alameda county study examined the nine-year mortality rates of residents in Oakland, California (Haan *et al.*, 1987). Those subjects living in a federally designated poverty area had around 70% higher all cause mortality (Haan *et al.*, 1987) which persisted despite adjustment for a wide array of potentially confounding variables, including socio-economic status, health practices, social networks and psychological factors. The development of multi-level modelling techniques now allows more formal empirical testing of area or 'contextual' effects. Results confirm that area characteristics such as deprivation, enable better prediction of physiological measures such as lung function (Jones and Duncan, 1995), long-term illness (Shouls *et al.*, 1996) and suicide (Congdon, 1996) than methods based solely on using individual risk factors.

Community cohesiveness

The town of Roseto, Pennsylvania contained an Italian-American community with strikingly low rates of IHD mortality reported for the decade

1955–65 compared to several neighbouring communities. Detailed investigation excluded the role of medical services and confounding by conventional risk factors. One possible explanation put forward to explain this unusual observation was that this immigrant population had a very stable social structure, strong family cohesion and a supportive community. However, amongst the younger generation, acculturation was evident and it was hypothesized that breakdown of traditional values would be accompanied by an increase in mortality. Recent temporal data provide support for this hypothesis. Over a 50-year period the protective effect of residing in Roseto has disappeared so that mortality for IHD is now no different from its neighbours (Egolf *et al.*, 1992; Figure 17.4).

The importance of strongly cohesive communities may only occur in special circumstances, such as first generation immigrants. More general attitudes to one's local area and community have, however, been found to be related to measures of anxiety and health. Sooman and Macintyre (1995) constructed an area assessment score based on residents, opinions on amenities, problems with area, poor reputation, neighbourliness, fear of crime and area satisfaction. This score significantly predicted both anxiety and self-assessed health after adjustment for individual socioeconomic status.

It is difficult to know exactly what sort of policies would enhance local community cohesion and satisfaction. Local empowerment is one method which forms an underlying principle behind the 'healthy cities' movement (Ashton, 1992). Engaging all the population is, however, a difficult endeavour. Putnam and colleagues (1993) demonstrate the possibility of measuring and promoting what they coined as 'civic' communities. Their work was aimed at understanding why local/regional governments in the north of Italy were more responsive to local needs than those based in the south. They defined measures of civic communities by civic engagement, political equality, social structures of co-operation and measures of trust and tolerance. For example, community participation in local elections, football matches, bird-watching clubs and readership of local papers were all measures of an engaged community. Not surprisingly areas with more civic engagement also had more responsive institutional structures, although it was not clear that this was directly a result of the local involvement.

Whilst politicians of all complexions often advocate decentralization and communitarianism, such rhetoric must also enable local communities to have real power over local structures and facilities. Local governments need to encourage community participation both in the democratic process and in raising local pride and activity.

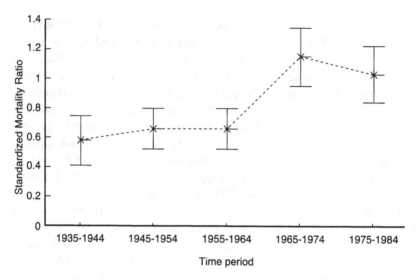

Figure 17.4. The 'Roseto effect' over 50 years standardized mortality ratios for myocardial infarction. (Data taken from Egolf *et al.*, 1992.)

General socio-economic and cultural factors

Income distribution

The risk of mortality is related to income levels. Individuals with lower income have a greater risk of all-cause mortality as well as an increased risk of IHD and accidental or violent death (Sorlie *et al.*, 1992; Diez-Roux *et al.*, 1995; Sorlie *et al.*, 1995; Backlund *et al.*, 1996; Davey-Smith *et al.*, 1996; Ettner, 1996). Backlund and colleagues (1996) illustrate the linear and *relative* nature of this relationship as increasing income is associated with progressively lower risk of mortality for both men between 25 and 44 years and those between 45 and 64 years (see Figure 17.5). For those aged over 65 years there appeared to be a ceiling effect while the worst mortality rates were experienced by the bottom 5% of the population. Improving the income of all subjects should reduce mortality but would not affect the relative differences between groups and hence inequalities. Alternatively, reducing the spread of the income distribution, the gap between those with the least and most income, should reduce this differential. Wilkinson has repeatedly demonstrated that in developed countries the effect of increasing GNP (Gross National Product) per capita is associated with diminishing health benefits (Wilkinson, 1994; this volume). The countries which demonstrate the longest life expectancy are not the wealthiest but those with the smallest spread of income. This is seen both cross-sectionally across countries and temporally within countries, those having the greatest

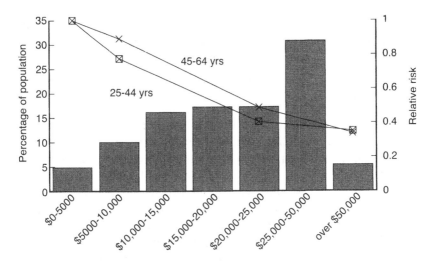

Figure 17.5. Income and mortality in the United States. Men between 25 and 64 years. (Data taken from Backlund *et al.* 1996.)

improvements in life expectancy showing either a diminution of relative poverty or less of an increase in the numbers who are poor. These relationships have also been seen within countries as well as between countries. Kaplan *et al.* (1996) have shown a correlation of 0.62 between all-cause mortality and the percentage of total household income received by the poorer 50% in each state. The correlation was even greater for homicides and violent crimes (0.74 and 0.70 respectively). Similarly, in the UK, where routine income data are not available, local authority areas that have a narrower spread of ward deprivation show lower mortality rates than those that are more unequal (Ben-Shlomo *et al.*, 1996).

There is little interventional data on the impact of social or fiscal policy on health. One study from Gary, Indiana in the USA, found that negative income tax did help reduce the proportion of low birth weight babies in a deprived area (Kehrer and Wolin, 1979). Other observational data from the USA demonstrate that whilst real median income levels increased in the Reagan years, the relative inequality of the population actually increased. This was mainly the effect of cutbacks in welfare programmes, social insurance and means-tested transfer programmes (Plotnick, 1993). In the UK, fiscal reforms have reduced taxation levels from 83% to 40% for the richest, whilst basic rate has declined from 33% to 25% since 1979. However, reductions in benefits and increases in levels of indirect taxation have disproportionately hit the poorer sections of society. Several fiscal options exist to reduce both absolute and relative poverty levels. These include a basic income scheme and negative income tax, which can ensure that everyone receives a basic minimum income. A graduated taxation scheme and abolition of a national insurance ceiling would help reduce income differentials and narrow population inequality.

Education

Many studies have shown a clear linear relationship between educational achievement and mortality, particularly from IHD and accidental or violent deaths, both at an ecological and at an individual level (Tyroler *et al.*, 1993; Diez-Roux *et al.*, 1995, see Macintyre, Chapter 2; Power and Matthews, Chapter 3). Often education is used as a proxy marker of socio-economic status instead of income or occupation, particularly in studies from the USA. Education and income may have different socio-cultural impacts on health. In the Atherosclerosis Risk in Community (ARIC) study, white men showed a bigger social differential for prevalence of coronary heart disease by education (odds ratio 3.8) whilst for black men, income was a better discriminator (odds ratio 3.4) (Diez-Roux *et al.*, 1995). Whilst education obviously is an important determinant of both occupation and income, it may operate through different mechanisms such as health knowledge, health behaviours, access to care and psycho-social mechanisms such as the ability to cope with life stressors. This may explain why some studies find an independent effect of education after adjustment for either income (Elo and Preston, 1996) or socio-economic status (Holme *et al.*, 1980).

There is evidence that educational interventions targeted at high risk populations may have long-term benefits (Hertzman and Wiens, 1996), although the outcomes for such studies have not usually included health measures. The Perry Preschool study was a trial comparing children allocated to either preschool or no preschool interventions (Schweinhart *et al.*, 1993, quoted in Hertzman and Weins, 1996). At age 27, those allocated to the active intervention were more likely to be earning more money, be a home-owner, a high school graduate and less likely to have had contact with social services or have had five or more arrests than the group who received no interventions. Evidence exists that intervention programmes in infancy, preschool and school-based can have a positive impact on cognitive development, social-emotional development and coping skills (Hertzman and Wiens, 1996). Such outcomes are likely to be associated with both adult health behaviours and risk of morbidity.

The role of medical interventions and the health service

It is generally accepted that medical care has made only a limited contribution to the improvements in mortality rates seen over the twentieth century (McKeown *et al.*, 1975; McKinlay *et al.*, 1989; Mackenbach *et al.*, 1990). However, recent advances in both surgical and medical interventions have led to a re-evaluation of medical services in terms of both preventing disease as well as reducing case fatality and improving quality of life. Bunker and colleagues (1995) have attempted to calculate the gains in life expectancy and quality of life associated with various interventions. They estimate that medical services in general add around 5 years of life expectancy, with the potential of another 2 or 2.5 years by extending access to effective treatments. Goldman and Cook (1984) estimated that 3.5% of the decline in IHD mortality could be attributed to coronary artery bypass grafting. By extending care to include surgery, medical treatments and coronary care units, it is estimated that life expectancy can be prolonged by an additional 1.2 years at a population level with a potential of an additional 6–8 months, with around a 55% improvement in quality of life (Bunker *et al.*, 1995). Whilst much improvement have similarly occurred for mortality from accidents, this is mostly related to non-medical interventions with medical interventions only adding an additional 1.5–2.0 months and a potential 3–4 months.

Given the limited but significant benefits of medical interventions, particularly for IHD, it is essential to ensure that access to effective treatment is equitable and based on need rather than socio-economic status. Unfortunately most studies, in particularly randomized controlled trials, do not explicitly address this issue and often fail to present results by relevant sub-groups. A recent re-analysis of the Multiple Risk Factor Intervention (MRFIT) trial clearly indicated an under-representation of poorer groups, common to most trials unless they are truly population-based, such as the Hypertension Detection and Follow-up Program (HDFP) trial (Hypertension Detection and Follow-up Program Cooperative Group, 1977). However, despite the selection biases, limited evidence suggests that improvements in diastolic blood pressure, smoking cessation, and LDL-cholesterol, seen under trial conditions, are very similar for both well educated and less educated subjects; education being used as a marker of socio-economic status (Cutler and Grandits, 1995). The HDFP provides even more compelling evidence that medical care can help eradicate mortality differences by the appropriate management of hypertension (Hypertension Detection and Follow-up Program Cooperative Group, 1987). Amongst the group who received conventional medical care (referred care) there was a two-fold mortality gradient based on whether the subject did or did not receive high school education. In contrast the special (stepped care) group showed almost non-existent gradients amongst both black and white subjects. Similarly the SHEP (Systolic Hypertension in the Elderly

Programme) anti-hypertension trial also found similar reductions in cardiovascular mortality for both educational groups with the less educated group showing larger benefits, although a formal test of interaction was not significant (Cutler and Grandits, 1995).

These data provide evidence on the efficacy of treatments but do not reflect the reality of day-to-day health care provision. Observational data consistently indicate that socio-economic status is related to the likelihood of receiving health interventions. This has been best documented in the USA, where the two-tier health care system ensures a large vulnerable segment of the population may not be able to afford major care expenditure (Hayward *et al.*, 1988). In the UK, it is assumed that a free health care system will not deter poorer individuals from treatment. Simulation models suggest that the UK health system does broadly provide equal treatment for equal need (Propper, 1994). However, inequities appear to exist both for receiving surgery for heart disease (Ben-Shlomo and Chaturvedi, 1995) and other common conditions (Chaturvedi and Ben-Shlomo, 1995). Men living in more affluent areas are more likely to receive coronary revascularisation surgery despite having less need as measured by mortality rates (Ben-Shlomo and Chaturvedi, 1995).

Differences in access are not inevitable and have not always been found, for example in Northern Ireland no differences were noted in access to coronary revascularisation surgery by area deprivation (Kee *et al.*, 1993). A recent report from Finland, similarly failed to find differences in the survival of diabetics by socio-economic status (Koskinen *et al.*, 1996). Equitable health services has been an important goal in Finnish health policy for decades. This suggests that health care purchasers must not only explicitly contract for equitable service provision but also take an active role in monitoring this both through routine activity data and clinical audit (Majeed *et al.*, 1994). The use of explicit guidelines, as in the stepped care approach used by the HDFP trail, will also help prevent extraneous demographic factors influencing the provision of health care.

Conclusions

Both health and social policies can reduce health inequalities. Most evidence suggests that the latter has a far greater potential impact, but is also less amenable to change due to its political complexity. Thus even with better evidence than currently exists, strong ideological barriers may prevent the most rational interventions ever being translated into policy recommendations.

Whilst diseases may come and go, Link and Phelan (1996) have argued that the fundamental social causes of disease will always depend on 'resources like money, power, prestige, and social connections that strongly influence people's ability to avoid risks and to minimise the consequences of disease once it occurs.' The ability to redistribute these more equitably within a society may be limited but it is clear that policy makers have a responsibility to at least attempt to tackle these factors. In 1848, the eminent physician Virchow noted the relationship between poor living conditions and typhus in Upper Silesia. He had no doubt that the solution to this problem lay in the political rather than medical realm by recommending that improvements in housing would only arise with 'full and unlimited democracy' (cited in Lynch, 1996). As he noted, 'do we not always find the diseases of the populace traceable to defects in society?' (cited in Amick *et al.*, 1995).

References

Amick, B. C., Levine, S., Tarlov, A. R. & Walsh, D. C. (1995). Introduction. In *Society and Health*, ed. B. C. Amick, S. Levine, A. R. Tarlov, & D. C. Walsh, 1st edn., pp. 3–17. New York: Oxford University Press.

Anderson, P. & Scott, E. (1992). Randomised controlled trial of general practitioner intervention in men with excessive alcohol consumption. *British Journal of Addiction* **87**, 891–900.

Arblaster, L., Entwistle, V., Lambert, M., Forster, M., Sheldon, T. & Watt, I. (1995). *Review of the Research on the Effectiveness of Health Service Interventions to Reduce Variations in Health*, part 1. York: CRD.

Ashton, J. (1992). *Healthy Cities.* Milton Keynes: Open University Press.

Backlund, E., Sorlie, P. & Johnson, N. J. (1996). The shape of the relationship between income and mortality in the United States: evidence from the National Longitudinal Mortality Study. *Annals of Epidemiology* **6**, 12–20.

Bartley, M. (1991). Health and labour force participation. *Journal of Social Policy* **20**, 327–64.

Bartley, M., Montgomery, S., Cook, D. & Wadsworth, M. (1996). Health and work insecurity in young men. In *Health and Social Organization: Towards a Health Policy for the 21st Century*, ed. D. Blane, E. Brunner & R. Wilkinson, pp. 255–71. London: Routledge.

Ben-Shlomo, Y. & Chaturvedi, N. (1995). Assessing equity in access to health care provision in the UK: does where you live affect your chances of getting a coronary artery bypass graft? *Journal of Epidemiology and Community Health* **49**, 200–4.

Ben-Shlomo, Y., Sheiham, A. & Marmot, M. (1991). Smoking and Health. In *British Social Attitudes: The 8th Report* ed. R. Jowell, L. Brook & B. Taylor, pp. 155–72. Aldershot: Dartmouth Publishing Company Limited.

Ben-Shlomo, Y., White, I. R. & Marmot, M. (1996). Does the variation in the socioeconomic characteristics of an area affect mortality? *British Medical Journal* **312**, 1013–14.

Blane, D., Davey-Smith, G. & Bartley, M. (1990). Social class differences in years of potential life lost: size, trends and principal causes. *British Medical Journal* **301**, 429–32.

Braddon, F. E. M., Wadsworth, M. E. J., Davies, J. M. C. & Cripps, H. A. (1988). Social and regional differences in food and alcohol consumptions and their measurement in a national birth cohort. *Journal of Epidemiology and Community Health* **42**, 341–9.

Bunker, J. P., Frazier, H. S. & Mosteller, F. (1995). The role of medical care in determining health: creating an inventory of benefits. In *Society and Health*, ed. B. C. Amick, S. Levine, A. R. Tarlov, & D. C. Walsh, pp. 304–41. New York: Oxford University Press.

Calhoun, H., Ben-Shlomo, Y., Dong, W. Bost, L. & Marmot, M. (1997). Ecological analysis of collectivity of alcohol consumption in England: importance of the average drinker. *British Medical Journal* **314**, 1164–8.

Carstairs, V. & Morris, R. (1991). *Deprivation and Health in Scotland.* Aberdeen: Aberdeen University Press.

Centerwell, B. S. (1984). Race, socioeconomic status, and domestic homicide, Atlanta, 1971–72. *American Journal of Public Health* **74**, 813–15.

Chaturvedi, N. & Ben-Shlomo, Y. (1995). From the surgery to the surgeon: does deprivation influence consultation and operation rates. *British Journal of General Practice* **45**, 127–31.

Congdon, P. (1996). The epidemiology of suicide in London. *Journal of the Royal Statistical Society* **159**, 515–33.

Cutler, J. A., & Grandits, G. (1995). What have we learned about socioeconomic status and cardiovascular disease from large trials? In *Report of the Conference on Socioeconomic Status and Cardiovascular Health and Disease: November 6–7, 1995*, ed. J. Stamler & H. P. Hazuda, pp. 145–52. Bethesda: National Institutes of Health.

Dahlgren, G. & Whitehead, M. (1991). *Policies and Strategies to Promote Social Equity in Health.* Stockholm: Institute for Future Studies.

Davey-Smith, G., Carroll, D. & Rankin, S. (1992). Socio-economic differentials in mortality: evidence from Glasgow graveyards. *British Medical Journal* **305**, 1554–7.

Davey-Smith, G. & Marmot, M. G. (1991). Trends in mortality in Britain: 1920–1986. *Annals of Nutrition and Metabolism* **35** (suppl 1), 53–63.

Davey-Smith, G., Neaton, J. D., Wentworth, D., Stamler, R. & Stamler, J. (1996). Socioeconomic differentials in mortality risk among men screened for the Multiple Risk Factor Intervention Trial: I. white men. *American Journal of Public Health* **86**, 486–96.

Department of Health. (1992). *The Health of the Nation: A Strategy for Health in England.* London: HMSO.

Diez-Roux, A. V., Nieto, F. J., Tyroler, H. A., Crum, L. D. & Szklo, M. (1995). Social inequalities and atherosclerosis. The atherosclerosis risk in communities study. *American Journal of Epidemiology* **141**, 960–72.

Doll, R., Peto, R., Wheatley, K., Gray, R. & Sutherland, I. (1994). Mortality in relation to smoking: 40 years' observations on male British doctors. *British Medical Journal* **309**, 901–11.

Eames, M., Ben-Shlomo, Y. & Marmot, M. G. (1993). Social deprivation and premature mortality: regional comparison across England. *British Medical Journal* **307**, 1097–102.

Ebrahim, S. & Davey-Smith, G. (1996). *Health Promotion in Older People for the Prevention of Coronary Heart Disease and Stroke*, London: Health Education Authority.

Edwards, G., Anderson, P., Babor, T. F. *et al.* (1994). *Alcohol Policy and the Public Good*. Oxford: Oxford University Press.

Egolf, B., Lasker, J., Wolf, S. & Potvin, L. (1992). The Roseto effect: a 50-year comparison of mortality rates. *American Journal of Public Health* **82**, 1089–92.

Elliott, P., Hills, M., Beresford, J., Kleinschmidt, I., Jolley, D., Pattendon, S., Rodrigues, L., Westlake, A. & Rose, G. (1992). Incidence of cancers of the larynx and lung near incinerators of waste solvents and oils in Great Britain. *Lancet* **339**, 854–7.

Elo, I. T. & Preston, S. H. (1996). Educational differentials in mortality: United States, 1979–85. *Social Science and Medicine* **42**, 47–57.

Ettner, S. L. (1996). New evidence on the relationship between income and health. *Journal of Health Economics* **15**, 67–85.

Evandrou, M., Falkingham, J., Le Grand, J. & Winter, D. (1992). Equity in health and social care. *Journal of Social Policy* **21**, 489–523.

Ferrie, J. E., Shipley, M. J., Marmot, M. G., Stansfeld, S. & Davey-Smith, G. D. (1995). Health effects of anticipation of job change and non-employment: longitudinal data from the Whitehall II study. *British Medical Journal* **311**, 1264–9.

Foster, P. (1996). Inequalities in health: what health systems can and cannot do. *Journal of Health Services Research and Policy* **1**, 179–82.

Goldman, L. & Cook, E. F. (1984). The decline in ischemic heart disease mortality rates. An analysis of the comparative effects of medical interventions and changes in lifestyle. *Annals of Internal Medicine* **101**, 825–36.

Graham, H. (1988). Women and smoking in the United Kingdom: the implications for health promotion. *Health Promotion* **3**, 371–82.

Haan, M., Kaplan, G. A. & Camacho, T. (1987). Poverty and health: prospective evidence from the Alameda County study. *American Journal of Epidemiology* **125**, 989–98.

Hayward, R. A., Shapiro, M. F., Freeman, H. E. & Corey, C. R. (1988). Inequities in health services among insured Americans. *New England Journal of Medicine* **318**, 1507–12.

Hertzman, C. & Wiens, M. (1996). Child development and long-term outcomes: a population health perspective and summary of successful interventions. *Social Science and Medicine* **43**, 1083–95.

Holme, I., Helgeland, A., Hjermann, I., Leren, P. & Lund-Larsen, P. G. (1980). Four-year mortality by some socio-economic indicators: the Oslo study. *Journal of Epidemiology and Community Health* **34**, 48–52.

Hypertension Detection and Follow-up Program Cooperative Group (1977). Race, education, and prevalence of hypertension. *American Journal of Epidemiology* **106**, 351–61.

Hypertension Detection and Follow-up Program Cooperative Group (1987). Education level and 5-year all-cause mortality in the HDFP. *Hypertension* **9**, 641–6.

Intersalt Cooperative Research Group (1988). Intersalt: an international study of electrolyte excretion and blood pressure. Results for 24-hour urinary sodium and potassium excretion. *British Medical Journal* **297**, 319–28.

Jones, K. & Duncan, C. (1995). Individuals and their ecologies: analysing the geography of chronic illness within a multilevel modelling framework. *Health and Place* **1**, 27–40.

Kaplan, G. A., Pamuk, E. R., Lynch, J. W., Cohen, R. D. & Balfour, J. L. (1996). Inequality in income and mortality in the United States: analysis of mortality and potential pathways. *British Medical Journal* **312**, 999–1003.

Karasek, R. A. & Theorell, T. (1990) *Healthy Work: Stress, Productivity and the Reconstruction of Working Life.* New York: Basic Books.

Kee, F., Gaffney, B., Currie, S. & O'Reilly, D. (1993). Access to coronary catheterisation: fair shares for all? *British Medical Journal* **307**, 1305–7.

Kehrer, B. H. & Wolin, C. M. (1979). Impact of low income maintenance on low birth weight: evidence from the Gary experiment. *The Journal of Human Resources* **14**, 434–62.

Koskinen, S. V. P., Martelin, T. P. & Valkonen, T. (1996). Socioeconomic differences in mortality among diabetic people in Finland: five-year follow up. *British Medical Journal* **313**, 975–8.

Krieger, N. (1994). Epidemiology and the web of causation: has anyone seen the spider? *Social Science and Medicine* **39**, 887–903.

Kristensen, P.S. (1989). Cardiovascular diseases under work environment: a critical review of the epidemiologic literature on chemical factors. *Scandinavian Journal of Work Environment and Health* **15**, 245–64.

Kunst, A. E. & Mackenbach, J. P. (1994). The size of mortality differences associated with educational level in nine industrialized countries. *American Journal of Public Health* **84**, 932–7.

Ling Tang, J., Law, M. & Wald, N. (1994). How effective is nicotine replacement in helping people to stop smoking? *British Medical Journal* **308**, 21–6.

Link, B. G. & Phelan, J. C. (1996). Editorial: understanding sociodemographic differences in health – the role of fundamental social causes. *American Journal of Public Health* **86**, 471–2.

Lynch, J. (1996). Social position and health. *Annals of Epidemiology* **6**, 21–3.

Mackenbach, J. (1995). Tackling inequalities in health. Great need for evidence based interventions. *British Medical Journal* **310**, 1152–3.

Mackenbach, J. P., Bouvier-Colle, M. H. & Jougla, E. (1990). 'Avoidable' mortality and health services: a review of aggregate data studies. *Journal of Epidemiology and Community Health* **44**, 106–11.

Majeed, F. A., Chaturvedi, N., Reading, R. & Ben-Shlomo, Y. (1994). Equity in the NHS: Monitoring and promoting equity in primary and secondary care. *British Medical Journal* **308**, 1426–9.

Marmot, M. G., Davey-Smith, G., Stansfeld, S., Patel, C., North, F., Head, J., White, I., Brunner, E. & Feeny, A. (1991). Health inequalities among British civil servants: the Whitehall II study. *Lancet* **337**, 1387–94.

Marmot, M. G. & McDowall, M. (1986). Mortality decline and widening social inequalities. *Lancet* **2**, 274–6.

McKeown, T., Record, R. G. & Turner, R. D. (1975). An interpretation of the decline of mortality in England and Wales during the twentieth century. *Population Studies* **29**, 391–421.

McKinlay, J. B., McKinlay, S. M. & Beaglehole, R. (1989). A review of the evidence concerning the impact of medical measures on recent mortality and morbidity in the United States. *International Journal of Health Services* **19**,

181–208.

Moser, K. A., Fox, A. J. & Jones, D. R. (1984). Unemployment and mortality in the OPCS Longitudinal Study. *Lancet* **2**, 1324–9.

Najman, J. M. (1993). Health and poverty: past, present and prospects for the future. *Social Science and Medicine* **36**, 157–66.

Novotny, T. E., Fiore, M. C., Hatziandreu, E. J., Giovino, G. A., Mills, S. L. & Pierce, J. P. (1990). Trends in smoking by age and sex, United States, 1974–1987: the implications for disease impact. *Preventive Medicine* **19**, 552–61.

OPCS (1989). *General Household Survey 1986*. OPCS Social Survey Division. Series GHS No. 16. London: HMSO.

Pamuk, E. R. (1985). Social class inequality in mortality from 1921 to 1972 in England and Wales. *Population Studies* **39**, 17–31.

Patrick, D. & Wickizer, T. M. (1995). Community and Health. In *Society and Health*, ed. B. C. Amick, S. Levine, A. R. Tarlov & D. C. Walsh, p. 374. New York: Oxford University Press.

Phillimore, P., Beattie, A. & Townsend, P. (1994). Widening inequality of health in northern England, 1981–1991. *British Medical Journal* **308**, 1125–8.

Pierce, J. P., Fiore, M. C., Novotny, T. E., Hatziandreu, E. J. & Davis, R. M. (1989). Trends in cigarette smoking in the United States. Projections to the year 2000. *Journal of the American Medical Association* **261**, 61–5.

Plotnick, R. D. (1993). Changes in poverty, income inequality, and the standard of living in the United States during the Reagan years. *International Journal of Health Services* **23**, 347–58.

Propper, C. (1994). Equity and the UK National Health Service: a review of the evidence. *The Economic and Social Review* **25**, 343–65.

Putnam, R. D., Leonardi, R. & Nanetti, R. Y. (1993). *Making Democracy Work: Civic Traditions in Modern Italy*. Princeton: Princeton University Press.

Rose, G. & Day, S. (1990). The population mean predicts the number of deviant individuals. *British Medical Journal* **301**, 1031–4.

Schnall, P. L., Landsbergis, P. A. & Baker, D. (1994). Job strain and cardiovascular disease. *Annual Review of Public Health* **15**, 381–411.

Shouls, S., Congdon, P. & Curtis, S. (1996). Modelling inequality in reported long term illness in the UK: combining individual and area characteristics. *Journal of Epidemiology and Community Health* **50**, 366–76.

Siegrist, J. & Matschinger, H. (1989). Restricted status control and cardiovascular risk. In *Stress, personal control and health*, ed. A. Steptoe & A. Appels, pp. 65–82. Chichester: Wiley.

Siegrist, J., Peter, R., Junge, A. *et al.* (1990). Low status control, high effort at work and ischemic heart disease: prospective evidence from blue-collar men. *Social Science and Medicine* **31**, 1127–34.

Siegrist, J., Peter, R., George, W. *et al.* (1991). Psychosocial and biobehavioral characteristics of hypertensive men with elevated atherogenic lipids. *Atherosclerosis* **86**, 211–18.

Siegrist, J. (1995). Stressful work, self-experience, and cardiovascular disease prevention. In *Behavioural Medicine Approaches to Cardiovascular Disease Prevention*, ed. K. Orth-Gomer & N. Schneiderman. New Jersey: Lawrence Erlbaum Associates.

Sloggett, A. & Joshi, H. (1994). Higher mortality in deprived areas: community or personal disadvantage. *British Medical Journal* **309**, 1470–4.

Sooman, A. & Macintyre, S. (1995). Health and perceptions of the local environment in socially contrasting neighbourhoods in Glasgow. *Health and Place* **1**, 15–26.

Sorlie, P. D., Backlund, E. & Keller, J. B. (1995). US mortality by economic, demographic, and social characteristics: the National Longitudinal Mortality Study. *American Journal of Public Health* **85**, 949–56.

Sorlie, P., Rogot, E., Anderson, R., Johnson, N. J. & Backlund, E. (1992). Black – white mortality differences by family income. *Lancet* **340**, 346–50.

Susser, M. & Susser, E. (1996). Choosing a future for epidemiology: II From black box to Chinese boxes and eco-epidemiology. *American Journal of Public Health* **86**, 674–7.

Theorell, T. (1995). The demand–control–support model for studying health in relation to the work environment: An Interactive Model. In *Behavioural Medicine Approaches to Cardiovascular Disease Prevention,* ed. K. Orth-Gomer & N. Schneiderman, New Jersey: Lawrence Erlbaum Associates.

Townsend, J. (1995). The burden of smoking. In *Tackling Inequalities in Health*, ed. M. Benzeval, K. Judge & M. Whitehead, p. 166. London: Kings Fund.

Townsend, J., Roderick, P. & Copper, J. (1994). Cigarette smoking by socioeconomic group, sex and age: effects of price, income and health publicity. *British Medical Journal* **309**, 923–7.

Tyroler, H. A., Wing, S. & Knowles, M. G. (1993). Increasing inequality in coronary heart disease mortality in relation to educational achievement profiles of places of residence, United States, 1962 to 1987. *Annals of Epidemiology* **3** (suppl), S51–4.

Uchitelle, L. & Kleinfield, N. R. (1996). The downsizing of America: paying the price in people. *Herald Tribune.*

Wallace, P., Cutler S. & Haines A. (1988). Randomised controlled trial of general practitioner intervention in patients with excessive alcohol consumption. *British Medical Journal* **297**, 663–8.

Walt, G. (1994). How far does research influence policy? *European Journal of Public Health* **4**, 233–5.

Wilkinson, R. G. (1994). The epidemiological transition: from material scarcity to social disadvantage? *Daedalus* **123**, 61–77.

Woods, R. & Williams, N. (1995). Must the gap widen before it can be narrowed? Long-term trends in social class mortality differentials. *Continuity and Change* **10**, 105–37.

Index

331